EXPERIMENTAL METHODS FOR ENGINEERS

Fourth Edition

J. P. Holman

Professor of Mechanical Engineering
Southern Methodist University

In collaboration with

W. J. Gajda, Jr.

Associate Professor of Electrical Engineering
University of Notre Dame

McGraw-Hill Book Company

New York   St. Louis   San Francisco   Auckland   Bogotá   Hamburg
Johannesburg   London   Madrid   Mexico   Montreal   New Delhi
Panama   Paris   São Paulo   Singapore   Sydney   Tokyo   Toronto

This book was set in Times Roman.
The editors were Rodger H. Klas and Susan Hazlett;
the production supervisor was Phil Galea.
New drawings were done by Danmark & Michaels, Inc.
The cover was designed by Rafael Hernandez.
R. R. Donnelley & Sons Company was printer and binder.

EXPERIMENTAL METHODS FOR ENGINEERS

234567890 DOCDOC 8987654

ISBN 0-07-029613-8

Library of Congress Cataloging in Publication Data

Holman, J. P. (Jack Philip)
    Experimental methods for engineers.

    Bibliography: p.
    Includes index.
    1. Engineering—Laboratory manuals.   2. Engineering
instruments.   I. Gajda, W. J.   II. Title.
TA152.H6   1984   620′.0028   83-12020
ISBN 0-07-029613-8

CONTENTS

Chapter 4   Basic Electrical Measurements and Sensing Devices   100

PREFACE

Experimental measurements can be vexatious, and a textbook about experimental methods cannot alleviate all the problems which are perplexing to the experimental engineer. Engineering education has placed as increased emphasis on the ability of an individual to perform a theoretical analysis of a problem. Experimental methods are not unimportant, but analytical studies have, at times, seemed to deserve more emphasis, particularly with the enormous computing power which is available. Laboratory work has also become more sophisticated in the modern engineering curricula. The conventional laboratory courses have consistently been modernized to include experiments with rather elaborate electronic instrumentation and microprocessor or computer-based data acquisition systems. Surprisingly enough, however, many engineering graduates do not seem capable of performing simple engineering measurements with acceptable precision. Furthermore, they are amazingly inept when asked the question: How *good* is the measurement?

This book represents a first survey of experimental methods for undergraduate students. As such, it covers a broad range of topics and may be lacking in depth on certain of these topics. In these instances the reader is referred to more detailed treatments in specialized monographs for further information.

It is important that engineers be able to perform successful experiments, and it is equally important that they know or be able to estimate the accuracy of their measurements. This book proposes to discuss a rather broad range of instruments and experimental measurement techniques. Strong emphasis is placed on problem solving, and the importance of accuracy, error, and uncertainty in experimental measurements is stressed throughout all the discussions. The book is generally suitable for a one-semester course at the junior or senior level in conventional engineering curricula. Such a course would be expected to devote one laboratory period a week to the material in this book. To supplement this work, it would be desirable to cover some of the text material in a lecture session. The lectures would be concerned with the principles of instrumentation, whereas the laboratory periods would afford the student an opportunity to use some of the devices discussed in this text and others which may be available to faculty planning the course. The particular experiments, or the instruments used in the laboratory periods, will depend on the facilities available and the objectives set by each curriculum. A mathematical background through ordinary differential equations is assumed for the developments, and it is expected that basic courses in thermodynamics, engineering mechanics, and electric-circuit theory will precede a course based on this text.

Whatever the course arrangement for which this text is applied, it is strongly recommended that the problems at the end of each chapter receive careful attention. These problems force the student to examine several instruments in order to determine their accuracy and the uncertainties which might result from faulty measurement techniques. In many instances the problems are very similar to numerical examples in the text. Other problems require the student to extend the text material through derivations, design of experiments, etc. The selection of problems for a typical course will depend, naturally, on the types of experiments and laboratory facilities available for use with the course.

A few remarks concerning the arrangement of the text material are in order. A brief presentation of all topics was desired, so that a rather broad range of experimental methods could be discussed within the framework of a book suitable in length for a one-semester course. Chapters 1 and 2 provide initial motivational remarks and some brief definitions of important terms common to all measurement systems. Next, a simple presentation of some of the principles of statistical data analysis is given in Chap. 3. Some of the concepts in Chap. 3 are used in almost every subsequent chapter in the book, particularly the concept of experimental uncertainty.

Chapter 4 discusses several simple electrical-measurement and amplifier circuits and the principles of operation of typical electric transducers. Many of these transducers are applicable to measurement problems described in later chapters. Chapters 5 and 6, concerning dimensional and pressure measurements, offer fairly conventional presentations of their subject matter, except that numerical examples and problems are included to emphasize the importance of experimental uncertainty in the various devices. Flow measurement is discussed in Chap. 7 in a rather conventional manner. A notable feature of this chapter is the section on flow-visualization techniques. Again, the examples and problems illustrate some of the advantages and shortcomings of the various experimental techniques. Chapter 8 is quite specific in its discussion of temperature-measurement devices. Strong emphasis is placed on the errors which may arise from conduction, convection, and radiation heat transfer between the temperature-sensing device and its thermal environment. Methods are given to correct for these effects.

Chapter 9 is brief but gives the reader an insight into the problems associated with transport-property measurements. The material in this chapter is dependent on the measurement techniques discussed in Chaps. 6, 7, and 8. It may be noted that the material in Chap. 9 could be dispersed through the three previous chapters and still achieve a balanced presentation; however, it was believed best to bring transport properties and thermal measurements into sharper focus by grouping them together in one chapter.

Static force, torque, and strain measurements are discussed in Chap. 10. The strain measurements are related to some elementary principles of experimental stress analysis, and the operation of the electrical-resistance strain gage is emphasized.

Some of the elementary principles of motion- and vibration-measurement devices are discussed in Chap. 11. Included in this presentation is a discussion of sound waves, sound-pressure level, and acoustic measurements. The inclusion of the acoustics material in Chap. 11 is somewhat arbitrary, since this material would be equally pertinent in Chap. 6.

Chapter 12 discusses thermal- and nuclear-radiation measurements, which are finding increasingly wide application in a number of industries. The presentation is brief, but some of the more important detection techniques are examined, and examples are given to illustrate the important principles. A short presentation of the statistics of counting illustrates the importance of background activity in nuclear-radiation detection. The thermal-radiation measurements are properly related to the material in Chap. 8.

Chapter 13 discusses some of the measurement techniques employed in air-pollution-control applications. Such measurements make use of the basic pressure, flow, and temperature measurement techniques discussed in Chaps. 6, 7, and 8. The importance of electronic data processing and its relation to the basic electrical measuring devices of Chap. 4 are discussed in a general way in Chap. 14. Because the fields of electronic microprocessors and data acquisition systems change so rapidly, the discussion follows a fairly general pattern.

Some remarks concerning units are in order. There is no question that the professional engineering community is moving rapidly toward the adoption of SI (metric) units wherever possible. Accordingly, the educational system must anticipate this movement and teach students to operate in SI. Courses in analysis (fluid mechanics, heat transfer, mechanics, etc.) can adopt SI as the primary system with appropriate conversions to the old English system. The experimental engineer may have a more cumbersome task. One does not buy new instruments, gages, or meters just to change scales to SI units. Common practice is to replace the old instrument only when it wears out or a better device becomes available. This means that the experimental engineer will be operating in a "bilingual" mode for many years. In addition, many field engineers do not *want* to change to SI and will not until they are forced to do so. Of course, companies heavily involved with export markets, like the auto industry, are already heavily into SI units. It is in recognition of this fact that I have chosen to mix SI and English units throughout the book, even though my personal desire is to change over to SI completely as soon as possible. There is another problem which the engineer must deal with. Some work is performed in metric units which are not in the SI system. Examples are the calorie unit for energy and the kilogram *force* per square centimeter for pressure. Alas, these are hazards to be encountered, and one must adapt as best as one can.

In this fourth edition updating citations have been added in various sections. New material on the correlation coefficient has been added to Chap. 3. The section on amplifiers in Chap. 4 has been completely rewritten, and the number of operational-amplifier configurations has been greatly expanded. A section on pH measurement has been added to Chap. 9. In Chap. 3 an alternate for uncertainty analysis is given which is particularly applicable to complicated data reduction involving computers. Other chapters involve tinkering here and there in an effort to improve the presentation and to respond to helpful suggestions from reviewers.

The report writing section of Chap. 2 has been expanded with a new section on graphical presentations. New material has also been added on oral presentations and management of meetings for experimental efforts. The rewite efforts of Dr. W. J. Gajda of the University of Notre Dame in Chaps. 4 and 14 have been largely retained, and I am still indebted to him for this careful integration of material with description of transducer characteristics.

Although several years have passed since the first edition was written and reviewed, the comments and helpful criticisms which were gathered at that time still exert a major influence on the book. The review of Professor S. J. Kline of Stanford University was particularly helpful, and the subsequent suggestions of W. S. Kut, W. L. Rogers, J. H. Dittfach, J. E. Blair, D. B. Marsland, and others are greatly appreciated. Partial sponsorship of the first edition was provided by Dr. C. A. Albritton and the Science Information Institute of Southern Methodist University, and their help is gratefully acknowledged. Finally, the insights gained by many lively discussions with Erich Soehngen have provided the spice, flavor, and appreciation for the importance of experimental measurements which, hopefully, are found in this text.

It is fairly obvious that a book of modest length cannot cover all details of experimental methods which will be of interest to the reader. For this reason, the following brief list of selected books and periodicals is provided for further study and reference. This list covers mainly those sources of *general* information on instruments and measurements. Techniques for measurement in any particular field will appear in those journals which regularly present research results in the field. Thus, the reader should consult such specialized references for information on new measurement techniques which may be employed from time to time.

J. P. Holman

LIST OF SELECTED
BOOKS AND PERIODICALS

Books

1. Ambrosius, E. E., R. D. Fellows, and A. D. Brickman: "Mechanical Measurement and Instrumentation," Ronald Press, New York, 1966.
2. Bartholomew, D.: "Electrical Measurements and Instrumentation," Allyn and Bacon, Inc., Boston, Mass., 1963.
3. Beckwith, T. G. and W. L. Buck: "Mechanical Measurements," 2d ed., Addison-Wesley Publishing Company, Inc., Reading, Mass., 1969.
4. Cady, W. M.; "Physical Measurements in Gas Dynamics and Combustion," Princeton University Press, Princeton, N.J., 1954.
5. Cerni, R. H., and L. E. Foster: "Instrumentation for Engineering Measurement," John Wiley & Sons, Inc., New York, 1962.
6. Considine, D. M. (ed.): "Process Instruments and Controls Handbook," McGraw-Hill Book Company, New York, 1957.
7. Cook, N. H. and E. Rabinowicz: "Physical Measurement and Analysis," Addison-Wesley Publishing Company, Inc., Reading, Mass., 1963.
8. Dally, J. W., and W. F. Riley: "Experimental Stress Analysis," 2d ed., McGraw-Hill Book Company, New York, 1978.
9. Doebelin, E. O.: "Measurement Systems: Application and Design," 2d ed., McGraw-Hill Book Company, New York, 1983.
10. Dove, R. C., and P. H. Adams: "Experimental Stress Analysis and Motion Measurement," Charles E. Merrill Books, Inc., Columbus, Ohio, 1964.
11. Eckman, D. P.: "Industrial Instrumentation," John Wiley & Sons, Inc., New York, 1950.
12. Ewing, G. W.: "Instrumental Methods of Chemical Analysis," 4th ed., McGraw-Hill Book Company, New York, 1975.
13. Frank, E.: "Electrical Measurement Analysis," McGraw-Hill Book Company, New York, 1959.
14. Keast, D. N.: "Measurements in Mechanical Dynamics," McGraw-Hill Book Company, New York, 1967.
15. Lion, K. S.: "Instrumentation in Scientific Research," McGraw-Hill Book Company, New York, 1959.
16. Schenck, H.: "Theories of Engineering Experimentation," 2d ed., McGraw-Hill Book Company, New York, 1968.
17. Tuve, G. L., and L. C. Domholdt: "Engineering Experimentation," McGraw-Hill Book Company, New York, 1966.

18. Lipson, C., and N. J. Sheth: "Statistical Design and Analysis of Engineering Experiments," McGraw-Hill, 1973.
19. Wilson, E. B., Jr.: "An Introduction to Scientific Research," McGraw-Hill Book Company, New York, 1952.
20. Benedict, R. P.: "Fundamentals of Temperature, Pressure, and Flow Measurements," John Wiley & Sons, Inc., New York, 1969.
21. Bragg, G. M.: "Principles of Experimentation and Measurement," Prentice-Hall, Inc., Englewood Cliffs, N.J., 1974.
22. Eckert, E. R. G., and R. J. Goldstein (eds.): "Measurements in Heat Transfer," McGraw-Hill Book Company, New York, 1976.

Periodicals

1. *Control Engineering*
2. *Industrial Laboratory* (USSR; English translation)
3. *Instruments and Control Systems*
4. *Instruments and Experimental Techniques* (USSR; English translation)
5. *Journal of the Instrument Society of America*
6. *Journal of Research of the National Bureau of Standards*
7. *Journal of Scientific Instruments* (Great Britain)
8. *Measurement Techniques* (USSR; English translation)
9. *The Review of Scientific Instruments*
10. *Transactions of the Instrument Society of America*
11. *Transactions of the Society of Instrument Technology* (Great Britain)

ONE

INTRODUCTION

There is no such thing as an easy experiment, nor is there any substitute for careful experimentation in many areas of basic research and applied product development. Because experimentation is so important in all phases of engineering, there is a very definite need for the engineer to be familiar with methods of measurement as well as analysis techniques for interpreting experimental data.

Experimental techniques have changed quite rapidly with the development of electronic devices for sensing primary physical parameters and for controlling process variables. In many instances more precision is now possible in the measurement of basic physical quantities through the use of these new devices. Further development in instrumentation techniques is certain to be very rapid because of the increasing demand for measurement and control of physical variables in a wide variety of applications. To meet this demand the engineer must be familiar with the basic principles of instrumentation and the ideas which govern instrumentation development and usage.

Obviously, a sound knowledge of many engineering principles is necessary to perform successful experiments; it is for this reason that experimentation is so difficult. To design the experiment, the engineer must be able to specify the physical variables to be investigated and the role they will play in later analytical work. Then, to design or procure the instrumentation for the experiment, the engineer must have a knowledge of the governing principles of a broad range of instruments. Finally, to analyze the data, the experimental engineer must have a combination of keen insight into the physical principles of the processes being investigated and a knowledge of the limitations of the data.

Research involves a combination of analytical and experimental work. The theoretician strives to explain or predict the results of experiments on the basis of analytical models which are in accordance with fundamental physical principles that have been well established over the years. When experimental data are encountered which do not fit into the scheme of existing physical theories, a skeptical eye is cast first at the experimental data and then at appropriate theories. In some cases the theories are modified or revised to take the results of the new experimental data into account, after being sure that the validity of the data has been ascertained. In any event, all physical theories must eventually rely upon experiment for verification.

Whether the research is of a basic or developmental character, the dominant role of experimentation is still present. The nuclear physicist must always test theories in the laboratory to be sure of their validity, just as the engineer who conducts research on a new electronic circuit or a new type of hydraulic flow system certainly must perform a significant number of experiments in order to establish the usefulness of the device. Physical experimentation is the ultimate test of most theories.

In many engineering applications certain basic physical phenomena are well known, and experience with devices using these phenomena is already available. Examples are vacuum tubes, solid-state electronics, flowmeters, and certain mechanisms. There will always arise, however, new uses for such devices in combination with other devices, e.g., a new type of amplifier, a new flow-control system, etc. In these cases the engineer must utilize all available experience with the previous devices in order to design the apparatus for a new application. No matter how reliable the information on which the design is based, the engineer must insist on thorough experimental tests of the new device before the design is finalized and production is initiated.

There is a whole spectrum of tests and experiments which an engineer may be called upon to perform. These might vary from a very crude test to determine the weight of a device to some exceedingly precise electronic measurements of nuclear radioactivity. Since the range of experiments is so broad, the experimental background of the engineer must be correspondingly diverse in order to function effectively in many experimental situations. Obviously, it is quite hopeless to expect any one person to operate at a high level of effectiveness in all areas of experimental work. The primary capability of a particular individual will necessarily be developed in an area of experimentation closely connected with that person's analytical and theoretical capability and related interests. The broader the interests of an individual, the more likely it is that that individual will develop broad experimental interests and capabilities.

In the past, there have been some engineers who were primarily experimentalists—those individuals who designed devices by trial and error with very little analytical work as a preliminary to the experimentation. There are some older areas of engineering where this technique will still prevail, primarily where years of experience have built up a background knowledge to rely upon. But, in new fields, more emphasis must be placed on a combination of theory and experimentation. To create a rather absurd example, we might cite the development of a rocket engine. It would be possible to build different sizes of rockets and test them until a lucky combination of design parameters was found; however, the cost would be prohibitive. The proper approach is one of test and theoretical study where experimental data are constantly evaluated and compared with theoretical estimates. New theories are formulated on the basis of the experimental measurements, and these theories help to guide further tests and the final design.

The engineer should know what to look for before beginning the experiments. The objective of the experiments will dictate the accuracy required, expense justified, and level of human effort necessary. A simple calibration check of a mercury-in-glass thermometer would be a relatively simple matter requiring a limited amount of equipment and time; however, the accurate measurement of the temperature of a high-speed gas stream at 1600°C (2912°F) would involve more thought and care. A test of an amplifier for a home music system might be less exacting than a test of an amplifier to be used as part of the electronic equipment in a satellite, and so on.

The engineer is not only interested in the measurement of physical variables but also concerned with their control. The two functions are closely related, however, because one must be able to measure a variable such as temperature or flow in order to control it. The accuracy of control is necessarily dependent on the accuracy of measurement. Hence, we see that a good knowledge of measurement techniques is necessary for the design of control systems. A detailed consideration of control systems is beyond the scope of our discussion, but the applicability of specific instruments and sensing devices to control systems will be indicated from time to time.

It is not enough for the engineer to be able to skillfully measure certain physical variables. For the data to have maximum significance the engineer must be able to specify the precise degree of accuracy with which a certain variable has been measured. To specify this accuracy the limitations of the apparatus must be understood, and full account must be taken of certain random and/or regular errors which may occur in the experimental data. Statistical techniques are available for analyzing data to determine expected errors and deviations from the true measurements. The engineer must be familiar with these techniques in order to analyze the data effectively.

All too frequently the engineer embarks on an experimental program in a stumbling blind-faith manner. Data are collected at random, many of which are not needed for later analysis. Certain ranges of operation are not investigated thoroughly enough, resulting in the collection of data which may have limited correlative value. The engineer must be sure to take enough data, but should not waste time and money by taking too many. The obvious point is that experiments should be carefully planned. Most experimentalists do indeed plan tests with respect to the range of certain variables that they will want to investigate. But they often neglect the fact that more data points may be necessary in certain ranges of operation than in others in order to ensure the same degree of accuracy in the final data evaluation. In other words, the anticipated methods of data analysis, statistical or otherwise, should be taken into account in planning the experiment, just as one would take into account certain variables in designing the physical size of the experimental apparatus. The engineer should always ask the question: How many data do I need to ensure that my data are not just the result of luck? We will have more to say about experiment planning throughout the book, and the reader should consider these opening remarks as only the initial motivation.

A few remarks concerning experimental research are in order at this point. It is very difficult to describe the atmosphere and technique of performing research. For, unlike standard performance testing where experiments are conducted according to some well-established procedure, in research there is seldom a clear-cut way of proceeding. Each problem is a different one, and if the research is worthwhile, it has not been attacked extensively before. This means that the engineer engaged in research must be prepared to face numerous experimental difficulties of varying complexities. Some desirable objectives of the research may have to be relaxed because of the unavailability of instrumentation to measure the variables involved. Many seemingly trivial details become significant problems before a new experimental apparatus is functioning properly. One of the most basic problems is that the engineer seldom gets to measure in a direct manner the variable he or she really wants. There are always corrections to apply to the measurements, and seldom do they fall into the category of "standard" corrections. One trivial detail piles on another until the whole experimental problem takes on a complexity which is usually not anticipated at the start of the research. Again we state the truism: there is no such thing as an *easy* experiment.

Neophyte experimentalists frequently assume that a certain experiment will be easy to perform. All they need do is hook up the apparatus, flip the switch, and out will come reams of significant data which will startle colleagues (or supervisors). They do not realize that one simple instrument may not work and thus spoil the experiment. Once this instrument is functioning properly, another may go bad, and so on. When the apparatus is functioning, neophytes are then tempted to take data at random without giving much consideration to the results that they will want to derive from the data. They try to solve all problems at once, and vary many parameters at the same time, so that little control is exerted on the data and it eventually becomes necessary to go back and do some of the work over. The important point, once again, is that careful planning is called for. In experimental research great care and patience will usually produce the best results in the *quickest* possible way.

The above remarks may appear discouraging to beginners for whom this book is written. On the contrary, they are intended to advise beginners so that they can avoid some of the more obvious pitfalls. And, even more important, the intent is to let the beginner know that some troubles are to be expected and that perseverance in the face of these troubles combined with intelligent planning will almost always lead to the desired results—accurate and meaningful data.

Certainly, not all experimental work is of a "research" nature. The majority of measurements are performed in routine practical industrial applications. Such measurements call for skill and communication ability as much as research because they may be very directly related to the profit and loss of a company.

The objective of the presentation in this book is to impart a broad knowledge of experimental methods and measurement techniques. To accomplish this objective a rather large number of instruments will be discussed from the standpoint of both theory of operation and specific functional characteristics. Emphasis will be placed on analytical calculations to familiarize the reader with important points in the theoretical development as well as in the descriptive information pertaining to operating characteristics. As a further means of emphasizing the discussions, the uncertainties which may arise in the various instruments are given particular attention.

The study of experimental methods is a necessary extension of all analytical subjects. A knowledge of the methods of verifying analytical work injects new life and vitality into the theories, and a clear understanding of the difficulties of experimental measurements creates a careful attitude in the theoretician which cannot be generated in any other way.

TWO

BASIC CONCEPTS

2-1 INTRODUCTION

In this chapter we seek to explain some of the terminology used in experimental methods and to show the generalized arrangement of an experimental system. We shall also discuss briefly the standards which are available and the importance of calibration in any experimental measurement. A major portion of the discussion on experimental errors is deferred until Chap. 3, and only the definition of certain terms is given here.

2-2 DEFINITION OF TERMS

We are frequently concerned with the *readability* of an instrument. This term indicates the closeness with which the scale of the instrument may be read; an instrument with a 12-in scale would have a higher readability than an instrument with a 6-in scale and the same range. The *least count* is the smallest difference between two indications that can be detected on the instrument scale. Both readability and least count are dependent on scale length, spacing of graduations, size of pointer (or pen if a recorder is used), and parallax effects.

The *sensitivity* of an instrument is the ratio of the linear movement of the pointer on the instrument to the change in the measured variable causing this motion. For example, a 1-mV recorder might have a 25-cm scale length. Its sensitivity would be 25 cm/mV, assuming that the measurement was linear all across the scale. For a digital instrument readout the term "sensitivity" does not have the same meaning because different scale factors can be applied with the push of a button. However, the manufacturer will usually specify the sensitivity for a certain scale setting, e.g., 100 nA on a 200-μA scale range for current measurement.

An instrument is said to exhibit *hysteresis* when there is a difference in readings depending on whether the value of the measured quantity is approached from above or below. Hysteresis may be the result of mechanical friction, magnetic effects, elastic deformation, or thermal effects.

The *accuracy* of an instrument indicates the deviation of the reading from a known input. Accuracy is usually expressed as a percentage of full-scale reading, so that a 100-kPa pressure gage having an accuracy of 1 percent would be accurate within ± 1 kPa over the entire range of the gage.

The *precision* of an instrument indicates its ability to reproduce a certain reading with a given accuracy. As an example of the distinction between precision and accuracy, consider the measurement of a known voltage of 100 volts (V) with a certain meter. Five readings are taken, and the indicated values are 104, 103, 105, 103, and 105 V. From these values it is seen that the instrument could not be depended on for an accuracy of better than 5 percent (5 V), while a precision of ± 1 percent is indicated, since the maximum deviation from the mean reading of 104 V is only 1 V. It may be noted that the instrument could be calibrated so that it could be used to dependably measure voltages within ± 1 V. This simple example illustrates an important point. Accuracy can be improved up to but not beyond the precision of the instrument by calibration. The precision of an instrument is usually subject to many complicated factors and requires special techniques of analysis, which will be discussed in Chap. 3.

We should alert the reader at this time to some data analysis terms which will appear in Chap. 3. Accuracy has already been mentioned as relating the deviation of an instrument reading from a *known* value. The deviation is called the *error*. In many experimental situations we may not have a known value with which to compare instrument readings, and yet we may feel fairly confident that the instrument is within a certain plus or minus range of the true value. In such cases, we say that the plus or minus range expresses the *uncertainty* of the instrument readings. Many experimentalists are not very careful in using the words "error" and "uncertainty." As we shall see in Chap. 3, uncertainty is the term that should be most often applied to instruments.

2-3 CALIBRATION

The calibration of all instruments is important, for it affords the opportunity to check the instrument against a known standard and subsequently to reduce errors in accuracy. Calibration procedures involve a comparison of the particular instrument with either (1) a primary standard, (2) a secondary standard with a higher accuracy than the instrument to be calibrated, or (3) a known input source. For example, a flowmeter might be calibrated by (1) comparing it with a standard flow-measurement facility of the National Bureau of Standards,

(2) comparing it with another flowmeter of known accuracy, or (3) direct calibration with a primary measurement such as weighing a certain amount of water in a tank and recording the time elapsed for this quantity to flow through the meter. In item 2, the keywords are "known accuracy." The meaning here is that the accuracy of the meter must be specified by a reputable source.

The importance of calibration cannot be overemphasized because it is calibration which firmly establishes the accuracy of the instruments. Rather than accept the reading of an instrument, it is usually best to make at least a simple calibration check to be sure of the validity of the measurements. Not even manufacturers' specifications or calibrations can always be taken at face value. Most instrument manufacturers are reliable; some, alas, are not. We shall be able to give more information on calibration methods throughout the book as various instruments and their accuracies are discussed.

2-4 STANDARDS

In order that investigators in different parts of the country and different parts of the world may compare the results of their experiments on a consistent basis, it is necessary to establish certain standard units of length, weight, time, temperature, and electrical quantities. The National Bureau of Standards has the primary responsibility for maintaining these standards in the United States.

The meter and the kilogram are considered fundamental units upon which, through appropriate conversion factors, the English system of length and mass is based. The standard meter is defined as the length of a platinum-iridium bar maintained at very accurate conditions at the International Bureau of Weights and Measures in Sèvres, France. Similarly, the kilogram is defined in terms of a platinum-iridium mass maintained at this same bureau. The conversion factors for the English and metric systems in the United States are fixed by law as

$$1 \text{ meter} = 39.37 \text{ inches}$$

$$1 \text{ pound-mass} = 453.59237 \text{ grams}$$

Secondary standards of length and mass are maintained at the National Bureau of Standards for calibration purposes. In 1960 the General Conference on Weights and Measures defined the standard meter in terms of the wavelength of the orange-red light of a krypton-86 lamp. The standard meter is thus

$$1 \text{ meter} = 1,650,763.73 \text{ wavelengths}$$

In the near future it is expected that the definition of the meter will be changed to the distance light travels in 1/299,792,458ths of a second. For the measurement, light from a helium-neon laser illuminates iodine which fluoresces at a highly stable frequency.

The inch is exactly defined as

$$1 \text{ inch} = 2.54 \text{ centimeters}$$

Standard units of time are established in terms of known frequencies of oscillation of certain devices. One of the simplest devices is a pendulum. A torsional vibrational system may also be used as a standard of frequency. The torsional system is widely used in clocks and watches. Ordinary 60-hertz (Hz) line voltage may be used as a frequency standard under certain circumstances. An electric clock uses this frequency as a standard because it operates from a synchronous electric motor whose speed depends on line frequency. A tuning fork is a suitable frequency source, as are piezoelectric crystals. Electronic oscillators may also be designed to serve as very precise frequency sources.

The fundamental unit of time, the second (s), has been defined in the past as $\frac{1}{86400}$ of a mean solar day. The solar day is measured as the time interval between two successive transits of the sun across a meridian of the earth. The time interval varies with location on the earth and time of year; however, the *mean solar day* for one year is constant. The solar year is the time required for the earth to make one revolution around the sun. The mean solar year is 365 days 5 h 48 min 48 s.

The above definition of the second is quite exact but is dependent on astronomical observations in order to establish the standard. In October 1967, the Thirteenth General Conference on Weights and Measures adopted a definition of the second as the duration of 9,192,631,770 periods of the radiation corresponding to the transition between the two hyperfine levels of the fundamental state of the atom of cesium-133, Ref. [7]. This standard can be readily duplicated in standards laboratories throughout the world. The estimated accuracy of this standard is 2 parts in 10^9.

Very precise frequency and time standards are broadcast by the National Bureau of Standards' radio station, WWV. This station furnishes standard time intervals, time signals, standard musical pitch, standard radio frequencies, and radio-propagation forecasts. The accuracy of the radio and audio frequencies as transmitted is better than 1 part in 100,000,000, and the time interval is accurate to 1 part in 100,000,000 $\pm$ 1 microsecond (μs).

Standard units of electrical quantities are derivable from the mechanical units of force, mass, length, and time. These units represent the absolute electrical units and differ slightly from the international system of electrical units established in 1948. A detailed description of the international system is given in Refs. [1] and [2]. The main advantage of this system is that it affords the establishment of a standard cell, the output of which may be directly related to the absolute electrical units. The conversion from the international system is established by the following relations:

$$1 \text{ international ohm} = 1.00049 \text{ absolute ohms}$$

$$1 \text{ international volt} = 1.000330 \text{ absolute volts}$$

$$1 \text{ international ampere} = 0.99835 \text{ absolute ampere}$$

Laboratory calibration is usually made with the aid of secondary standards such as standard cells for voltage sources and standard resistors as standards of comparison for measurement of electrical resistance.

Table 2-1 Primary points for the International Practical Temperature Scale of 1968

	Temperature	
Point Normal pressure = 14.6959 psia = 1.0132×10^5 Pa	°C	°F
Triple point of equilibrium hydrogen	−259.34	−434.81
Boiling point of equilibrium hydrogen at $\frac{25}{76}$ normal pressure	−256.108	−428.99
Normal boiling point (1 atm) of equilibrium hydrogen	−252.87	−423.17
Normal boiling point of neon	−246.048	−410.89
Triple point of oxygen	−218.789	−361.82
Normal boiling point of oxygen	−182.962	−297.33
Triple point of water	0.01	32.018
Normal boiling point of water	100	212.00
Normal freezing point of zinc	419.58	787.24
Normal freezing point of silver	961.93	1763.47
Normal freezing point of gold	1064.43	1947.97

Table 2-2 Secondary fixed points for the International Practical Temperature Scale of 1968

Point	Temperature, °C
Triple point, normal H_2	−259.194
Boiling point, normal H_2	−252.753
Triple point, Ne	−248.595
Triple point, N_2	−210.002
Boiling point, N_2	−195.802
Sublimation point, CO_2 (normal)	−78.476
Freezing point, Hg	−38.862
Ice point	0
Triple point, phenoxibenzene	26.87
Triple point, benzoic acid	122.37
Freezing point, In	156.634
Freezing point, Bi	271.442
Freezing point, Cd	321.108
Freezing point, Pb	327.502
Boiling point, Hg	356.66
Boiling point, S	444.674
Freezing point, Cu-Al eutectic	548.23
Freezing point, Sb	630.74
Freezing point, Al	660.74
Freezing point, Cu	1084.5
Freezing point, Ni	1455
Freezing point, Co	1494
Freezing point, Pd	1554
Freezing point, Pt	1772
Freezing point, Rh	1963
Freezing point, Ir	2447
Freezing point, W	3387

An absolute temperature scale was proposed by Lord Kelvin in 1854. It forms the basis for thermodynamic calculations. This absolute scale is so defined that particular meaning is given to the second law of thermodynamics when this temperature scale is used. The International Practical Temperature Scale of 1968 [2] furnishes an experimental basis for a temperature scale which approximates as closely as possible the absolute thermodynamic temperature scale. In the international scale, 11 primary points are established as shown in Table 2-1. Secondary fixed points are also established as given in Table 2-2. In addition to the fixed points, precise procedures are established for interpolating between these points. These interpolation procedures are given in Table 2-3.

Both the Fahrenheit (°F) and Celsius (°C) temperature scales are in wide use, and the experimentalist must be able to work in either. The absolute Fahrenheit scale is called the Rankine (°R) scale, while absolute Celsius has been designated the Kelvin (K) scale. The relationship between these scales is as follows:

$$K = {}^\circ C + 273.15$$

$$^\circ R = {}^\circ F + 459.67$$

$$^\circ F = \tfrac{9}{5}{}^\circ C + 32.0$$

The absolute thermodynamic temperature scale is discussed by Holman [4].

Table 2-3 Interpolation procedures for International Practical Temperature Scale of 1968

Range, °C	Procedure
−259.34 to 0	Platinum resistance thermometer with cubic polynomial coefficients determined from calibration at fixed points, using four ranges
0 to 630.74	Platinum resistance thermometer with second-degree polynomial coefficients determined from calibration at three fixed points in the range
630.74 to 1064.43	Standard platinum–platinum rhodium (10%) thermocouple with second-degree polynomial coefficients determined from calibration at antimony, silver, and gold points
Above 1064.43	Temperature defined by: $$\frac{J_t}{J_{Au}} = \frac{e^{C_2/\lambda(T_{Au} + T_0)} - 1}{e^{C_2/\lambda(T + T_0)} - 1}$$ J_t, J_{Au} = radiant energy emitted per unit time, per unit area, and per unit wavelength at wavelength λ, at temperature T, and gold-point temperature T_{Au}, respectively $C_2 = 1.438$ cm − K $T_0 = 273.16$ K λ = wavelength

2-5 DIMENSIONS AND UNITS

Despite strong emphasis in the professional engineering community on standardizing units with an international system, a variety of instruments will be in use for many years, and an experimentalist must be conversant with the units which appear on the gages and readout equipment. The main difficulties arise in mechanical and thermal units, because electrical units have been standardized for some time. The SI (Système International d'Unités) set of units will eventually prevail, and we shall express examples and problems in this system as well as the English system employed in the United States for many years.

One must be careful not to confuse the meaning of the terms "units" and "dimensions." A *dimension* is a physical variable used to specify the behavior or nature of a particular system. For example, the length of a rod is a dimension of the rod. In like manner, the temperature of a gas may be considered one of the thermodynamic dimensions of the gas. When we say the rod is so many meters long, or the gas has a temperature of so many degrees Celsius, we have given the units with which we choose to measure the dimension. In our development we shall use the dimensions

$$L = \text{length}$$

$$M = \text{mass}$$

$$F = \text{force}$$

$$\tau = \text{time}$$

$$T = \text{temperature}$$

All the physical quantities used may be expressed in terms of these fundamental dimensions. The units to be used for certain dimensions are selected by somewhat arbitrary definitions which usually relate to a physical phenomenon or law. For example, Newton's second law of motion may be written

Force ~ time rate of change of momentum

$$F = k \frac{d(mv)}{d\tau}$$

where k is the proportionality constant. If the mass is constant,

$$F = kma \tag{2-1}$$

where the acceleration is $a = dv/d\tau$. Equation (2-1) is usually written

$$F = \frac{1}{g_c} ma \tag{2-2}$$

with $1/g_c = k$. Equation (2-2) is used to define our systems of units for mass, force, length, and time. Some typical systems of units are

1. 1 pound-force will accelerate 1 pound-mass 32.174 feet per second squared.
2. 1 pound-force will accelerate 1 slug-mass 1 feet per second squared.

3. 1 dyne-force will accelerate 1 gram-mass 1 centimeter per second squared.
4. 1 newton (N) force will accelerate 1 kilogram-mass 1 meter per second squared.
5. 1 kilogram-force will accelerate 1 kilogram-mass 9.80665 meter per second squared.

The kilogram-force is sometimes given the designation *kilopond* (kp).

Since Eq. (2-2) must be dimensionally homogeneous, we shall have a different value of the constant g_c for each of the unit systems in items 1 to 5 above. These values are

1. $g_c = 32.174$ lbm $\cdot$ ft/lbf $\cdot$ s^2
2. $g_c = 1$ slug $\cdot$ ft/lbf $\cdot$ s^2
3. $g_c = 1$ g $\cdot$ cm/dyn $\cdot$ s^2
4. $g_c = 1$ kg $\cdot$ m/N $\cdot$ s^2
5. $g_c = 9.80665$ kgm $\cdot$ m/kgf $\cdot$ s^2

It matters not which system of units is used so long as it is consistent with the above definitions.

Work has the dimensions of a product of force times a distance. Energy has the same dimensions. The units for work and energy may be chosen from any of the systems used above. Thus the units of work and energy for each of the above systems are

1. lbf $\cdot$ ft
2. lbf $\cdot$ ft
3. dyn $\cdot$ cm $= 1$ erg
4. N $\cdot$ m $= 1$ joule (J)
5. kgf $\cdot$ m $= 9.80665$ J

In addition, we may use the units of energy which are based on thermal phenomena:

1 British thermal unit (Btu) will raise 1 pound-mass of water 1 degree Fahrenheit at 68°F.
1 calorie (cal) will raise 1 gram of water 1 degree Celsius at 20°C.
1 kilocalorie will raise 1 kilogram of water 1 degree Celsius at 20°C.

The conversion factors for the various units of work and energy are

$$1 \text{ Btu} = 778.16 \text{ lbf} \cdot \text{ft}$$

$$1 \text{ Btu} = 1055 \text{ J}$$

$$1 \text{ kcal} = 4182 \text{ J}$$

$$1 \text{ lbf} \cdot \text{ft} = 1.356 \text{ J}$$

$$1 \text{ Btu} = 252 \text{ cal}$$

Additional conversion factors are given in the Appendix.

The weight of a body is defined as the force exerted on the body as a result of the acceleration of gravity. Thus

$$W = \frac{g}{g_c} m \tag{2-3}$$

where W is the weight and g is the acceleration of gravity. Note that the weight of a body has the dimensions of a force. We now see why systems 1 and 5 above were devised; 1 lbm will weigh 1 lbf at sea level, and 1 kgm will weigh 1 kgf.

Unfortunately, *all* the above unit systems are used in various places throughout the world. While the foot-pound force, pound-mass, second, degree Fahrenheit, Btu system is still widely used in the United States, there is increasing impetus to institute the SI units as a worldwide standard. In this system, the fundamental units are meter, newton, kilogram mass, second, and degree Celsius; a "thermal" energy unit is not used; i.e., the joule (N · m) becomes the energy unit used throughout. The watt (J/s) is the unit of power in this system. In SI the concept of g_c is not normally used, and the newton is defined as

$$1 \text{ newton} = 1 \text{ kilogram-meter per second squared} \tag{2-4}$$

Even so, one should keep in mind the physical relation between force and mass as expressed by Newton's second law of motion.

Despite the present writer's personal enthusiasm for complete changeover to the SI system, the fact is that a large number of engineering practitioners still use English units and will continue to do so for some time to come. Therefore, we shall use both English and SI units in this text so that the reader can operate effectively in the expected industrial environment.

Table 2-4 lists the basic and supplementary SI units, while Table 2-5 gives a list of derived SI units for various physical quantities. The SI system also specifies standard multiplier prefixes, as shown in Table 2-6. For example, 1 atm pressure

Table 2-4 Basic and supplemental SI units

Quantity	Unit	Symbol
Basic units		
Length	meter	m
Mass	kilogram	kg
Time	second	s
Electric current	ampere	A
Temperature	kelvin	K
Luminous intensity	candela	cd
Supplemental units		
Plane angle	radian	rad
Solid angle	steradian	sr

Table 2-5 Derived SI units

Quantity	Name(s) of unit	Unit symbol or abbreviation, where differing from basic form	Unit expressed in terms of basic or supplementary units
Area	square meter		m^2
Volume	cubic meter		m^3
Frequency	hertz, cycle per second	Hz	s^{-1}
Density, concentration	kilogram per cubic meter		kg/m^3
Velocity	meter per second		m/s
Angular velocity	radian per second		rad/s
Acceleration	meter per second squared		m/s^2
Angular acceleration	radian per second squared		rad/s^2
Volumetric flow rate	cubic meter per second		m^3/s
Force	newton	N	$kg \cdot m/s^2$
Surface tension	newton per meter, joule per square meter	N/m, J/m^2	kg/s^2
Pressure	newton per square meter, pascal	N/m^2, Pa	$kg/m \cdot s^2$
Viscosity, dynamic	newton-second per square meter poiseuille	$N \cdot s/m^2$, Pl	$kg/m \cdot s$
Viscosity, kinematic; diffusivity; mass conductivity	meter squared per second		m^2/s
Work, torque, energy, quantity of heat	joule, newton-meter, watt-second	J, $N \cdot m$, $W \cdot s$	$kg \cdot m^2/s^2$
Power, heat flux	watt, joule per second	W, J/s	$kg \cdot m^2/s^2$
Heat flux density	watt per square meter	W/m^2	kg/s^3
Volumetric heat release rate	watt per cubic meter	W/m^3	$kg/m \cdot s^3$
Heat transfer coefficient	watt per square meter degree	$W/m^2 \cdot deg$	$kg/s^3 \cdot deg$
Latent heat, enthalpy (specific)	joule per kilogram	J/kg	m^2/s^2
Heat capacity (specific)	joule per kilogram degree	$J/kg \cdot deg$	$m^2/s^2 \cdot deg$
Capacity rate	watt per degree	W/deg	$kg \cdot m^2/s^3 \cdot deg$
Thermal conductivity	watt per meter degree	$W/m \cdot deg$, $J \cdot m/s \cdot m^2 \cdot deg$	$kg \cdot m/s^3 \cdot deg$
Mass flux, mass flow rate	kilogram per second		kg/s
Mass flux density, mass flow rate per unit area	kilogram per square meter-second		$kg/m^2 \cdot s$
Mass-transfer coefficient	meter per second		m/s
Quantity of electricity	coulomb	C	$A \cdot s$
Electromotive force	volt	V, W/A	$kg \cdot m^2/A \cdot s^3$
Electric resistance	ohm	Ω, V/A	$kg \cdot m^2/A^2 \cdot s^3$
Electric conductivity	ampere per volt meter	$A/V \cdot m$	$A^2 \cdot s^3/kg \cdot m^3$
Electric capacitance	farad	F, $A \cdot s/V$	$A^3 \cdot s^4/kg \cdot m^2$
Magnetic flux	weber	Wb, $V \cdot s$	$kg \cdot m^2/A \cdot s^2$
Inductance	henry	H, $V \cdot s/A$	$kg \cdot m^2/A^2 \cdot s^2$
Magnetic permeability	henry per meter	H/m	$kg \cdot m/A^2 \cdot s^2$
Magnetic flux density	tesla, weber per square meter	T, Wb/m^2	$kg/A \cdot s^2$
Magnetic field strength	ampere per meter		A/m
Magnetomotive force	ampere		A
Luminous flux	lumen	lm	$cd \cdot sr$
Luminance	candela per square meter		cd/m^2
Illumination	lux, lumen per square meter	lx, lm/m^2	$cd \cdot sr/m^2$

Table 2-6 Standard prefixes and multiples in SI units

Multiples and submultiples	Prefixes	Symbols	Pronunciations
10^{12}	tera	T	tĕr′ȧ
10^{9}	giga	G	jĭ′ gȧ
10^{6}	mega	M	mĕg′ȧ
10^{3}	kilo	k	kĭl′ô
10^{2}	hecto	h	hek′tô
10	deka	da	dĕk′ȧ
10^{-1}	deci	d	dĕs′ĭ
10^{-2}	centi	c	sĕn′tĭ
10^{-3}	milli	m	mĭl ĭ
10^{-6}	micro	μ	mī′ krô
10^{-9}	nano	n	năn′ô
10^{-12}	pico	p	pē′ cô
10^{-15}	femto	f	fĕm′ tô
10^{-18}	atto	a	ăt′ tô

is 1.0132×10^{5} N/m^{2}, which could be written 1 atm = 0.10132 MN/m^{2}. Conversion factors are given in the Appendix as well as in sections of the text which discuss specific measurement techniques for pressure, energy flux, etc.

2-6 THE GENERALIZED MEASUREMENT SYSTEM

Most measurement systems may be divided into three parts:

1. A detector-transducer stage, which detects the physical variable and performs either a mechanical or an electrical transformation to convert the signal into a more usable form. In the general sense, a transducer is a device that transforms one physical effect into another. In most cases, however, the physical variable is transformed into an electric signal because this is the form of signal that is most easily measured.
2. Some intermediate stage, which modifies the direct signal by amplification, filtering, or other means so that a desirable output is available.
3. A final or terminating stage, which acts to indicate, record, or control the variable being measured.

As an example of a measurement system, consider the measurement of a low-voltage signal at a low frequency. The detector in this case may be just two wires and possibly a resistance arrangement, which are attached to appropriate terminals. Since we want to indicate or record the voltage, it may be necessary to perform some amplification. The amplification stage is then stage 2 designated above. The final stage of the measurement system may be either a voltmeter or a recorder that operates in the range of the output voltage of the amplifier. In actuality, an electronic voltmeter is a measurement system like the one described here. The amplifier and the readout voltmeter are contained in one cabinet, and various switches enable the user to change the range of the instrument by varying the input conditions to the amplifier.

Consider the simple bourdon-tube pressure gage shown in Fig. 2-1. This gage offers another example of the generalized measurement system. In this case the bourdon tube is the detector-transducer stage because it converts the pressure signal into a mechanical displacement of the tube. The intermediate stage consists of the gearing arrangement, which amplifies the displacement of the end of the tube so that a relatively small displacement at that point produces as much as three-quarters of a revolution of the center gear. The final indicator stage consists of the pointer and the dial arrangement, which, when calibrated with known pressure inputs, gives an indication of the pressure signal impressed on the bourdon tube. A schematic diagram of the generalized measurement system is shown in Fig. 2-2.

When a control device is used for the final measurement stage, it is necessary to apply some feedback signal to the input signal to accomplish the control objectives. The control stage compares the signal representing the measured vari-able with some other signal in the same form representing the assigned value the measured variable *should have*. The assigned value is given by a predetermined setting of the controller. If the measured signal agrees with the predetermined setting, then the controller does nothing. If the signals do not agree, the controller

Figure 2-1 Bourdon-tube pressure gage as the generalized measurement system.

Figure 2-2 Schematic of the generalized measurement system.

issues a signal to a device, which acts to alter the value of the measured variable. This device can be many things, depending on the variable which is to be controlled. If the measured variable is the flow rate of a fluid, the control device might be a motorized valve placed in the flow system. If the measured flow rate is too high, then the controller would cause the motorized valve to close, thereby reducing the flow rate. If the flow rate were too low, the valve would be opened. Eventually the operation would cease when the desired flow rate was achieved. The control feedback function is indicated in Fig. 2-2.

It is very important to realize that the accuracy of control cannot be any better than the accuracy of the measurement of the control variable. Therefore, one must be able to measure a physical variable accurately before one can hope to control the variable. In the flow system mentioned above, the most elaborate controller could not control the flow rate any more closely than the accuracy with which the primary sensing element measures the flow. We shall have more to say about some simple control systems in a later chapter. For the present we want to emphasize the importance of the measurement system in any control setup.

The overall schematic of the generalized measurement system is quite simple, and, as one might suspect, the difficult problems are encountered when suitable devices are sought to fill the requirements for each of the "boxes" on the schematic diagram. Most of the remaining chapters of the book are concerned with the types of detectors, transducers, modifying stages, etc., that may be used to fill these boxes.

2-7 BASIC CONCEPTS IN DYNAMIC MEASUREMENTS

A *static* measurement of a physical quantity is performed when the quantity is not changing with time. The deflection of a beam under a constant load would be a static deflection. However, if the beam were set in vibration, the deflection would vary with time, and the measurement process might be more difficult. Measurements of flow processes are much easier to perform when the fluid is in a nice steady state, and become progressively more difficult to perform when rapid changes with time are encountered.

Many experimental measurements are taken under such circumstances that ample time is available for the measurement system to reach steady state, and hence one need not be concerned with the behavior under non-steady-state conditions. In many other situations, however, it may be desirable to determine the behavior of a physical variable over a period of time. Sometimes the time interval is short, and sometimes it may be rather extended. In any event, the measurement problem usually becomes more complicated when the transient characteristics of a system need to be considered. In this section we wish to discuss some of the more important characteristics and parameters applicable to a measurement system under dynamic conditions.

To initiate the discussion let us consider a simple spring-mass damper system as shown in Fig. 2-3. We might consider this as a simple mechanical-measurement system where $x_1(t)$ is the input displacement variable which acts through the spring-mass damper arrangement to produce an output displacement $x_2(t)$. Both x_1 and x_2 vary with time. Suppose we wish to find $x_2(t)$, knowing $x_1(t)$, m, k, and the damping constant c. We assume that the damping force is proportional to velocity, so that the differential equation governing the system is obtained from Newton's second law of motion as

$$k(x_1 - x_2) + c\left(\frac{dx_1}{dt} - \frac{dx_2}{dt}\right) = m\frac{d^2x_2}{dt^2} \tag{2-5}$$

Written in another form,

$$m\frac{d^2x_2}{dt^2} + c\frac{dx_2}{dt} + kx_2 = c\frac{dx_1}{dt} + kx_1 \tag{2-6}$$

Now suppose that $x_1(t)$ is the harmonic function

$$x_1(t) = x_0 \cos \omega_1 t \tag{2-7}$$

where x_0 is the amplitude of the displacement and ω_1 is the frequency.

We might imagine this simple vibrational system as being similar to a simple spring scale. The mass of the scale is m, the spring inside the scale is represented

Figure 2-3 Simple spring-mass damper system.

by the spring constant k, and whatever mechanical friction may be present is represented by c. We are subjecting the scale to an oscillating-displacement function and wish to know how the body of the scale will respond; i.e., we want to know $x_2(t)$. We might imagine that the spring scale is shaken by hand. When the oscillation $x_1(t)$ is very slow, we would note that the scale body very nearly follows the applied oscillation. When the frequency of the oscillation is increased, the scale body will react more violently until, at a certain frequency called the *natural frequency*, the amplitude of the displacement of the spring will take on its maximum value and could be *greater* than the amplitude of the impressed oscillation $x_1(t)$. The smaller the value of the damping constant, the larger the maximum amplitude at the natural frequency. If the impressed frequency is increased further, the amplitude of the displacement of the spring body will decrease rather rapidly. The reader may conduct an experiment to verify this behavior using a simple spring-mass system shaken by hand.

Clearly, the displacement function $x_2(t)$ depends on the frequency of the impressed function $x_1(t)$. We say that the system *responds* differently depending on the input frequency, and the overall behavior is designated as the *frequency response* of the system.

A simple experiment with the spring-mass system will show that the displacement of the mass is not in *phase* with the impressed displacement; i.e., the maximum displacement of the mass does not occur at the same time as the maximum displacement of the impressed function. This phenomenon is described as *phase shift*. We could solve Eq. (2-6) and determine the detailed characteristics of $x_2(t)$, including the frequency response and phase-shift behavior. However, we shall defer the solution for this system until Chap. 11, where vibration measurements are discussed. At this point in our discussion we have used this example because it is easy to visualize in a physical sense. Now, we shall consider a practical application, which involves a transformation of a *force-input function* into a displacement function. We shall present the solution to this problem because it shows very clearly the nature of frequency response and phase shift while emphasizing a practical system that might be used for a transient force or pressure measurement.

This system is shown in Fig. 2-4. The forcing function

$$F(t) = F_0 \cos \omega_1 t \tag{2-8}$$

is impressed on the spring-mass system, and we wish to determine the displacement of the mass $x(t)$ as a function of time. The differential equation for the system is

$$m \frac{d^2x}{dt^2} + c \frac{dx}{dt} + kx = F_0 \cos \omega_1 t \tag{2-9}$$

Equation (2-9) has the solution

$$x = \frac{(F_0/k) \cos (\omega_1 t - \phi)}{\{[1 - (\omega_1/\omega_n)^2]^2 + [2(c/c_c)(\omega_1/\omega_n)]^2\}^{1/2}} \tag{2-10}$$

$$m\frac{d^2x}{dt^2} + c\frac{dx}{dt} + kx = F_0 \cos \omega_1 t$$

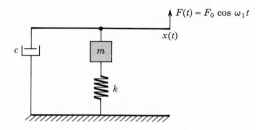

Figure 2-4 Spring-mass damper system subjected to a force input.

where

$$\phi = \tan^{-1}\frac{2(c/c_c)(\omega_1/\omega_n)}{1 - (\omega_1/\omega_n)^2} \tag{2-11}$$

$$\omega_n = \sqrt{\frac{k}{m}} \tag{2-12}$$

$$c_c = 2\sqrt{mk} \tag{2-13}$$

ϕ is called the *phase angle*, ω_n is the natural frequency, and c_c is called the *critical damping coefficient*.

The ratio of output to input amplitude $x_0/(F_0/k)$, where x_0 is the amplitude of the motion given by

$$x_0 = \frac{F_0/k}{\{[1 - (\omega_1/\omega_n)^2]^2 + [2(c/c_c)(\omega_1/\omega_n)]^2\}^{1/2}} \tag{2-14}$$

is plotted in Fig. 2-5 to show the frequency response of the system, and the phase angle ϕ is plotted in Fig. 2-6 to illustrate the phase-shift characteristics. From these graphs we make several observations:

1. For low values of c/c_c the amplitude is very nearly constant up to a frequency ratio of about 0.3.
2. For large values of c/c_c (overdamped systems) the amplitude is reduced substantially.
3. The phase-shift characteristics are a strong function of damping ratio for all frequencies.

We might say that the system has good *linearity* for low damping ratios and up to a frequency ratio of 0.3 since the amplitude is essentially constant in this range.

While this brief discussion has been concerned with a simple mechanical system, we may remark that similar frequency and phase-shift characteristics are exhibited by electrical and thermal systems as well, and whenever time-varying measurements are made, due consideration must be given to these characteristics. Ideally, we should like to have a system with a linear frequency response over all

Figure 2-5 Frequency response of the system in Fig. 2-4.

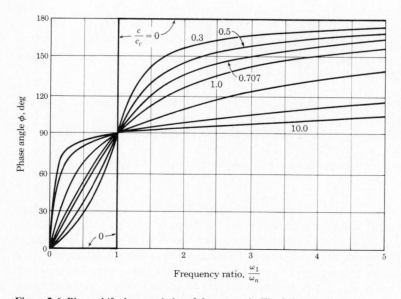

Figure 2-6 Phase-shift characteristics of the system in Fig. 2-4.

ranges and with zero phase shift, but this is never completely attainable, although a certain instrument may be linear over a range of operation in which we are interested so that the behavior is good enough for the purposes intended. There are methods of providing compensation for the adverse frequency-response characteristics of an instrument, but these methods represent an extensive subject in themselves and cannot be discussed here. We shall have something to say about the dynamic characteristics of specific instruments in subsequent chapters.

2-8 SYSTEM RESPONSE

We have already discussed the meaning of frequency response and observed that in order for a system to have good response it must treat all frequencies the same within the range of application so that the ratio of output-to-input amplitude remains the same over the frequency range desired. We say that the system has *linear frequency response* if it follows this behavior.

Amplitude response pertains to the ability of the system to react in a linear way to various input amplitudes. In order for the system to have *linear amplitude response*, the ratio of output-to-input amplitude should remain constant over some specified range of input amplitudes. When this linear range is exceeded, the system is said to be *overdriven*, as in the case of a voltage amplifier where too high an input voltage is used.

We have already noted the significance of phase-shift response and its relation to frequency response. Phase shift is particularly important where complex waveforms are concerned because severe distortion may result if the system has poor phase-shift response.

Suppose a *step* or instantaneous input signal is applied to a system. In general there will be a slight delay in the output response, and this delay is called the *rise time* or *delay* of the system. This phenomenon is illustrated in Fig. 2-7. The *slew rate* is the maximum rate of change the system can handle.

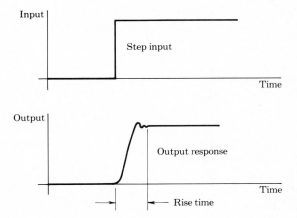

Figure 2-7 Effect of rise time on output response to a step input.

(a)

t/RC

(b)

Figure 2-8 Capacitor discharge through a resistance. (a) Schematic; (b) plot of voltage.

Many systems exhibit an exponential decay type of behavior. When a capacitor is discharged through a resistance as shown in Fig. 2-8, the voltage varies with time according to

$$\frac{E(t)}{E_0} = e^{-(1/RC)t} \tag{2-15}$$

where R is the value of the external resistance and C is the capacitance. The voltage across the capacitor as a function of time is $E(t)$, and the initial voltage is E_0. Some types of thermal systems also display this same kind of response. The temperature of a hot block of metal allowed to cool in a room varies with approximately the same kind of relation as shown in Eq. (2-15). For systems which have this kind of behavior we may note that when

$$t = RC \tag{2-16}$$

the ratio E/E_0 has a value of 36.8 percent of its initial value. Since the terms of

Eq. (2-16) have the units of time, they are described as the *time constant* of the system. This meaning of time constant is carried into all types of systems—even those which do not have a simple behavior like Eq. (2-15). Thus, the time constant is usually taken as the time for the system to attain a state of 63.2 percent of the difference between the initial and steady-state values. In the capacitance system we speak of voltage, in a thermal system we speak of temperature, and in some mechanical systems we might speak of velocity or displacement as the physical variables that change with time. The time constant would then be related to the initial and steady-state values of these variables. A plot of the voltage decay is shown in Fig. 2-8b, illustrating the position of the time constant.

2-9 DISTORTION

Suppose a harmonic function of a complicated nature, i.e., composed of many frequencies, is transmitted through the mechanical system of Figs. 2-3 and 2-4. If the frequency spectrum of the incoming waveform were sufficiently broad, there

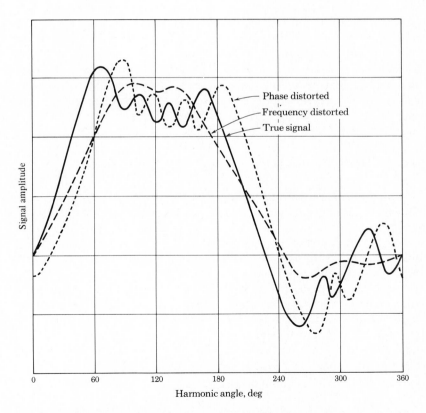

Figure 2-9 Effects of frequency response and phase-shift response on complex waveform.

would be different amplitude and phase-shift characteristics for each of the input-frequency components, and the output waveform might bear little resemblance to the input. Thus, as a result of the frequency-response characteristics of the system, distortion in the waveform would be experienced. Distortion is a very general term that may be used to describe the variation of a signal from its true form. Depending on the system, the distortion may result from either poor frequency response or poor phase-shift response. In electronic devices various circuits are employed to reduce distortion to very small values. For mechanical systems the dynamic response characteristics are not as easily controlled and remain a subject for further development. The effects of poor frequency and phase-shift response on a complex waveform are illustrated in Fig. 2-9.

2-10 IMPEDANCE MATCHING

In many experimental setups it is necessary to connect various items of electrical equipment in order to perform the overall measurement objective. When connections are made between electrical devices, proper care must be taken to avoid impedance mismatching. The input impedance of a two-terminal device may be illustrated as in Fig. 2-10. The device behaves as if the internal resistance R_i were connected in series with the internal voltage source E. The connecting terminals for the instrument are designated as A and B, and the open-circuit voltage presented at these terminals is the internal voltage E. Now, if an external load R is connected to the device and the internal voltage E remains constant, the voltage presented at the output terminals A and B will be dependent on the value of R. The potential presented at the output terminals is

$$E_{AB} = E \frac{R}{R + R_i} \tag{2-17}$$

The larger the value of R, the more closely the terminal voltage approaches the internal voltage E. Thus, if the device is used as a voltage source with some internal impedance, the external impedance (or load) should be large enough that the voltage is essentially preserved at the terminals. Or, if we wish to measure the internal voltage E, the impedance of the measuring device connected to the terminals should be large compared with the internal impedance.

Figure 2-10 Two-terminal device with internal impedance R_i.

Now suppose that we wish to deliver power from the device to the external load R. The power is given by

$$P = \frac{E_{AB}^2}{R} \tag{2-18}$$

We ask for the value of the external load that will give the maximum power for a constant internal voltage E and internal impedance R_i. Equation (2-18) is rewritten

$$P = \frac{E^2}{R} \left(\frac{R}{R + R_i} \right)^2 \tag{2-19}$$

and the maximizing condition

$$\frac{dP}{dR} = 0 \tag{2-20}$$

is applied. There results

$$R = R_i \tag{2-21}$$

That is, the maximum amount of power may be drawn from the device when the impedance of the external load just matches the internal impedance. This is the essential principle of impedance matching in electric circuits.

Clearly, the internal impedance and external load of a complicated electronic device may contain inductive and capacitative components that will be important in ac transmission and dissipation. Nevertheless, the basic idea is the same. The general principles of matching, then, are that the external impedance should match the internal impedance for maximum energy transmission (minimum attenuation), and the external impedance should be large compared with the internal impedance when a measurement of internal voltage of the device is desired. It is this latter principle that makes an electronic voltmeter essential for measurement of voltages in electronic circuits. The electronic voltmeter has a very high internal impedance (greater than 10 MΩ), so that little current is drawn and the voltage presented to the terminals of the instrument is not altered appreciably by the measurement process. Many such voltmeters today operate on digital principles but all have the capability of very high input impedance.

Impedance-matching problems are usually encountered in electrical systems but can be important in mechanical systems as well. We might, for example, imagine the simple spring-mass system of the previous section as a mechanical transmission system. From the curves describing the system behavior it is seen that frequencies below a certain value are transmitted through the system; i.e., the force is converted to displacement with little attenuation. Near the natural frequency undesirable amplification of the signal is performed, and above this frequency severe attenuation is present. We might say that this system exhibits a behavior characteristic of a variable impedance that is frequency-dependent.

When it is desired to transmit mechanical motion through a system, the natural-frequency and damping characteristics must be taken into account so that good "matching" is present. The problem is an impedance-matching situation, although it is usually treated as a subject in mechanical vibrations.

2-11 EXPERIMENT PLANNING

The key to success in experimental work is to ask continually: What am I looking for? Why am I measuring this—does the measurement really answer any of my questions? What does the measurement tell me? These questions may seem rather elementary, but they should be asked frequently throughout the progress of any experimental program. Some particular questions which should be asked in the initial phases of experiment planning are:

1. What primary variables shall be investigated?
2. What *control* must be exerted on the experiment?
3. What ranges of the primary variables will be necessary to describe the phenomena under study?
4. How many data points should be taken in the various ranges of operation to ensure good sampling of data considering instrument accuracy and other factors? (See Chap. 3.)
5. What instrument accuracy is required for each measurement?
6. If a dynamic measurement is involved, what frequency response must the instruments have?
7. Are the instruments available commercially, or must they be constructed especially for the particular experiment?
8. What safety precautions are necessary if some kind of hazardous operation is involved in the experiment?
9. What financial resources are available to perform the experiment, and how do the various instrument requirements fit into the proposed budget?
10. What provisions have been made for recording the data?

The importance of control in any experiment should always be recognized. The physical principle, apparatus, or device under investigation will dictate the variables which must be controlled carefully. For example, a heat-transfer test of a particular apparatus might involve some heat loss to the surrounding air in the laboratory where the test equipment is located. Consequently, it would be wise to maintain (control) the surrounding temperature at a reasonably constant value. If one run is made with the room temperature at 90°F and another at 50°F, large unwanted effects may occur in the measurements. Or suppose a test is to be made of the effect of cigarette smoke on the eating habits of mice. Clearly, we would want to control the concentration of smoke inhaled by the mice and also observe another group of mice which were not exposed to cigarette smoke at all. All other environmental variables should be the same if we are to establish the effect of the cigarette smoke on eating habits.

In the case of the heat-transfer test we make a series of measurements of the characteristics of a device under certain specified operating conditions—no comparison with other devices is made. For the smoke test with mice it is necessary to measure the performance of the mice under specified conditions and *also* to compare this performance with the performance of another group under different controlled conditions. For the heat-transfer test we establish an absolute measurement of performance, but for the mice a *relative performance* is all that can be ascertained. We have chosen two diverse examples of absolute and relative experiments, but the lesson is clear. Whenever a comparison test is performed to establish relative performance, control must be exerted over more than one experimental setup in order for the comparison to be significant.

It would seem obvious that very careful provisions should be made to record the data and all ideas and observations concerned with the experiment. Yet, many experimenters record data and important sketches on pieces of scratch paper or in such a disorganized manner that they may be lost or thrown away. In some experiments the readout instrument is a recording type so that a record is automatically obtained and there is little chance for loss. For many experiments, however, visual observations must be made and values recorded on an appropriate data sheet. This data sheet should be very carefully planned so that it may subsequently be used, if desired, for data reduction. Frequently, much time may be saved in the reduction process by eliminating unnecessary transferal of data from one sheet to another. If a computer is to be used for data reduction, then the primary data sheet should be so designed that the data may be easily transferred to the input device of the computer. Even with digital readout systems the printout must be carefully labeled, either in the machine programming or by hand.

A bound notebook should be maintained to record sketches and significant observations of an unusual character which may occur during both the planning and the execution stages of the experiment. The notebook is also used for recording thoughts and observations of a theoretical nature as the experiment progresses. Upon the completion of the experimental program the well-kept notebook forms a clear and sequential record of the experiment planning, observations during the experiment, and, where applicable, correspondence of important observations with theoretical predictions. Every experimenter should get into the habit of keeping a good notebook. In some cases the output of various instrument transducers will be fed directly to a data acquisition system and minicomputer which processes the data and furnishes a printed output of the desired results. Then the notebook becomes the repository of important sketches and listings of computer programs, as well as documentation which may be needed later to analyze the data or repeat the experiment. Sample printouts or computer-generated graphs may be taped in the notebook along with program documentation.

As a summary of our remarks on experimental planning we present the generalized experimental procedure given in Table 2-7. This procedure is, of

Table 2-7 Generalized experimental procedure

1. *a.* Establish the need for the experiment.
   b. Establish the optimum budgetary, manpower, and time requirements, including time sequencing of the project. Modify scope of the experiment to *actual* budget, manpower, and time schedule which is allowable.

2. Begin detail planning for the experiment, clearly establish objectives of experiment (verify performance of production model, verify theoretical analysis of particular physical phenomenon, etc.). If experiments are similar to those of previous investigators, be sure to make use of experience of the previous workers. *Never* overlook the possibility that the work may have been done before and reported in the literature.

3. Continue planning by performing the following steps:
   a. Establish the primary variables which must be measured (force, strain, flow, pressure, temperature, etc.).
   b. Determine as nearly as possible the accuracy which may be required in the primary measurements and the number of such measurements which will be required for proper data analysis.
   c. Set up data reduction calculations *before* conducting the experiments to be sure that adequate data are being collected to meet the objectives of the experiment.
   d. Analyze the possible errors in the anticipated results *before* the experiments are conducted so that modifications in accuracy requirements on the various measurements may be changed if necessary.

4. Select instrumentation for the various measurements to match the anticipated accuracy requirements. Modify the instrumentation to match budgetary limitations if necessary.

5. Collect a few data points and conduct a preliminary analysis of these data to be sure that the experiment is going as planned.

6. Modify the experimental apparatus and/or procedure in accordance with the findings in item 5.

7. Collect the bulk of experimental data and analyze the results.

8. Organize, discuss, and publish the findings and results of the experiments, being sure to include information pertaining to all items 1 to 7, above.

course, a flexible one, and the reader should consider the importance of each item in relation to the entire experimental program. Notice particularly item 1*a*. The engineer should give careful thought to the *need* for the experiment. Perhaps, after some sober thinking, the engineer will decide that a previously planned experiment is really not necessary at all and that the desired information could be found from an analytical study or from the results of experiments which have already been conducted. ***Do not take this item lightly.*** A great amount of money is wasted by individuals who rush into a program only to discover later that the experiments were unnecessary for their own particular purposes.

The Role of Uncertainty Analysis in Experiment Planning

Items 3*b* and *d* in Table 2-7 note the need to perform preliminary analyses of experimental uncertainties in order to effect a proper selection of instruments and to design the apparatus to meet the overall goals of the experiment. These items are worthy of further amplification.

Recall our previous comments about the terms accuracy, error, and uncertainty. We noted that many persons use the term "error" when "uncertainty" is the proper nomenclature. As promised before, we shall clarify this matter in Chap. 3.

In Chap. 3 we shall see how one goes about estimating uncertainty in an experimental measurement. For now, let us consider how these estimates can aid our experiment planning. It is clear that certain variables we wish to measure are set by the particular experimental objectives, but there may be several choices open in the method we use to measure these variables. An electric-power measurement could be performed, for example, by measuring current and voltage and taking the product of these variables. The power might also be calculated by measuring the voltage drop across a known resistor, or possibly through some calorimetric determination of the heat dissipated from a resistor. The choice of the method used can be made on the basis of an uncertainty analysis, which indicates the relative accuracy of each method. A flow measurement might be performed by sensing the pressure drop across an obstruction meter, or possibly by counting the number of revolutions of a turbine placed in the flow (see Chap. 7). In the first case the overall uncertainty depends on the accuracy of a measurement of pressure differential and other variables, such as flow area, while in the second case the overall uncertainty depends on the accuracy of counting and a time determination.

The point is that a careful uncertainty analysis during the experiment planning period may enable the investigator to make a better selection of instruments for the program. Briefly, then, an uncertainty analysis enters into the planning phase with the following approximate steps:

1. Once the variables to be measured have been established, several alternative measurement techniques are selected.

Figure 2-11 (*a*) Preliminary stages of experiment planning; (*b*) intermediate stages of experiment planning; (*c*) final stages of experimental program.

2. An uncertainty analysis is performed on each measurement technique, taking into account the estimated accuracies of the instruments that will actually be used.
3. The different measurement techniques are then compared on the basis of cost, availability of instrumentation, ease of data collection, and calculated uncertainty. The technique with the least uncertainty is clearly the most desirable from an experimental-accuracy standpoint, but it may be too expensive. Frequently, however, the investigator will find that cost is not a strong factor and that the technique with the smallest uncertainty (within reason) is as easy to perform as some other less accurate method.

Figure 2-11 divides the procedure of Table 2-7 into a graphical pattern of preliminary, intermediate, and final stages of an experimental program. The feedback blocks in these diagrams are very important because they illustrate the need to continuously retrace one's steps and modify the program in accordance with the most current information that is available.

2-12 REPORT WRITING

Many books have been written on the subject of report writing, and the author has no intention of competing with these works. The importance of good report writing and data presentation cannot be overemphasized. No matter how good an experiment or how brilliant a discovery, it is worthless unless the information is communicated to other people. Some more specific remarks will now be made concerning item 8 in Table 2-7.

Very good advice for those engaged in writing or speaking activities is contained in the aphorism:

> A man who uses a great many words to express his meaning is like a bad marksman who instead of aiming a single stone at an object takes up a handful and throws at it in hopes he may hit.—*Samuel Johnson*

In other words, say what you have to say and then shut up! Be succinct but not laconic. When graphs or tables will present the idea clearly, use them, but do not also include a wordy explanation which tells the reader what can be plainly seen by careful inspection of the graph.

Third-person past tense is generally accepted as the most formal grammatical style for technical reports, and it is seldom incorrect to use such a style. In some instances, first person may be employed in order to emphasize a point or to stress the fact that a statement is primarily the opinion of the writer. The usual scientific writing style is also in the passive mode. Examples of the two styles are:

Third person Equation (5) is recommended for the final correlation in accordance with the limitations of the data as discussed above.

First person We (I) recommend Eq. (5) for the final correlation in accordance with the limitations of the data presented in our discussion above.

In the first-person statement, the writer is making the recommendation on a much more personal basis than in the third-person statement. The selection of the proper statement is a matter of idiom, which depends on many factors, including consideration of the person(s) who will read the report. For a formal paper in a scientific journal, the third-person statement might be preferable, whereas the first-person usage might be desirable in an engineering report to an individual.

Engineers could learn a few lessons from politicians and business executives. Have you ever noticed the difference in writing style in the annual reports of corporations? If the earnings situation is good, the point is brought forth very quickly and very clearly. Wordiness is avoided in the presentation of good points so that the reader will not miss them. When the profit statement is poor, the discussion becomes more circumlocutious and clear graphical or tabular presentations are seldom employed. The point is that a writer should make sure that the strong points of a report are not buried in excess words. The degree to which weak points are submerged depends upon the audience for which the report is intended and the circumstances under which the report is written.

The above comments should not be misconstrued. The engineer should not bury bad data in a presentation in hopes that no one will notice! On the other hand, one should recognize that some engineering reports are written for lay persons, and it is extremely important for the presentation not to give conflicting or confusing results which may be misinterpreted. Of course, a paper for publication in the technical or scientific literature should be completely objective. The purpose of these brief paragraphs has been to illustrate the fact that engineers may, in *some circumstances*, find it necessary to do a bit of selling in their writing. Such writing requires a great amount of skill and experience in order to be effective.

Be specific if you have something to be specific about. Consider the following two statements:

1. An analysis of the experimental data showed that the average deviation from the theoretical values was less than 1 percent. Uncertainties in the primary data were shown to account for a deviation of 0.5 percent. In view of this excellent agreement between Eq. (42) and the experimental data, this relation is believed to be an adequate representation of the physical phenomena and is recommended for calculation purposes.
2. The experimental data are in good agreement with the theoretical development. In view of this favorable comparison the assumptions pertaining to the derivation are verified.

Note the difference between the two statements. The first statement is quite specific and leaves the reader with a feeling of confidence in the experimental data and theoretical analysis. Upon examining the second statement the reader will immediately ask: How good is "good"; how favorable is "favorable"? The author has seen such statements applied to experimental data that differed from theoretical values by such a large factor that the veracity of the writer might be questioned by a careful reader.

A brief, general procedure for report writing might take the following form:

1. Make a written outline of the report with as much detail as possible.
2. Let the outline "cool" for a period of time while you direct your thoughts to other matters.
3. Go back to the outline and make whatever changes you feel are necessary.
4. Write the report in rough draft form as quickly as possible.
5. Let the report cool for a period of time, preferably a week or so.
6. Go back and make corrections on your draft. You will probably find that you did not say things quite the way you would like to, did not include as much information as you wanted to, or made several stupid mistakes in some of the data analysis.
7. If possible, have a colleague scrutinize the report carefully. This person should be one whose competence you respect.
8. Consider your colleague's comments very carefully, and rewrite the report in its final form.

Once a draft of the report is available, corrections must be made; Table 2-8 shows the accepted nomenclature for proofreader's marks. It is a good idea to get in the habit of using these standard marks.

Of course, the objective of a report is to communicate the ideas and information gained in the experimental work. The care and skill with which the discussion and conclusions are drawn will determine the overall success of the report.

The above remarks on report writing can be considered as general ones which may be applicable to the broad range of reports, papers, or monographs the engineer will be called upon to write. Of course, a carefully constructed outline is always the best starting point and will help the writer to make sure that all pertinent information is included. To aid in this outline construction we now give a brief discussion of different report elements. Not all these elements will be included in every report, and the writer must choose those appropriate to the audience.

Front Matter

Front matter includes the title page, with author affiliation(s), sponsor of report activity (if any), table of contents, list of nomenclature, list of tables, list of figures, preface (if any), and a letter of transmittal if required.

In many cases the letter of transmittal will be a simple interoffice memo for the company. For a major report to a project sponsor the letter may be more elaborate.

Table 2-8 Standard proofreader's marks

⤳	Delete	*le*	lowercase Word
⤳	Delete and close up	*cap.*	Capital letter
⊓	Quad (one em) space	*sc*	SMALL CAPITAL LETTER
⊔	Move down	*bf*	**Boldface type**
⊓	Move up	*ital*	*Italic type*
⊏ ⊏	Move to left	*rom*	Roman type
⊐	Move to right	*wf*	Wrong font
eq #	Equalize spacing between words	#	Insertspace
✕	Broken letter	◡	Close up
¶	Begin a new paragraph	9	Turn letter
no ¶	No new paragraph	⊙ .	Period
stet	Let type stand as set	∧ ,	Comma
⑦	Verify or supply information	⋁ '	Apostrophe
tr	Transpose letters or marked words	⋁/⋁ " "	Quotation marks
⑨	Spell out abbrev. or ⑦	j/ ;	Semicolon
⌣	Push space down	⊙ :	Colon
=	Straighten type	?/ ?	Question mark
//	Align type	!/ !	Exclamation mark
run in	Run in material	=/ -	Hyphen
	on same line	— En dash
bu	Change (x/y) to built-up fraction	— Em dash
sh	Change (x/y) to shilling fraction	— Two-em dash
⟨	Set s as subscript	(/)/ ( )	Parentheses
⟨	Set s as exponent	[/]/ [ ]	Brackets

A variety of schemes are used for organizing the table of contents. Many authors like to number every paragraph, thereby creating many subheadings. A technical paper for a journal would use no *numbered* headings and would not include a table of contents. Most book publishers do not use more than one subheading in each chapter, i.e., chapter 6 might have sections 6-1, 6-2, 6-3, etc., but *not* 6-1-1, 6-1-2, and so on.

Nomenclature lists are normally required for both journal articles and comprehensive technical reports. They may not be necessary in brief reports for a limited audience. Lists of tables and figures are frequently included in large reports, but are never used in papers for journal publication. Inclusion of a preface is optional in reports: it is normally used only in monographs (books).

Abstracts

Someone stated an old rule for army methods of communication: Tell them what you are going to tell them; tell them; and then tell them what you told them. The abstract should attempt to accomplish the first objective in a very short, succinct format, without mathematical formulations. It should tell what was done, and the conclusions which resulted from the work. Keep in mind that there are many people who will read *only* the abstract of a report because of heavy demands on their time. This is especially true of managers who must review a large number of reports on a broad range of subjects. A well-written abstract becomes particularly useful for these people, and may arouse their interest in examining the report in more detail.

Introduction

In some reports the introduction will take the place of a preface. In a very extensive report or book, both a preface and an introduction may be employed, while in brief reports neither may be used. The purpose of the introduction is to lay the groundwork for the more detailed discussions in the body of the report. Sometimes a review of previous work is given in the introduction.

Background and Previous Work

A survey of the literature may be appropriate and can be included under this title. It could also be included in an introduction.

Theoretical Presentation(s)

In some reports a large section will be devoted to development of theoretical information applicable to the subject. This section enables the reader to understand the implications of the experimental work and aids in proper interpretation of the data.

Experimental Apparatus and Procedure

Sufficient information must be supplied on the apparatus and experimental procedures for the reader to understand what was done. If the report is concerned with research and new knowledge, a rather detailed discussion of the apparatus may be necessary. If test results are being reported in accordance with standard procedures (ASME, ASTM, etc.), then the appropriate procedures can be cited without furnishing details in the report.

Results of Experiments

One normally will include a separate section in the report to give results of the experiments in a form which is consistent with the needs of the intended audience. Clear tabular and graphical presentations should be used as much as possible to conserve words in the report. Of course, some verbal discussion of the graphs and tables must be given, but such discussion should focus the reader's attention on salient features of the data, not just recite numbers or parameters which are obvious upon inspection. In other words the written and graphical presentations should be complementary so that the reader's time is conserved.

Interpretation of Results

Once the experimental results have been presented in a clear form, the author has a responsibility to interpret the results in the light of the theoretical presentation and the work of others in the same general subject area. In this section the background, theoretical presentation, and experimental results are all brought together to lead the reader to the conclusions of the study.

Conclusions

By the time the reader has reached this section, most of the conclusions of the work should already have been drawn. The object of the conclusion *section* is to collect all the important results and interpretations in clear summary form. This is the section which tells the reader what was covered in the body of the report. There will be many readers who will read only the abstract or conclusion sections of a report, so it is especially important that they be carefully written.

References and Bibliographies

Most technical reports require references to the work of others. References should be cited when a work was used in writing the report. For example, the statement:

Cheesewright [1] discusses turbulent natural convection . . .

with a citation:

Cheesewright, R.: Turbulent Natural Convection from a Vertical Plane Surface, *J. Heat Trans.*, vol. 90, p. 1, 1968.

is a valid reference citation. On the other hand, a standard textbook in heat transfer might have general information on free convection and yet not be employed as a specific reference for the report. It is common practice to list such sources under the title of Bibliography. Normally, only extensive works like textbooks or monographs have separate bibliographies. Because of the extra space consumed, bibliographies are almost never used in papers for journal publication.

Some latitude is available in the manner of literature citation. The above reference could be listed without its title, i.e.,

Cheesewright, R., *J. Heat Trans.*, vol. 90, p. 1, 1968.

Though such citations may be accepted by various journals, and do conserve space, they should be avoided whenever possible because the absence of the article title is very annoying to a reader. The object of a report is to communicate. A procedure which impedes communication is to be shunned.

Graphical Presentations

Most engineering reports include some form of graphical presentation; a simple plot of raw data, a statistical distribution of data, correlation of data, and perhaps a comparison of experimental data with analytical predictions. Regardless of the purpose of the graph, there are certain elements of good style which should be common to all.

1. Label the graph. Make sure all coordinates are labeled consistently, i.e., don't use all uppercase on one coordinate and mixed upper- and lowercase on the other.
2. If numerical values of variables are graphed, make sure that each coordinate has scale markings and that each label for a variable includes the *units* for that variable. Try to maintain the same unit system for all variables, i.e., do not mix English and SI units.
3. Recognize that some graphs may be reduced in size later because of space limitations. This is particularly true for those to be published in technical papers or symposia volumes. If any reduction is anticipated, the size of the labels and unit specifications on the graph must be increased so that they will be legible in their eventual format. For graphs prepared on computers this will be an easy matter to arrange.
4. The purpose of graphs is to convey information. Sometimes this is best accomplished by plotting several curves on one graph; sometimes not. For example, suppose that six pressure traverses are taken along the length of a wind tunnel for six different flow rates. In this case it would probably be best to include all six sets of data on one graph with each curve clearly labeled. A use of six *separate* graphs would not communicate as well and would certainly consume more space. On the other hand, suppose a heated plate is placed in the tunnel and heat transfer data collected for the same six flow rates. It would not be prudent to present these data on the same graph as the pressure data because the display would be too cluttered. A second graph would be in order. There is no hard rule to apply in these circumstances. Common sense usually works fine.
5. There is a class of presentations called *G* and *A* graphs (for *G*enerals and *A*dmirals). They are usually broad-brush and never loaded with a lot of data points or complicated mathematical expressions. The purpose of these graphs is to present general trends without precise data points, and they are normally employed for readers or viewers not very familiar with the subject. The person

responsible for this kind of presentation should normally seek professional artistic help.

Miscellaneous Helpful Hints

As a further aid to the writer we give the following bits of advice which do not fall in the categories of other sections on report writing.

1. Always give great attention to the audience for which the report is written.
2. Set up a list of symbols to be used in the report before starting to write. Then be sure to use these symbols consistently as the report is written. *This is not an insignificant item.* Many inexperienced writers make the mistake of switching symbols for a physical quantity in different sections of the report. Such practice can be very confusing to a reader and is not easily detected in proofreading. Even an experienced technical typist will have difficulty catching such mistakes. Be particularly careful about switching between uppercase and lowercase letters for the same variable.
3. In choosing symbols to be used in a report try to follow the practice used in standard textbooks or technical journals in the field. Don't go off and invent new symbols which will make the report hard for the reader to understand. As simple as this advice may sound, it is violated over and over again, even by experienced writers.
4. Once writing is started, write the report as if you had to make it correct on the first draft. Take your time. This will pay off later. Of course, unless you are very good, or very lucky, changes will have to be made, but they usually can be minimized with this approach. Avoid the procedure followed by some persons of assembling a draft consisting of notes or comments in the form of incomplete sentences. You will just have to go back and rewrite the whole thing and must still go through a correction stage. It is an old saying but still appropriate: *Try to do it right the first time.*
5. Be kind to the typist or word-processor operator. Be careful with handwriting so there is no confusion between English and Greek symbols. Here are some specific suggestions:
   (a) Use block letters for uppercase: viz.,

   U, not $\mathcal{U}$;
   E, not $\mathcal{E}$; etc.

   (b) Use script letters for lowercase: viz.,

   u, not $\cup$ or μ (mistaken for Greek "mu")
   v, not Y or v (mistaken for Greek "nu")
   k, not K (mistaken for Greek "kappa")
   h, not h
   p, not P (mistaken for Greek "rho")
   w, not W (mistaken for Greek "omega")

   (c) Be sure to make the "tails" very definite on Greek "eta" (η) and "mu" (μ).
   (d) Put "flag" to left on Greek "nu" (γ) and to right on lowercase "vee" (v).

2-13 WORD PROCESSORS AND COMPUTERS

The number of word processors available on the market ranges in the scores, with new models appearing almost daily. Two basic types of output are available: (1) dot matrix, as in a computer output and (2) letter quality, as from a daisy-wheel printer. Output of Greek and mathematical symbols is sometimes problematical because typing elements must be changed; however, we may expect this barrier to be overcome soon. Ordinary electronic typewriters with large memory storage can classify as word processors and are available at a low price.

No matter how simple or elaborate the word processor, information must be fed in by a human. This person may be the author (experimentalist) or someone who takes the author's write-up and prepares the report draft. If the report is highly technical and mathematical the input person obviously must have experience and skill in handling such material. The advantage of the word processor is that whole sections may be displayed on a screen and corrections or insertions made very easily.

As an example of the advances being made we can note that several technical textbooks have been prepared by their authors on word processors and furnished to the publishers on magnetic disks ready for processing into final book format. In addition, many technical journals now use elaborate machines to go directly from author's edited manuscript to camera-ready copy in a matter of a few minutes.

Despite the power available in modern word-processing equipment, brief, simple reports can frequently be handled quite expeditiously, and economically, by an author preparing a handwritten or typed draft for a good technical typist.

2-14 ORAL PRESENTATIONS

There are many times when an oral presentation must be made of the results of the experiments. Sometimes the persons attending this presentation will have copies of a written report and sometimes not. It is usually best to plan the presentation on the assumption that the attendees do *not* have the report or that, if they have it, they will not read the report until later. With this in mind the presentation can be planned using the following guidelines.

1. Determine the audience for the presentation. Are they likely to be familiar with background information on your topic? If not, then you should include enough remarks to emphasize the importance of the work.
2. Determine the allotted time for the presentation and *plan to stay within this time limit*. If there are to be questions or discussion then the formal presentation should end before the overall time limit is reached.
3. Use slides or overhead view graphs for the presentation. If a report has been prepared with good graphics the slides can be made from figures in the report. For less formal presentations some blackboard work may be in order.

4. Avoid the use of complicated mathematical relations. If such relations are used in the report, present only summary comments which will enable the listener to understand what was done, and for what reason, but without seeing the details.
5. Don't leave the audience hanging. Be sure to summarize the results and conclusions with some clear, concise statements.
6. In the question-and-answer session some questioners tend to speak softly and cannot be heard by other people in the audience. It is a good idea to quickly repeat the question, particularly if you are using a microphone, so that everyone will understand your response.
7. Finally, a courteous "thank you" is in order at the close of the presentation.

2-15 PLANNING SESSIONS AND CONFERENCES

Seldom does an experiment involve only one person. In many cases a very large number of people will be involved, and it will be necessary to get everyone together in some kind of conference. It is not an overstatement to say that a great amount of time is wasted in conferences because the sessions are not carefully planned or administered. The following remarks should provide some help in making more efficient use of the time devoted to conferences or meetings.

1. Establish the *need* for the meeting, and tell the attendees in advance. There can be many purposes for the conference, such as to brief all persons on the overall scope of the project, to come to a decision on equipment purchases, to reconcile experimental data, to divide and assign individual responsibilities, etc. The important point is that the purpose(s) be established in advance of the meeting. Don't just get everyone together "to have a meeting."
2. Decide in advance *who* should attend the meeting. Don't involve people needlessly and waste their time. For example, you don't need anyone from the computer center to decide on a pump purchase, but you may want to consult some experienced shop people. On the other hand, a meeting to discuss data reduction and analysis might properly involve some computer people.
3. Schedule the meeting and anticipate the total time needed. *Plan to stay within that time limit.*
4. If the meeting is to be rather long, establish an agenda with a certain amount of time allotted for each item. Be reasonable in setting these time limits. If there are five agenda items this does not mean that each will consume one-fifth of the total meeting time. The idea is to *schedule* discussion. Don't let it just happen.
5. Have action items to be accomplished, such as assignment of responsibilities either on an individual or subgroup basis, time schedules for reporting, or other specific tasks.
6. If follow-up meetings are necessary, schedule a time for that meeting which fits the schedule of all participants.

7. As in a report, it is a good idea to summarize the results of the meeting at the close in case there has been any misunderstandings on someone's part.

The purpose of the above remarks has not been to overly complicate the task of "holding a meeting." There is some skill to the task, though. The present writer has seen a skillful organizer accomplish more in a one-hour session than some blunderer might achieve in an all-day meeting.

PLANNING PROJECTS

1. Make an experimental plan for conducting a series of tests on a home stereo amplifier. Consider the following:
   (a) What measurements must be performed?
   (b) What measures of performance must be obtained?
   (c) What kind of instruments are needed, and what accuracy is necessary?
   (d) Make an outline for a report to be written based on the tests. In this outline show what information will be presented in each section of the report.
2. Make an experimental plan for testing a low-speed moped having an engine displacement of about 50 cc and a top speed of 35 mi/h. Consider the same factors as in project 1.
3. Make an experimental plan to test a home chain saw. Consider the same factors as in project 1.
4. Make an experimental plan to test a radar speed-measurement device used by the state police to detect highway speeders. Consider the same factors as in project 1.

REVIEW QUESTIONS

2-1 What is meant by sensitivity; accuracy; precision?

2-2 Why is instrument calibration necessary?

2-3 Why are standards necessary?

2-4 What is meant by frequency response?

2-5 Describe the meaning of phase shift.

2-6 Define time constant.

2-7 What kind of impedance matching is desired for (a) maximum power transmission and (b) minimum influence on the output of the system?

2-8 Why is a literature survey important in the preliminary stages of experiment planning?

2-9 Why is an uncertainty analysis important in the preliminary stages of experiment planning?

2-10 How can an uncertainty analysis help to reduce overall experimental uncertainty?

PROBLEMS

2-1 Consider an ordinary mercury-in-glass thermometer as a measurement system, and indicate which parts of the thermometer correspond to the boxes in the diagram of Fig. 2-2.

2-2 A thermometer is used for the range of 200 to 400°F, and it is stated that its accuracy is one-quarter of 1 percent. What does this mean in terms of temperature?

2-3 A sinusoidal forcing function is impressed on the system in Fig. 2-4. The natural frequency is 100 Hz, and the damping ratio c/c_c is 0.7. Calculate the amplitude ratio and time lag of the system for an input frequency of 40 Hz. (The time lag is the time interval between the maximum force input and maximum displacement output.)

2-4 For a natural frequency of 100 Hz and a damping ratio of 0.7, compute the input-frequency range for which the system in Fig. 2-4 will have an amplitude ratio of 1.00 ± 0.01.

2-5 A thermometer is initially at a temperature of 70°F and is suddenly placed in a liquid which is maintained at 300°F. The thermometer indicates 200 and 270°F after time intervals of 3 and 5 s, respectively. Estimate the time constant for the thermometer.

2-6 Plot the power output of the circuit in Fig. 2-10 as a function of R/R_i. Assume R_i and E as constants and show the plot in dimensionless form; i.e., use PR_i/E^2 as the ordinate for the curve.

2-7 A steel scale is graduated in increments of $\frac{1}{32}$ in. What is the readability and least count of such a scale?

2-8 A 10-μF capacitor is charged to a potential of 100 V. At time zero it is discharged through a 1-MΩ resistor. What is the time constant for this system?

2-9 The two-terminal device shown in Fig. 2-10 has an internal resistance of 5000 Ω. A meter with an impedance of 20,000 Ω is connected to the output to perform a voltage measurement. What is the percent error in determination of the internal voltage?

2-10 The device in Prob. 2-9 has an internal voltage of 100 V. Calculate the power output for the loading conditions indicated. What would be the maximum power output? What power output would result for a load resistance of 1000 Ω?

2-11 A 2-g mass is suspended from a simple spring. The deflection caused by this mass is 0.5 cm. What is the natural frequency of the system?

2-12 A 10-lbm turntable is placed on rubber supports such that a deflection of 0.25 in results from the weight. Calculate the natural frequency of such a system.

2-13 A resistance-capacitance system is to be designed with a time constant of 1 ms. Specify several combinations of R and C to accomplish this, and based on commercially available items obtain the cost of each combination.

2-14 What is the minimum amplitude reduction to be expected for the system in Prob. 2-12 for frequencies greater than 20, 40, and 60 Hz?

2-15 A small tank contains 10 liters of water at 20°C which is allowed to discharge from an opening in the side at the initial rate of 6 liters/h. The discharge rate is directly proportional to the volume of water remaining in the tank. Calculate an equivalent time constant for this system.

2-16 Pressure is measured in units of lbf/in^2 in the English system of units. Derive factors to convert to units of N/m^2 (Pa) and kilopond/cm^2.

2-17 One gallon equals 231 in^3. Derive a conversion factor to convert automobile fuel economy from miles/gallon to kilometers/liter.

2-18 A unit for viscosity in the English system is lbf $\cdot$ s/ft^2. Determine a factor to convert this to kg/m $\cdot$ s.

2-19 A unit for specific heat in the SI system is kJ/kg $\cdot$ °C. Derive a factor to convert this to Btu/lbm $\cdot$ °F, and to kcal/g $\cdot$ °C.

2-20 Derive a factor to convert density from g/m^3 to slugs/ft^3.

2-21 An English unit for thermal conductivity is Btu/hr $\cdot$ ft $\cdot$ °F. Derive a factor to convert to ergs/cm $\cdot$ °C.

2-22 A unit of kinematic viscosity in the metric system is the Stoke (St), defined as 1.0 cm^2/s. Derive a factor to convert to ft^2/s.

2-23 The SI unit for heat generation is W/m^3. Derive a factor to convert to Btu/h $\cdot$ ft^3.

2-24 A unit for dynamic viscosity in the metric system is the Poise (P) = 1.0 dyn $\cdot$ s/cm^2. Derive a factor to convert to lbm/h $\cdot$ ft.

REFERENCES

1. Silsbee, F. B.: Extension and Dissemination of the Electrical and Magnetic Units by the National Bureau of Standards, *Natl. Bur. Std. (U.S.), Circ.* 531, 1952.
2. International Practical Temperature Scale of 1968, *Metrologia*, vol. 5, no. 2, April 1969.
3. Eshbach, O. N.: "Handbook of Engineering Fundamentals," 2d ed., John Wiley & Sons, Inc., New York, 1952.
4. Holman, J. P.: "Thermodynamics," 3d ed., McGraw-Hill Book Company, New York, 1980.
5. Constant, F. W.: "Theoretical Physics," vol. 2, Addison-Wesley Publishing Company, Inc., Reading, Mass., 1958.
6. Stimson, H. F.: The International Temperature Scale of 1948, *J. Res. Natl. Bur. Std.* (paper 1962), vol. 42, p. 211, March 1949.
7. *NBS Tech. News Bull.*, vol. 52, no. 1, p. 10, January 1968.
8. Beckwith, T. G., and N. L. Buck: "Mechanical Measurements," 2d ed., Addison-Wesley Publishing Company, Inc., Reading, Mass., 1969.
9. Schenck, H., Jr.: "An Introduction to the Engineering Research Project," McGraw-Hill Book Company, New York, 1969.
10. Doebelin, E. O.: "Measurement Systems: Analysis and Design," 2d ed., McGraw-Hill Book Company, New York, 1981.
11. Time Standards, *Instr. Control Systems*, p. 87, October 1965.
12. Turabian, K. L.: "A Manual for Writers," 4th ed., University of Chicago Press, Chicago, 1973.
13. Wilson, E. B.: "An Introduction to Scientific Research," McGraw-Hill Book Company, New York, 1952.

THREE

ANALYSIS OF EXPERIMENTAL DATA

3-1 INTRODUCTION

Some form of analysis must be performed on all experimental data. The analysis may be a simple verbal appraisal of the test results, or it may take the form of a complex theoretical analysis of the errors involved in the experiment and matching of the data with fundamental physical principles. Even new principles may be developed in order to explain some unusual phenomenon. Our discussion in this chapter will consider the analysis of data to determine errors, precision, and general validity of experimental measurements. The correspondence of the measurements with physical principles is another matter, quite beyond the scope of our discussion. Some methods of graphical data presentation will also be discussed. The interested reader should consult the monograph by Wilson [4] for many interesting observations concerning correspondence of physical theory and experiment.

The experimentalist should always know the validity of data. The automobile test engineer must know the accuracy of the speedometer and gas gage in order to express the fuel-economy performance with confidence. A nuclear engineer must know the accuracy and precision of many instruments just to make some simple radioactivity measurements with confidence. In order to specify the performance of an amplifier, an electrical engineer must know the accuracy with which the appropriate measurements of voltage, distortion, etc., have been conducted. Many considerations enter into a final determination of the validity of the results of experimental data, and we wish to present some of these considerations in this chapter.

Errors will creep into all experiments regardless of the care which is exerted. Some of these errors are of a random nature, and some will be due to gross blunders on the part of the experimenter. Bad data due to obvious blunders may be discarded immediately. But what of the data points that just "look" bad? We cannot throw out data because they do not conform with our hopes and expectations unless we see something obviously wrong. If such "bad" points fall outside the range of normally expected random deviations, they may be discarded on the basis of some consistent statistical data analysis. The keyword here is "consistent." The elimination of data points must be consistent and should not be dependent on human whims and bias based on what "ought to be." In many instances it is very difficult for the individual to be consistent and unbiased. The pressure of a deadline, disgust with previous experimental failures, and normal impatience all can influence rational thinking processes. However, the competent experimentalist will strive to maintain consistency in the primary data analysis. Our objective in this chapter is to show how one may go about maintaining this consistency.

3-2 CAUSES AND TYPES OF EXPERIMENTAL ERRORS

In this section we present a discussion of some of the types of errors that may be present in experimental data and begin to indicate the way these data may be handled. First, let us distinguish between single-sample and multisample data.

Single-sample data are those in which some uncertainties may not be discovered by repetition. Multisample data are obtained in those instances where enough experiments are performed so that the reliability of the results can be assured by statistics. Frequently, cost will prohibit the collection of multisample data, and the experimenter must be content with single-sample data and prepared to extract as much information as possible from such experiments. The reader should consult Refs. [1] and [4] for further discussions on this subject, but we state a simple example at this time. If one measures pressure with a pressure gage and a single instrument is the only one used for the entire set of observations, then some of the error that is present in the measurement will be sampled only once no matter how many times the reading is repeated. Consequently, such an experiment is a single-sample experiment. On the other hand, if more than one pressure gage is used for the same total set of observations, then we might say that a multisample experiment has been performed. The *number* of observations will then determine the success of this multisample experiment in accordance with accepted statistical principles.

An experimental error is an experimental error. If the experimenter knew what the error was, he or she would correct it and it would no longer be an error. In other words, the real errors in experimental data are those factors that are always vague to some extent and carry some amount of uncertainty. Our task is to determine just how uncertain a particular observation may be and to devise a consistent way of specifying the uncertainty in analytical form. A reasonable definition of experimental uncertainty may be taken as the *possible value* the error may have. This uncertainty may vary a great deal depending upon the circumstances of the experiment. Perhaps it is better to speak of experimental uncertainty instead of experimental error because the magnitude of an error is always uncertain. Both terms are used in practice, however, so the reader should be familiar with the meaning attached to the terms and the ways that they relate to each other.

It is very common for people to speak of experimental errors when the correct terminology should be "uncertainty." Because of this common usage, we ask that the reader accept the faulty semantics when they occur and view each term in its proper context.

At this point we may mention some of the types of errors that may cause uncertainty in an experimental measurement. First, there can always be those gross blunders in apparatus or instrument construction, which may invalidate the data. Hopefully, the careful experimenter will be able to eliminate most of these errors. Second, there may be certain *fixed errors*, which will cause repeated readings to be in error by roughly the same amount but for some unknown reason. These fixed errors are sometimes called *systematic errors*. Third, there are the *random errors*, which may be caused by personal fluctuations, random electronic fluctuations in the apparatus or instruments, various influences of friction, etc. These random errors usually follow a certain statistical distribution, *but not always*. In many instances it is very difficult to distinguish between fixed errors and random errors.

The experimentalist may sometimes use theoretical methods to estimate the magnitude of a fixed error. For example, consider the measurement of the temperature of a hot gas stream flowing in a duct with a mercury-in-glass thermometer. It is well known that heat may be conducted from the stem of the thermometer, out of the glass body, and into the surroundings. In other words, the fact that part of the thermometer is exposed to the surroundings at a temperature different from the gas temperature to be measured may influence the temperature of the stem of the thermometer. There is a heat flow from the gas to the stem of the thermometer, and, consequently, the temperature of the stem must be lower than that of the hot gas. Therefore, the temperature we read on the thermometer is not the true temperature of the gas, and it will not make any difference how many readings are taken—we shall always have an error resulting from the heat-transfer condition of the stem of the thermometer. This is a *fixed* error, and its magnitude may be estimated with theoretical calculations based upon known thermal properties of the gas and the glass thermometer.

3-3 ERROR ANALYSIS ON A COMMONSENSE BASIS

We have already noted that it is somewhat more explicit to speak of experimental uncertainty rather than experimental error. Suppose that we have satisfied ourselves with the uncertainty in some basic experimental measurements, taking into consideration such factors as instrument accuracy, competence of the people using the instruments, etc. Eventually, the primary measurements must be combined to calculate a particular result that is desired. We shall be interested in knowing the uncertainty in the final result due to the uncertainties in the primary measurements. This may be done by a commonsense analysis of the data, which may take many forms. One rule of thumb that could be used is that the error in the result is equal to the maximum error in any parameter used to calculate the result. Another commonsense analysis would combine all the errors in the most detrimental way in order to determine the maximum error in the final result. Consider the calculation of electric power from

$$P = EI$$

where E and I are measured as

$$E = 100 \text{ V} \pm 2 \text{ V}$$

$$I = 10 \text{ A} \pm 0.2 \text{ A}$$

The nominal value of the power is $100 \times 10 = 1000$ W. By taking the worst possible variations in voltage and current, we could calculate

$$P_{max} = (100 + 2)(10 + 0.2) = 1040.4 \text{ W}$$

$$P_{min} = (100 - 2)(10 - 0.2) = 960.4 \text{ W}$$

Thus, using this method of calculation, the uncertainty in the power is $+4.04$ percent, -3.96 percent. It is quite unlikely that the power would be in error by these amounts because the voltmeter variations would probably not correspond with the ammeter variations. When the voltmeter reads an extreme "high," there is no reason why the ammeter must also read an extreme "high" at that particular instant; indeed, this combination is most unlikely.

The simple calculation applied to the electric-power equation above is a useful way of inspecting experimental data to determine what errors *could* result in a final calculation; however, the test is too severe and should be used only for rough inspections of data. It is significant to note, however, that if the results of the experiments appear to be in error by *more* than the amounts indicated by the above calculation, then the experimenter had better examine the data more closely. In particular, the experimenter should look for certain fixed errors in the instrumentation, which may be eliminated by applying either theoretical or empirical corrections.

As another example we might conduct an experiment where heat is *added* to a container of water. If our temperature instrumentation should indicate a *drop* in temperature of the water, our good sense would tell us that something is wrong and the data point(s) should be thrown out. No sophisticated analysis procedures are necessary to discover this kind of error.

The term "common sense" has many connotations and means different things to different people. In the brief example given above, it is intended as a quick and expedient vehicle, which may be used to examine experimental data and results for gross errors and variations. In subsequent sections we shall present methods for determining experimental uncertainties in a more precise manner.

3-4 UNCERTAINTY ANALYSIS

A more precise method of estimating uncertainty in experimental results has been presented by Kline and McClintock [1]. The method is based on a careful specification of the uncertainties in the various primary experimental measurements. For example, a certain pressure reading might be expressed as

$$p = 100 \text{ kN/m}^2 \pm 1 \text{ kN/m}^2$$

When the plus or minus notation is used to designate the uncertainty, the person making this designation is stating the degree of accuracy with which he or she *believes* the measurement has been made. We may note that this specification is in itself uncertain because the experimenter is naturally uncertain about the accuracy of these measurements.

If a very careful calibration of an instrument has been performed recently, with standards of very high precision, then the experimentalist will be justified in assigning a much lower uncertainty to measurements than if they were performed with a gage or instrument of unknown calibration history.

To add a further specification of the uncertainty of a particular measurement, Kline and McClintock propose that the experimenter specify certain odds for the uncertainty. The above equation for pressure might thus be written

$$p = 100 \text{ kN/m}^2 \pm 1 \text{ kN/m}^2 \text{ (20 to 1)}$$

In other words, the experimeter is willing to bet with 20 to 1 odds that the pressure measurement is within ± 1 kN/m^2. It is important to note that the specification of such odds can *only* be made by the experimenter based on the total laboratory experience.

Suppose a set of measurements is made and the uncertainty in each measurement may be expressed with the same odds. These measurements are then used to calculate some desired result of the experiments. We wish to estimate the uncertainty in the calculated result on the basis of the uncertainties in the primary measurements. The result R is a given function of the independent variables $x_1, x_2, x_3, \ldots, x_n$. Thus,

$$R = R(x_1, x_2, x_3, \ldots, x_n) \tag{3-1}$$

Let w_R be the uncertainty in the result and $w_1, w_2, \ldots, w_n$ be the uncertainties in the independent variables. If the uncertainties in the independent variables are all given with the same odds, then the uncertainty in the result having these odds is given in Ref. [1] as

$$w_R = \left[\left(\frac{\partial R}{\partial x_1} w_1 \right)^2 + \left(\frac{\partial R}{\partial x_2} w_2 \right)^2 + \cdots + \left(\frac{\partial R}{\partial x_n} w_n \right)^2 \right]^{1/2} \tag{3-2}$$

If this relation is applied to the electric power relation of the previous section, the expected uncertainty is 2.83 percent instead of 4.04 percent.

Example 3-1 The resistance of a certain size of copper wire is given as

$$R = R_0[1 + \alpha(T - 20)]$$

where $R_0 = 6\ \Omega \pm 0.3$ percent is the resistance at 20°C, $\alpha = 0.004°C^{-1} \pm 1$ percent is the temperature coefficient of resistance, and the temperature of the wire is $T = 30 \pm 1°C$. Calculate the resistance of the wire and its uncertainty.

SOLUTION The nominal resistance is

$$R = (6)[1 + (0.004)(30 - 20)] = 6.24\ \Omega$$

The uncertainty in this value is calculated by applying Eq. (3-2). The various terms are:

$$\frac{\partial R}{\partial R_0} = 1 + \alpha(T - 20) = 1 + (0.004)(30 - 20) = 1.04$$

$$\frac{\partial R}{\partial \alpha} = R_0(T - 20) = (6)(30 - 20) = 60$$

$$\frac{\partial R}{\partial T} = R_0\alpha = (6)(0.004) = 0.024$$

$$w_{R_0} = (6)(0.003) = 0.018\ \Omega$$

$$w_\alpha = (0.004)(0.01) = 4 \times 10^{-5}\ °C^{-1}$$

$$w_T = 1°C$$

Thus, the uncertainty in the resistance is

$$w_R = [(1.04)^2(0.018)^2 + (60)^2(4 \times 10^{-5})^2 + (0.024)^2(1)^2]^{1/2}$$
$$= 0.0305\ \Omega \text{ or } 0.49\%$$

Particular notice should be given to the fact that the uncertainty propagation in the result w_R predicted by Eq. (3-2) depends on the squares of the uncertainties in the independent variables w_n. This means that if the uncertainty in one variable is significantly larger than the uncertainties in the other variables, say, by a factor of 5 or 10, then it is the largest uncertainty that predominates and the others may probably be neglected.

To illustrate, suppose there are three variables with a product of sensitivity and uncertainty $[(\partial R/\partial x)w_x]$ of magnitude 1, and one variable with a magnitude of 5. The uncertainty in the result would be

$$(5^2 + 1^2 + 1^2 + 1^2)^{1/2} = \sqrt{28} = 5.29$$

The importance of this brief remark concerning the relative magnitude of uncertainties is evident when one considers the design of an experiment, procurement of instrumentation, etc. Very little is gained by trying to reduce the "small" uncertainties. Because of the square propagation it is the "large" ones that predominate, and any improvement in the overall experimental result must be achieved by improving the instrumentation or experimental technique connected with these relatively large uncertainties. In the examples and problems that follow, both in this chapter and throughout the book, the reader should always note the relative effect of uncertainties in primary measurements on the final result.

In Sec. 2-11 (Table 2-7) the reader was cautioned to examine possible experimental errors *before* the experiment is conducted. Equation (3-2) may be used very effectively for such analysis, as we shall see in the sections and chapters that follow. A further word of caution may be added here. It is equally as unfortunate to overestimate uncertainty as to underestimate it. An underestimate gives false security, while an overestimate may make one discard important results, miss a real effect, or buy much too expensive instruments. The purpose of this chapter is to indicate some of the methods for obtaining reasonable estimates of experimental uncertainty.

In the previous discussion of experimental planning we noted that an uncertainty analysis may aid the investigator in selecting alternative methods to measure a particular experimental variable. It may also indicate how one may improve the overall accuracy of a measurement by attacking certain critical variables in the measurement process. The next three examples illustrate these points.

Example 3-2 A resistor has a nominal stated value of 10 $\Omega \pm 1$ percent. A voltage is impressed on the resistor, and the power dissipation is to be calculated in two different ways: (1) from $P = E^2/R$ and (2) from $P = EI$. In (1) only a voltage measurement will be made, while both current and voltage will be measured in (2). Calculate the uncertainty in the power determination in each case when the measured values of E and I are:

$$E = 100 \text{ V} \pm 1\% \quad \text{(for both cases)}$$

$$I = 10 \text{ A} \pm 1\%$$

Figure Example 3-2 Power measurement across a resistor.

SOLUTION The schematic is shown in the accompanying figure. For the first case we have

$$\frac{\partial P}{\partial E} = \frac{2E}{R} \qquad \frac{\partial P}{\partial R} = -\frac{E^2}{R^2}$$

and we apply Eq. (3-2) to give

$$w_P = \left[\left(\frac{2E}{R} \right)^2 w_E^2 + \left(-\frac{E^2}{R^2} \right)^2 w_R^2 \right]^{1/2} \qquad (a)$$

Dividing by $P = E^2/R$ gives

$$\frac{w_P}{P} = \left[4 \left(\frac{w_E}{E} \right)^2 + \left(\frac{w_R}{R} \right)^2 \right]^{1/2} \qquad (b)$$

Inserting the numerical values for uncertainty,

$$\frac{w_P}{P} = [4(0.01)^2 + (0.01)^2]^{1/2} = 2.236\%$$

For the second case we have

$$\frac{\partial P}{\partial E} = I \qquad \frac{\partial P}{\partial I} = E$$

and after similar algebraic manipulation, we obtain

$$\frac{w_P}{P} = \left[\left(\frac{w_E}{E} \right)^2 + \left(\frac{w_I}{I} \right)^2 \right]^{1/2} \qquad (c)$$

Inserting the numerical values of uncertainty,

$$\frac{w_P}{P} = [(0.01)^2 + (0.01)^2]^{1/2} = 1.414\%$$

Thus, the second method of power determination provides considerably less uncertainty than the first method, even though the primary uncertainties in each quantity are the same. In this example the utility of the uncertainty analysis is that it affords the individual a basis for *selection of a measurement method* to produce a result with less uncertainty.

Example 3-3 The power measurement in Example 3-2 is to be conducted by measuring voltage and current across the resistor with the circuit shown in the accompanying figure. The voltmeter has an internal resistance R_m, and the value of R is known only approximately. Calculate the nominal value of the power dissipated in R and the uncertainty for the following conditions:

$$R = 100 \ \Omega \qquad \text{(not known exactly)}$$
$$R_m = 1000 \ \Omega \pm 5\%$$
$$I = 5 \ \text{A} \pm 1\%$$
$$E = 500 \ \text{V} \pm 1\%$$

Figure Example 3-3 Effect of meter impedance on measurement.

SOLUTION A current balance on the circuit yields

$$I_1 + I_2 = I$$

$$\frac{E}{R} + \frac{E}{R_m} = 1$$

and

$$I_1 = I - \frac{E}{R_m} \qquad (a)$$

The power dissipated in the resistor is

$$P = EI_1 = EI - \frac{E^2}{R_m} \qquad (b)$$

The nominal value of the power is thus calculated as

$$P = (500)(5) - \frac{500^2}{1000} = 2250 \text{ W}$$

In terms of known quantities the power has the functional form $P = f(E; I, R_m)$, and so we form the derivatives

$$\frac{\partial P}{\partial E} = I - \frac{2E}{R_m} \qquad \frac{\partial P}{\partial I} = E$$

$$\frac{\partial P}{\partial R_m} = \frac{E^2}{R_m^2}$$

The uncertainty for the power is now written as

$$w_P = \left[\left(I - \frac{2E}{R_m} \right)^2 w_E^2 + E^2 w_I^2 + \left(\frac{E^2}{R_m^2} \right)^2 w_{R_m}^2 \right]^{1/2} \qquad (c)$$

Inserting the appropriate numerical values gives

$$w_P = \left[\left(5 - \frac{1000}{1000} \right)^2 5^2 + (25 \times 10^4)(25 \times 10^{-4}) + \left(25 \times \frac{10^4}{10^6} \right)^2 (2500) \right]^{1/2}$$

$$= [16 + 25 + 6.25]^{1/2}(5)$$

$$= 34.4 \text{ W}$$

or

$$\frac{w_P}{P} = \frac{34.4}{2250} = 1.53\%$$

In order of influence on the final uncertainty in the power we have

1. Uncertainty of current determination
2. Uncertainty of voltage measurement
3. Uncertainty of knowledge of internal resistance of voltmeter

There are other conclusions we can draw from this example. The relative influence of the experimental quantities on the overall power determination is noted above. But this listing may be a bit misleading in that it implies that the uncertainty of the meter impedance does not have a large effect on the final uncertainty in the power determination. This results from the fact that $R_m \gg R$ ($R_m = 10R$). If the meter impedance were lower, say, 200 Ω, we would find that it was a dominant factor in the overall uncertainty. For a very *high* meter impedance there would be little influence, even with a very inaccurate knowledge of the exact value of R_m. Thus, we are led to the simple conclusion that we need not worry too much about the precise value of the internal impedance of the meter as long as it is very large compared with the resistance we are measuring the voltage across. This fact should influence *instrument selection* for a particular application.

Example 3-4 A certain obstruction-type flowmeter (orifice, venturi, nozzle), shown in the accompanying figure, is used to measure the flow of air at low velocities. The relation describing the flow rate is

$$\dot{m} = CA \left[\frac{2g_c p_1}{R T_1} (p_1 - p_2) \right]^{1/2} \tag{a}$$

where C is an empirical-discharge coefficient, A is the flow area, p_1 and p_2 are the upstream and downstream pressures, T_1 is the upstream temperature, and

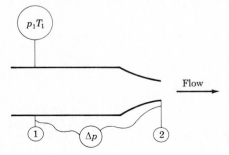

Figure Example 3-4 Uncertainty in a flowmeter.

R is the gas constant for air. Calculate the percent uncertainty in the mass flow rate for the following conditions:

$$C = 0.92 \pm 0.005 \quad \text{(from calibration data)}$$

$$p_1 = 25 \text{ psia} \pm 0.5 \text{ psia}$$

$$T_1 = 70°\text{F} \pm 2°\text{F} \qquad T_1 = 530°\text{R}$$

$$\Delta p = p_1 - p_2 = 1.4 \text{ psia} \pm 0.005 \text{ psia} \qquad \text{(measured directly)}$$

$$A = 1.0 \text{ in}^2 \pm 0.001 \text{ in}^2$$

SOLUTION In this example the flow rate is a function of several variables, each subject to an uncertainty.

$$\dot{m} = f(C, A, p_1, \Delta p, T_1) \tag{b}$$

Thus, we form the derivatives:

$$\frac{\partial \dot{m}}{\partial C} = A \left(\frac{2g_c p_1}{RT_1} \Delta p \right)^{1/2}$$

$$\frac{\partial \dot{m}}{\partial A} = C \left(\frac{2g_c p_1}{RT_1} \Delta p \right)^{1/2}$$

$$\frac{\partial \dot{m}}{\partial p_1} = 0.5 CA \left(\frac{2g_c}{RT_1} \Delta p \right)^{1/2} p_1^{-1/2} \tag{c}$$

$$\frac{\partial \dot{m}}{\partial \Delta p} = 0.5 CA \left(\frac{2g_c p_1}{RT_1} \right)^{1/2} \Delta p^{-1/2}$$

$$\frac{\partial \dot{m}}{\partial T_1} = -0.5 CA \left(\frac{2g_c p_1}{R} \Delta p \right)^{1/2} T_1^{-3/2}$$

The uncertainty in the mass flow rate may now be calculated by assembling these derivatives in accordance with Eq. (3-2). Designating this assembly as Eq. (c) and then dividing by Eq. (a) gives

$$\frac{w_{\dot{m}}}{\dot{m}} = \left[\left(\frac{w_C}{C} \right)^2 + \left(\frac{w_A}{A} \right)^2 + \frac{1}{4} \left(\frac{w_{p_1}}{p_1} \right)^2 + \frac{1}{4} \left(\frac{w_{\Delta p}}{\Delta p} \right)^2 + \frac{1}{4} \left(\frac{w_{T_1}}{T_1} \right)^2 \right]^{1/2} \tag{d}$$

We may now insert the numerical values for the quantities to obtain the percent uncertainty in the mass flow rate.

$$\frac{w_{\dot{m}}}{\dot{m}} = \left[\left(\frac{0.005}{0.92} \right)^2 + \left(\frac{0.001}{1.0} \right)^2 + \frac{1}{4} \left(\frac{0.5}{25} \right)^2 + \frac{1}{4} \left(\frac{0.005}{1.4} \right)^2 + \frac{1}{4} \left(\frac{2}{530} \right)^2 \right]^{1/2}$$

$$= [29.5 \times 10^{-6} + 1.0 \times 10^{-6} + 1.0 \times 10^{-4} + 3.19 \times 10^{-6}$$

$$+ 3.57 \times 10^{-6}]^{1/2}$$

$$= [1.373 \times 10^{-4}]^{1/2} = 1.172\% \tag{e}$$

The main contribution to uncertainty is the p_1 measurement with its basic uncertainty of 2 percent. Thus, to improve the overall situation the accuracy of this measurement should be attacked first. In order of influence on the flow-rate uncertainty, we have

1. Uncertainty in p_1 measurement (± 2 percent)
2. Uncertainty in value of C
3. Uncertainty in determination of T_1
4. Uncertainty in determination of Δp
5. Uncertainty in determination of A

By inspecting Eq. (*e*) we see that the first two items make practically the whole contribution to uncertainty. The value of the uncertainty analysis in this example is that it shows the investigator how to improve the overall measurement accuracy of this technique. First, obtain a more precise measurement of p_1. Then try to obtain a better calibration of the device, i.e., a better value of C. In Chap. 7 we shall see how values of the discharge coefficient C are obtained.

3-5 EVALUATION OF UNCERTAINTIES FOR COMPLICATED DATA REDUCTION

We have seen in the preceding discussion and examples how uncertainty analysis can be a useful tool to examine experimental data. In many cases data reduction is a rather complicated affair and is often performed with a computer routine written specifically for the task. A small adaptation of the routine can provide for direct calculation of uncertainties without resorting to an analytical determination of the partial derivatives in Eq. (3-2). We still assume that this equation applies, although it could involve several computational steps. We also assume that we are able to obtain estimates by some means of the uncertainties in the primary measurements, i.e., w_1, w_2, etc.

Suppose a set of data is collected in the variables $x_1, x_2, \ldots, x_n$ and a result calculated. At the same time one may perturb the variables by $\Delta x_1, \Delta x_2$, and so on, and calculate new results. We would have

$$R(x_1) = R(x_1, x_2, \ldots, x_n)$$
$$R(x_1 + \Delta x_1) = R(x_1 + \Delta x_1, x_2, \ldots, x_n)$$

$$R(x_2) = R(x_1, x_2, \ldots, x_n)$$
$$R(x_2 + \Delta x_2) = R(x_1, x_2 + \Delta x_2, \ldots, x_n)$$

For small enough values of Δx the partial derivatives can be well approximated by

$$\frac{\partial R}{\partial x_1} \simeq \frac{R(x_1 + \Delta x_1) - R(x_1)}{\Delta x_1}$$

$$\frac{\partial R}{\partial x_2} \simeq \frac{R(x_2 + \Delta x_2) - R(x_2)}{\Delta x_2}$$

and these values could be inserted in Eq. (3-2) to calculate the uncertainty in the result.

At this point we must again alert the reader to the ways uncertainties or errors of instruments are normally specified. Suppose a pressure gage is available and the manufacturer states that it is accurate within ± 1.0 percent. This statement normally refers to *percent of full scale*. So a gage with a range of 0 to 100 kPa would have an uncertainty of ± 10 percent when reading a pressure of only 10 kPa. Of course, this means that the uncertainty in the calculated result, either as an absolute value or percentage, can vary widely depending on the range of operation of instruments used to make the primary measurements. The above procedure can be used to advantage in complicated data-reduction schemes.

A very full description of this technique and many other considerations of uncertainty analysis are given by Moffat [7].

Example 3-5 Calculate the uncertainty of the wire resistance in Example 3-1 using the technique of this section.

SOLUTION In Example 3-1 we have already calculated the nominal resistance as $6.24 \, \Omega$. We now perturb the three variables R_0, α, and T by small amounts to evaluate the partial derivatives. We shall take

$$\Delta R_0 = 0.01 \qquad \Delta \alpha = 1 \times 10^{-5} \qquad \Delta T = 0.1$$

Then

$$R(R_0 + \Delta R_0) = (6.01)[1 + (0.004)(30 - 20)] = 6.2504$$

and the derivative is approximated as

$$\frac{\partial R}{\partial R_0} \approx \frac{R(R_0 + \Delta R_0) - R}{\Delta R_0} = \frac{6.2504 - 6.24}{0.01} = 1.04$$

or the same result as in Example 3-1. Similarly,

$$R(\alpha + \Delta \alpha) = (6.0)[1 + (0.00401)(30 - 20)] = 6.2406$$

$$\frac{\partial R}{\partial \alpha} \approx \frac{R(\alpha + \Delta \alpha) - R}{\Delta \alpha} = \frac{6.2406 - 6.24}{1 \times 10^{-5}} = 60$$

$$R(T + \Delta T) = (6)[1 + (0.004)(30.1 - 20)] = 6.2424$$

$$\frac{\partial R}{\partial T} \approx \frac{R(T + \Delta T) - R}{\Delta T} = \frac{6.2424 - 6.24}{0.1} = 0.24$$

All the derivatives are the same as in Example 3-1 so the uncertainty in R would be the same, or 0.0305 Ω.

3-6 STATISTICAL ANALYSIS OF EXPERIMENTAL DATA

We shall not be able to give an extensive presentation of the methods of statistical analysis of experimental data; we may only indicate some of the more important methods currently employed. First, it is important to define some pertinent terms.

When a set of readings of an instrument is taken, the individual readings will vary somewhat from each other, and the experimenter is usually concerned with the *mean* of all the readings. If each reading is denoted by x_i and there are n readings, the *arithmetic mean* is given by

$$x_m = \frac{1}{n} \sum_{i=1}^{n} x_i \tag{3-3}$$

The *deviation* d_i for each reading is defined by

$$d_i = x_i - x_m \tag{3-4}$$

We may note that the average of the deviations of all the readings is zero since

$$\bar{d}_i = \frac{1}{n} \sum_{i=1}^{n} d_i = \frac{1}{n} \sum_{i=1}^{n} (x_i - x_m)$$

$$= x_m - \frac{1}{n}(nx_m) = 0 \tag{3-5}$$

The average of the absolute values of the deviations is given by

$$|\bar{d}_i| = \frac{1}{n} \sum_{i=1}^{n} |d_i| = \frac{1}{n} \sum_{i=1}^{n} |x_i - x_m| \tag{3-6}$$

Note that this quantity is not necessarily zero.

The *standard deviation* or *root mean square deviation* is defined by

$$\sigma = \left[\frac{1}{n} \sum_{i=1}^{n} (x_i - x_m)^2 \right]^{1/2} \tag{3-7}$$

and the square of the standard deviation σ^2 is called the *variance*. This is sometimes called the *population* or *biased* standard deviation because it strictly applies only when a large number of samples is taken to describe the population.

In many circumstances the engineer will not be able to collect as many data points as necessary to describe the underlying population. Generally speaking, it is desirable to have at least 20 measurements in order to obtain reliable estimates of standard deviation and general validity of the data. For small sets of data an *unbiased* or *sample standard deviation* is defined by

$$\sigma = \frac{\sum\limits_{i=1}^{n} (x_i - x_m)^2}{n - 1}^{1/2} \tag{3-8}$$

Note that the factor $n - 1$ is used instead of n as in Eq. (3-7). The sample or unbiased standard deviation should be used when the underlying population is not known. However, when comparisons are made against a known population or standard, Eq. (3-7) is the proper one to use for standard deviation. An example would be the calibration of a voltmeter against a known voltage source.

There are other kinds of mean values of interest from time to time in statistical analysis. The *median* is the value that divides the data points in half. For example, if measurements made on five production resistors give 10, 12, 13, 14, and 15 kΩ, the median value would be 13 kΩ. The *arithmetic* mean, however, would be

$$R_m = \frac{10 + 12 + 13 + 14 + 15}{5} = 12.8 \text{ k}\Omega$$

In some instances it may be appropriate to divide data into quartiles and deciles also. So, when we say that a student is in the upper quartile of the class we mean that that student's grade is among the top 25 percent of all students in the class.

Sometimes it is appropriate to use a *geometric mean* when studying phenomena which grow in proportion to their size. This would apply to certain biological processes and to growth rates in financial resources. The geometric mean is defined by

$$x_g = [x_1 \cdot x_2 \cdot x_3 \cdots x_n]^{1/n} \tag{3-9}$$

As an example of the use of this concept, consider the 5-year record of a mutual fund investment:

Year	Asset value	Rate of increase over previous year
1	1000	
2	890	0.89
3	990	1.1124
4	1100	1.1111
5	1250	1.1364

The average growth rate is therefore

$$\text{Average growth} = [(0.89)(1.1124)(1.1111)(1.1364)]^{1/4}$$

$$= 1.0574$$

To see that this is indeed a valid average growth rate we can observe that

$$(1000)(1.0574)^4 = 1250$$

Example 3-6 The following readings are taken of a certain physical length. Compute the mean reading, standard deviation, variance, and average of the absolute value of the deviation, using the " biased " basis.

Reading	x, cm
1	5.30
2	5.73
3	6.77
4	5.26
5	4.33
6	5.45
7	6.09
8	5.64
9	5.81
10	5.75

SOLUTION The mean value is given by

$$x_m = \frac{1}{n} \sum_{i=1}^{n} x_i = \tfrac{1}{10}(56.13) = 5.613 \text{ cm}$$

The other quantities are computed with the aid of the following table:

Reading	$d_i = x_i - x_m$	$(x_i - x_m)^2 \times 10^2$
1	−0.313	9.797
2	0.117	1.369
3	1.157	133.865
4	−0.353	12.461
5	−1.283	164.609
6	−0.163	2.657
7	0.477	22.753
8	0.027	0.0729
9	0.197	3.881
10	0.137	1.877

$$\sigma = \left[\frac{1}{n} \sum_{i=1}^{n} (x_i - x_m)^2 \right]^{1/2} = [(\tfrac{1}{10}(3.533)]^{1/2} = 0.5944 \text{ cm}$$

$$\sigma^2 = 0.3533 \text{ cm}^2$$

$$|\bar{d}_i| = \frac{1}{n} \sum_{i=1}^{n} |d_i| = \frac{1}{n} \sum_{i=1}^{n} |x_i - x_m|$$

$$= \tfrac{1}{10}(4.224) = 0.4224 \text{ cm}$$

Example 3-7 Calculate the best estimate of standard deviation for the data of Example 3-6 based on the "sample" or unbiased basis.

Solution The calculation gives

$$\sigma = \left[\frac{1}{10 - 1} (3.536) \right]^{1/2} = (0.3929)^{1/2} = 0.627 \text{ cm}$$

Suppose an "honest" coin is flipped a large number of times. It will be noted that after a large number of tosses heads will be observed about the same number of times as tails. If one were to consistently bet on either heads or tails, the best one could hope for would be a break-even proposition over a long period of time. In other words, the *frequency of occurrence* is the same for both heads or tails for a very large number of tosses. It is common knowledge that a few tosses of a coin, say 5 or 10, may not be a break-even proposition, as a large number of tosses would be. This observation illustrates the fact that frequency of occurrence of an event may be dependent on the total number of events which are observed.

The *probability* that one will get a head when flipping an unweighted coin is $\frac{1}{2}$, regardless of the number of times the coin is tossed. The probability that a tail will occur is also $\frac{1}{2}$. The probability that either a head or a tail will occur is $\frac{1}{2} + \frac{1}{2}$ or unity. (We ignore the possibility that the coin will stand on edge.) *Probability* is a mathematical quantity that is linked to the frequency with which a certain phenomenon occurs after a large number of tries. In the case of the coin, it is the number of times heads would be expected to result in a large number of tosses divided by the total number of tosses. Similarly, the toss of an unloaded die results in the occurence of each side one-sixth of the time. Probabilities are expressed in values of less than one, and a probability of unity corresponds to certainty. In other words, if the probabilities for all possible events are added, the result must be unity. For separate events, the probability that one of the events will occur is the sum of the individual probabilities for the events. For a die, the probability that any one side will occur is $\frac{1}{6}$. The probability for one of three given sides is $\frac{1}{6} + \frac{1}{6} + \frac{1}{6}$, or $\frac{1}{2}$, and so on.

Suppose two dice are thrown and we wish to know the probability that both will display a 6. The probability for a 6 on a single die is $\frac{1}{6}$. By a short listing of the possible arrangements that the dice may have, it can be seen that there can be 36 possibilities and that the desired result of two 6's represents only one of these possibilities. Thus, the probability is $\frac{1}{36}$. For a throw of 7 or 11, there are 6 possible ways of getting a 7; thus the probability of getting a 7 is $\frac{6}{36}$ or $\frac{1}{6}$. There are only 2 ways of getting an 11; thus the probability is $\frac{2}{36}$ or $\frac{1}{18}$. The probability of getting *either* a 7 or an 11 is $\frac{6}{36} + \frac{2}{36}$.

If several *independent* events occur at the same time such that each event has a probability p_i, the probability that all events will occur is given as the product of the probabilities of the individual events. Thus, $p = \prod p_i$, where the $\prod$ designates a product. This rule could be applied to the problem of determining the probability of a double 6 in the throw of two dice. The probability of getting a 6 on each die is $\frac{1}{6}$, and the total probability is therefore $(\frac{1}{6})(\frac{1}{6})$, or $\frac{1}{36}$. This reasoning could not be applied to the problem of obtaining a 7 on the two dice because the number on each die is not *independent* of the number on the other die, since a 7 can be obtained in more than one way.

As a final example we ask what the chances are of getting a royal flush in the first five cards drawn off the top of the deck. There are 20 suitable possibilities for the first draw (4 suits, 5 possible cards per suit) out of a total of 52 cards. On the second draw we have fixed the suit so that there are only 4 suitable cards out of the 51 remaining. There are three suitable cards on the third draw, two on the fourth, and only one on the fifth draw. The total probability of drawing the royal flush is thus the product of the probabilities of each draw, or

$$\frac{20}{52} \times \frac{4}{51} \times \frac{3}{50} \times \frac{2}{49} \times \frac{1}{48} = \frac{1}{649,740}$$

In the above discussion we have seen that the probability is related to the number of ways a certain event may occur. In this case we are assuming that all events are equally likely, and hence the probability that an event will occur is the number of ways the event may occur divided by the number of possible events. Our primary concern is the application of probability and statistics to the analysis of experimental data. For this purpose we need to discuss next the meaning and use of *probability distributions*. We shall be concerned with a few particular distributions that are directly applicable to experimental data analysis.

3-7 PROBABILITY DISTRIBUTIONS

Suppose we toss a horseshoe some distance x. Even though we make an effort to toss the horseshoe the same distance each time, we would not always meet with success. On the first toss the horseshoe might travel a distance x_1, on the second toss a distance of x_2, and so on. If one is a good player of the game, there would be more tosses which have an x distance equal to that of the objective. Also, we would expect fewer and fewer tosses for those x distances which are further and further away from the target. For a large number of tosses, the probability that it will travel a distance is obtained by dividing the number traveling this distance by the total number of tosses. Since each x distance will vary somewhat from other x distances, we might find it advantageous to calculate the probability of a toss landing in a certain increment of x between x and $x + \Delta x$. When this calculation is made, we might get something like the situation shown in Fig. 3-1. For a good player, the maximum probability is expected to surround the distance x_m designating the position of the target.

The curve shown in Fig. 3-1 is called a *probability distribution*. It shows how the probability of success in a certain event is distributed over the distance x. Each value of the ordinate $p(x)$ gives the probability that the horseshoe will land between and $x + \Delta x$, where Δx is allowed to approach zero. We might consider the deviation from x_m as the error in the throw. If the horseshoe player has good

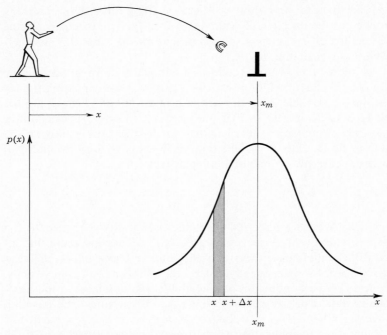

Figure 3-1 Distribution of throws for a " good " horseshoes player.

aim, large errors are less likely than small errors. The area under the curve is unity since it is certain that the horseshoe will land somewhere.

We should note that more than one variable may be present in a probability distribution. In the case of the horseshoes player, a person might throw the object an exact distance of x_m and yet to one side of the target. The sideways distance is another variable, and a large number of throws would have some distribution in this variable as well.

A particular probability distribution is the *binomial distribution*. This distribution gives the number of successes n out of N possible independent events when each event has a probability of success p. The probability that n events will succeed is given in Ref. [2] as

$$p(n) = \frac{N!}{(N-n)!\,n!}\, p^n(1-p)^{N-n} \tag{3-10}$$

It will be noted that the quantity $(1 - p)$ is the probability of failure of each independent event.

Example 3-8 An unweighted coin is flipped three times. Calculate the probability of getting zero, one, two, or three heads in these tosses.

SOLUTION The binomial distribution applies in this case. The probability of getting a head on each throw is $p = \frac{1}{2}$ and $N = 3$, while n takes on the values 0, 1, 2, and 3. The probabilities are calculated as

$$p(0) = \frac{3!}{(3!)(0!)} \left(\frac{1}{2}\right)^0 \left(\frac{1}{2}\right)^3 = \frac{1}{8}$$

$$p(1) = \frac{3!}{(2!)(1!)} \left(\frac{1}{2}\right)^1 \left(\frac{1}{2}\right)^2 = \frac{3}{8}$$

$$p(2) = \frac{3!}{(1!)(2!)} \left(\frac{1}{2}\right)^2 \left(\frac{1}{2}\right)^1 = \frac{3}{8}$$

$$p(3) = \frac{3!}{(0!)(3!)} \left(\frac{1}{2}\right)^3 \left(\frac{1}{2}\right)^0 = \frac{1}{8}$$

Now suppose that the number of possible independent events N is very large and the probability of occurrence of each p is very small. The calculation of the probability of n successes out of the N possible events using Eq. (3-10) would be most cumbersome because of the size of the numbers. The limit of the binomial distribution as $N \to \infty$ and $p \to 0$, such that

$$Np = a = \text{const}$$

is called the *Poisson distribution* and is given by

$$p_a(n) = \frac{a^n e^{-a}}{n!} \tag{3-11}$$

The Poisson distribution is applicable to the calculation of the decay of radioactive nuclei, as we shall see in a subsequent chapter. It may be shown that the standard deviation of the Poisson distribution is

$$\sigma = \sqrt{a} \tag{3-12}$$

We have noted that a probability distribution like Fig. 3-1 is obtained when we observe frequency of occurrence over a large number of observations. When a limited number of observations is made and the raw data plotted, we call the plot a *histogram*. For example, the following distribution of throws might be observed for a horseshoes player:

Distance from target, cm	Number of throws
0–10	5
10–20	15
20–30	13
30–40	11
40–50	9
50–60	8
60–70	10
70–80	6
80–90	7
90–100	5
100–110	5
110–120	3
Over 120	2
Total	99

Number of throws

Distance from target, cm

Figure 3-2 Histogram with $\Delta x = 10$ cm.

Number of throws

Distance from target, cm

Figure 3-3 Histogram with $\Delta x = 20$ cm.

Figure 3-4 Cumulative frequency diagram.

These data are plotted in Fig. 3-2 using increments of 10 cm in Δx. The same data are plotted in Fig. 3-3 using a Δx of 20 cm. The *relative frequency*, or fraction of throws in each Δx increment, could also be used to convey the same information. A *cumulative frequency* diagram could be employed for these data, as shown in Fig. 3-4. If this figure had been constructed on the basis of a very large number of throws, then we could appropriately refer to the ordinate as the probability that the horseshoe will land within a distance x of the target.

3-8 THE GAUSSIAN OR NORMAL ERROR DISTRIBUTION

Suppose an experimental observation is made and some particular result recorded. We know (or would strongly suspect) that the observation has been subjected to many random errors. These random errors may make the final reading either too large or too small, depending on many circumstances which are unknown to us. Assuming that there are many small errors that contribute to the final error and that each small error is of equal magnitude and equally likely to be positive or negative, the *gaussian* or *normal error distribution* may be derived. If the measurement is designated by x, the gaussian distribution gives the probability that the measurement will lie between x and $x + dx$ and is written

$$P(x) = \frac{1}{\sigma\sqrt{2\pi}} e^{-(x - x_m)^2/2\sigma^2} \tag{3-13}$$

In this expression, x_m is the mean reading and σ is the standard deviation. Some may prefer to call $P(x)$ the *probability density*. The units of $P(x)$ are those of $1/x$ since these are the units of $1/\sigma$. A plot of Eq. (3-13) is given in Fig. 3-5. Note that the most probable reading is x_m. The standard deviation is a measure of the width of the distribution curve; the larger the value of σ, the flatter the curve and hence the larger the expected error of all the measurements. Equation (3-13) is normalized so that the total area under the curve is unity. Thus,

$$\int_{-\infty}^{+\infty} P(x)\,dx = 1.0 \tag{3-14}$$

At this point we may note the similarity between the shape of the normal error curve and the expected experimental distribution for tossing horseshoes as shown in Fig. 3-1. This is what we would expect, because the good horseshoes player's throws will be bunched around the target. The better the player is at the game, the more closely the throws will be grouped around the mean and the more probable will be the mean distance x_m. Thus, in the case of the horseshoes player, a smaller standard deviation would mean a larger percentage of "ringers."

We may quickly anticipate the next step in the analysis as one of trying to determine the precision of a set of experimental measurements through an application of the normal error distribution. One may ask: But how do you know

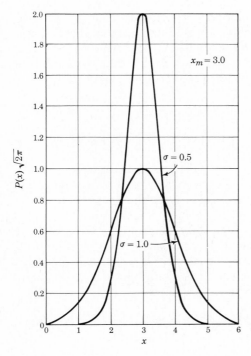

Figure 3-5 The gaussian or normal error distribution for two values of the standard deviation.

that the assumptions pertaining to the derivation of the normal error distribution apply to experimental data? The answer is that for sets of data where a large number of measurements are taken, experiments indicate that the measurements do indeed follow a distribution like that shown in Fig. 3-5 when the experiment is under control. If an important parameter is not controlled, one gets just scatter, i.e., no sensible distribution at all. Thus, as a matter of experimental verification the gaussian distribution is believed to represent the *random* errors in an adequate manner for a properly controlled experiment.

By inspection of the gaussian distribution function of Eq. (3-13) we see that the maximum probability occurs at $x = x_m$, and the value of this probability is

$$P(x_m) = \frac{1}{\sigma\sqrt{2\pi}} \tag{3-15}$$

It is seen from Eq. (3-15) that smaller values of the standard deviation produce larger values of the maximum probability, as would be expected in an intuitive sense. $P(x_m)$ is sometimes called a *measure of precision* of the data because it has a larger value for smaller values of the standard deviation.

We next wish to examine the gaussian distribution to determine the likelihood that certain data points will fall within a specified deviation from the mean of all the data points. The probability that a measurement will fall within a certain range x_1 of the mean reading is

$$P = \int_{x_m - x_1}^{x_m + x_1} \frac{1}{\sigma\sqrt{2\pi}} e^{-(x - x_m)^2/2\sigma^2} \, dx \tag{3-16}$$

Making the variable substitution,

$$\eta = \frac{x - x_m}{\sigma}$$

Equation (3-16) becomes

$$P = \frac{1}{\sqrt{2\pi}} \int_{-\eta_1}^{+\eta_1} e^{-\eta^2/2} \, d\eta \tag{3-17}$$

where

$$\eta_1 = \frac{x_1}{\sigma} \tag{3-18}$$

Values of the gaussian normal error function

$$\frac{1}{\sqrt{2\pi}} e^{-\eta^2/2}$$

and integrals of the gaussian function corresponding to Eq. (3-17) are given in Tables 3-1 and 3-2.

If we have a sufficiently large number of data points, the error for each point should follow the gaussian distribution and we can determine the probability that certain data fall within a specified deviation from the mean value. Example 3-9 illustrates the method of computing the chances of finding data points within one or two standard deviations from the mean. Table 3-3 gives the chances for certain deviations from the mean value of the normal-distribution curve.

Table 3-1 Values of the gaussian normal error distribution

Values of the function $(1/\sqrt{2\pi})e^{-\eta^2/2}$ for different values of the argument η. (Each figure in the body of the table is preceded by a decimal point.)

η	0.00	0.01	0.02	0.03	0.04	0.05	0.06	0.07	0.08	0.09
0.0	39894	39892	39886	39876	39862	39844	39822	39797	39767	39733
0.1	39695	39654	39608	39559	39505	39448	39387	39322	39253	39181
0.2	39104	39024	38940	38853	38762	38667	38568	38466	38361	38251
0.3	38139	38023	37903	37780	37654	37524	37391	37255	37115	36973
0.4	36827	36678	36526	36371	36213	36053	35889	35723	35553	35381
0.5	35207	35029	34849	34667	34482	34294	34105	33912	33718	33521
0.6	33322	33121	32918	32713	32506	32297	32086	31874	31659	31443
0.7	31225	31006	30785	30563	30339	30114	29887	29658	29430	29200
0.8	28969	28737	28504	28269	28034	27798	27562	27324	27086	26848
0.9	26609	26369	26129	25888	25647	25406	25164	24923	24681	24439
1.0	24197	23955	23713	23471	23230	22988	22747	22506	22265	22025
1.1	21785	21546	21307	21069	20831	20594	20357	20121	19886	19652
1.2	19419	19186	18954	18724	18494	18265	18037	17810	17585	17360
1.3	17137	16915	16694	16474	16256	16038	15822	15608	15395	15183
1.4	14973	14764	14556	14350	14146	13943	13742	13542	13344	13147
1.5	12952	12758	12566	12376	12188	12001	11816	11632	11450	11270
1.6	11092	10915	10741	10567	10396	10226	10059	09893	09728	09566
1.7	09405	09246	09089	08933	08780	08628	08478	08329	08183	08038
1.8	07895	07754	07614	07477	07341	07206	07074	06943	06814	06687
1.9	06562	06438	06316	06195	06077	05959	05844	05730	05618	05508
2.0	05399	05292	05186	05082	04980	04879	04780	04682	04586	04491
2.1	04398	04307	04217	04128	04041	03955	03871	03788	03706	03626
2.2	03547	03470	03394	03319	03246	03174	03103	03034	02965	02898
2.3	02833	02768	02705	02643	02582	02522	02463	02406	02349	02294
2.4	02239	02186	02134	02083	02033	01984	01936	01888	01842	01797
2.5	01753	01709	01667	01625	01585	01545	01506	01468	01431	01394
2.6	01358	01323	01289	01256	01223	01191	01160	01130	01100	01071
2.7	01042	01014	00987	00961	00935	00909	00885	00861	00837	00814
2.8	00792	00770	00748	00727	00707	00687	00668	00649	00631	00613
2.9	00595	00578	00562	00545	00530	00514	00499	00485	00470	00457
3.0	00443									
3.5	008727									
4.0	0001338									
4.5	0000160									
5.0	000001487									

Table 3-2 Integrals of the gaussian normal error function

Values of the integral $(1/\sqrt{2\pi}) \int_0^{\eta_1} e^{-\eta^2/2} \, d\eta$ are given for different values of the argument η_1. It may be observed that

$$\frac{1}{\sqrt{2\pi}} \int_{-\eta_1}^{+\eta_1} e^{-\eta^2/2} \, d\eta = 2 \, \frac{1}{\sqrt{2\pi}} \int_0^{\eta_1} e^{-\eta^2/2} \, d\eta$$

The values are related to the error function since

$$\text{erf } \eta_1 = \frac{1}{\sqrt{\pi}} \int_{-\eta_1}^{+\eta_1} e^{-\eta^2} \, d\eta$$

so that the tabular values are equal to $\frac{1}{2} \text{erf} (\eta_1/\sqrt{2})$. (Each figure in the body of the table is preceded by a decimal point.)

η_1	0.00	0.01	0.02	0.03	0.04	0.05	0.06	0.07	0.08	0.09
0.0	00000	00399	00798	01197	01595	01994	02392	02790	03188	03586
0.1	03983	04380	04776	05172	05567	05962	06356	06749	07142	07355
0.2	07926	08317	08706	09095	09483	09871	10257	10642	11026	11409
0.3	11791	12172	12552	12930	13307	13683	14058	14431	14803	15173
0.4	15554	15910	16276	16640	17003	17364	17724	18082	18439	18793
0.5	19146	19497	19847	20194	20450	20884	21226	21566	21904	22240
0.6	22575	22907	23237	23565	23891	24215	24537	24857	25175	25490
0.7	25804	26115	26424	26730	27035	27337	27637	27935	28230	28524
0.8	28814	29103	29389	29673	29955	30234	30511	30785	31057	31327
0.9	31594	31859	32121	32381	32639	32894	33147	33398	33646	33891
1.0	34134	34375	34614	34850	35083	35313	35543	35769	35993	36214
1.1	36433	36650	36864	37076	37286	37493	37698	37900	38100	38298
1.2	38493	38686	38877	39065	39251	39435	39617	39796	39973	40147
1.3	40320	40490	40658	40824	40988	41198	41308	41466	41621	41774
1.4	41924	42073	42220	42364	42507	42647	42786	42922	43056	43189
1.5	43319	43448	43574	43699	43822	43943	44062	44179	44295	44408
1.6	44520	44630	44738	44845	44950	45053	45154	45254	45352	45449
1.7	45543	45637	45728	45818	45907	45994	46080	46164	46246	46327
1.8	46407	46485	46562	46638	46712	46784	46856	46926	46995	47062
1.9	47128	47193	47257	47320	47381	47441	47500	47558	47615	47670
2.0	47725	47778	47831	47882	47932	47962	48030	48077	48124	48169
2.1	48214	48257	48300	48341	48382	48422	48461	48500	48537	48574
2.2	48610	48645	48679	48713	48745	48778	48809	48840	48870	48899
2.3	48928	48956	48983	49010	49036	49061	49086	49111	49134	49158
2.4	49180	49202	49224	49245	49266	49286	49305	49324	49343	49361
2.5	49379	49296	49413	49430	49446	49461	49477	49492	49506	49520
2.6	49534	49547	49560	49573	49585	49598	49609	49621	49632	49643
2.7	49653	49664	49674	49683	49693	49702	49711	49720	49728	49736
2.8	49744	49752	49760	49767	49774	49781	49788	49795	49801	49807
2.9	49813	49819	49825	49831	49836	49841	49846	49851	49856	49861
3.0	49865									
3.5	4997674									
4.0	4999683									
4.5	4999966									
5.0	4999997133									

Table 3-3 Chances for deviations from mean value of normal-distribution curve

Deviation	Chance of results falling within specified deviation
$\pm 0.6745\sigma$	1–1
σ	2.15–1
2σ	21–1
3σ	369–1

Example 3-9 Calculate the probabilities that a measurement will fall within one, two, and three standard deviations of the mean value, and compare them with the values in Table 3-3.

SOLUTION We perform the calculation using Eq. (3-17) with $\eta_1 = 1$, 2, and 3. The values of the integral may be obtained from Table 3-2. We observe that

$$\int_{-\eta_1}^{+\eta_1} e^{-\eta^2/2} \, d\eta = 2 \int_0^{\eta_1} e^{-\eta^2/2} \, d\eta$$

so that

$$P(1) = (2)(0.34134) = 0.6827$$

$$P(2) = (2)(0.47725) = 0.9545$$

$$P(3) = (2)(0.49865) = 0.9973$$

Using the odds given in Table 3-3, we would calculate the probabilities as

$$P(1) = \frac{2.15}{2.15 + 1} = 0.6827$$

$$P(2) = \frac{21}{21 + 1} = 0.9545$$

$$P(3) = \frac{369}{369 + 1} = 0.9973$$

It is a rare circumstance indeed when an experimenter does not find that some of the data points look bad and out of place in comparison with the bulk of the data. The experimenter is therefore faced with the task of deciding if these points are the result of some gross experimental blunder and hence may be neglected or if they represent some new type of physical phenomenon that is peculiar to a certain operating condition. The engineer cannot just throw out those points that do not fit with expectations—there must be some consistent basis for elimination.

Suppose n measurements of a quantity are taken and n is large enough that we may expect the results to follow the gaussian error distribution. This distribution may be used to compute the probability that a given reading will deviate a

Table 3-4 Chauvenet's criterion for rejecting a reading

Number of readings, n	Ratio of maximum acceptable deviation to standard deviation, d_{max}/σ
3	1.38
4	1.54
5	1.65
6	1.73
7	1.80
10	1.96
15	2.13
25	2.33
50	2.57
100	2.81
300	3.14
500	3.29
1,000	3.48

certain amount from the mean. We would not expect a probability much smaller than $1/n$ because this would be unlikely to occur in the set of n measurements. Thus, if the probability for the observed deviation of a certain point is less than $1/n$, a suspicious eye would be cast at that point with an idea toward eliminating it from the data. Actually, a more restrictive test is usually applied to eliminate data points. It is known as *Chauvenet's criterion* and specifies that a reading may be rejected if the probability of obtaining the particular deviation from the mean is less than $1/2n$. Table 3-4 lists values of the ratio of deviation to standard deviation for various values of n according to this criterion.

In applying Chauvenet's criterion to eliminate dubious data points, one first calculates the mean value and standard deviation using all data points. The deviations of the individual points are then compared with the standard deviation in accordance with the information in Table 3-4 (or by a direct application of the criterion), and the dubious points are eliminated. For the final data presentation a new mean value and standard deviation are computed with the dubious points eliminated from the calculation. Note that Chauvenet's criterion might be applied a second or third time to eliminate additional points; but this practice is unacceptable, and only the first application may be used.

Example 3-10 Using Chauvenet's criterion, test the data points of Example 3-6 for possible inconsistency. Eliminate the questionable points and calculate a new standard deviation for the adjusted data.

SOLUTION The best estimate of the standard deviation is given in Example 3-7 as 0.627 cm. We first calculate the ratio d_i/σ and eliminate data points in accordance with Table 3-4.

Reading	d_i/σ
1	0.499
2	0.187
3	1.845
4	0.563
5	2.046
6	0.260
7	0.761
8	0.043
9	0.314
10	0.219

In accordance with Table 3-4, we may eliminate only point number 5. When this point is eliminated, the new mean value is

$$x_m = \tfrac{1}{9}(51.80) = 5.756 \text{ cm}$$

The new value of the standard deviation is now calculated with the following table:

Reading	$d_i = x_i - x_m$	$(x_i - x_m)^2 \times 10^2$
1	−0.456	20.7936
2	−0.026	0.0676
3	1.014	102.8196
4	−0.496	24.602
6	−0.306	9.364
7	0.334	11.156
8	−0.116	1.346
9	0.054	0.292
10	−0.006	0.0036

$$\sigma = \left[\frac{1}{n-1} \sum_{i=1}^{n} (x_i - x_m)^2 \right]^{1/2} = \left[\tfrac{1}{8}(1.7044) \right]^{1/2}$$

$$= (0.213)^{1/2} = 0.4615 \text{ cm}$$

Thus, by the elimination of the one point, the standard deviation has been reduced from 0.627 to 0.462 cm. This is a 26.5 percent reduction.

3-9 PROBABILITY GRAPH PAPER

We have seen that the normal error distribution offers a means for examining experimental data for statistical consistency. In particular, it enables us to eliminate questionable readings with the Chauvenet criterion and thus obtain a better estimate of the standard deviation and mean reading. If the distribution of random errors is not *normal*, then this elimination technique will not apply. It is to our advantage, therefore, to determine if the data are following a normal

Reading

Figure 3-6 Probability graph paper.

distribution before making too many conclusions about the mean value, variances, etc. Specially constructed probability graph paper is available for this purpose and may be purchased from a technical drawing shop. The paper uses the coordinate system shown in Fig. 3-6. The ordinate has the percent of readings at or below the value of the abscissa, and the abscissa is the value of a particular reading. The ordinate spacings are arranged so that the gaussian-distribution curve will plot as a straight line on the graph. In addition, this straight line will intersect the 50 percent ordinate at an abscissa equal to the arithmetic mean of the data.

Thus, to determine if a set of data points is distributed normally, we plot the data on probability paper and see how well they match with the theoretical straight line. It is to be noted that the largest reading cannot be plotted on the graph because the ordinate does not extend to 100 percent. In assessing the validity of the data we should not place as much reliance on the points near the upper and lower ends of the curve since they are closer to the "tails" of the probability distribution and are thus less likely to be valid.

Example 3-11 The following data are collected for a certain measurement. Plot the data on probability paper and comment on the normality of the distribution.

Reading	x_i, cm
1	4.62
2	4.69
3	4.86
4	4.53
5	4.60
6	4.65
7	4.59
8	4.70
9	4.58
10	4.63
	$\sum x_i = 46.45$

From these data the mean value is calculated as

$$x_m = \tfrac{1}{10} \sum x_i = \tfrac{1}{10}(46.45) = 4.645 \text{ cm}$$

The data are plotted in the accompanying figure indicating a reasonably normal distribution. It should be noted that the straight line crosses the 50 percent ordinate at a value of approximately $x = 4.62$, which is not in agreement with the calculated value of x_m. Note that point 3, $x = 4.86$, does not appear on the plot since it would represent the 100 percent ordinate.

Figure Example 3-10

3-10 THE CHI-SQUARE TEST OF GOODNESS OF FIT

In the previous discussion we have noted that random experimental errors would be expected to follow the gaussian distribution, and the examples illustrated the method of calculating the probability of occurrence of a particular experimental determination. We might ask how it is known that the random errors or deviations do approximate a gaussian distribution. In general, we may ask how we can determine if experimental observations match some particular expected distribution for the data. As a simple example, consider the tossing of a coin. We would like to know if a certain coin is "honest," i.e., unweighted toward either heads or tails. If the coin is unweighted, then heads should occur half the time and tails should occur half the time. But suppose we do not want to take the time to make thousands of tosses to get a frequency distribution of heads and tails for a large number of tosses. Instead, we toss the coin a few times and wish to infer from these few tosses whether the coin is unweighted or weighted. Common sense tells us not to expect exactly six heads and six tails out of, say, twelve tosses. But how much deviation from this arrangement could we tolerate and still expect the coin to be unweighted? The chi-square test of goodness of fit is a suitable way of answering this question. It is based on a calculation of the quantity chi squared, defined by

$$\chi^2 = \sum_{i=1}^{n} \frac{[(\text{observed value})_i - (\text{expected value})_i]^2}{(\text{expected value})_i} \tag{3-19}$$

where n is the number of cells or groups of observations. The expected value is the value which would be obtained if the measurements matched the expected distribution perfectly.

The chi-square test may be applied to check the validity of various distributions. Calculations have been made [2] of the probability that the actual measurements match the expected distribution, and these probabilities are given in Table 3-5. In this table, F represents the number of degrees of freedom in the measurements and is given by

$$F = n - k \tag{3-20}$$

where n is the number of cells and k is the number of imposed conditions on the expected distribution. A plot of the chi-square function is given in Fig. 3-7.

While we initiated the discussion on the chi-square test in terms of random errors following the gaussian distribution, the test is an important tool for testing any expected experimental distribution. In other words, we might use the test to analyze random errors or to check the adherence of certain data to an expected

Number of degrees of freedom

Figure 3-7 The chi-square function.

Table 3-5 Chi-squared. P is the probability that the value in the table will be exceeded for a given number of degrees of freedom F†

$F = $ \ P	0.995	0.990	0.975	0.950	0.900	0.750	0.500	0.250	0.100	0.050	0.025	0.010	0.005
1	0.0^4393	0.0^3157	0.0^3982	0.0^2393	0.0158	0.102	0.455	1.32	2.71	3.84	5.02	6.63	7.88
2	0.0100	0.0201	0.0506	0.103	0.211	0.575	1.39	2.77	4.61	5.99	7.38	9.21	10.6
3	0.0717	0.115	0.216	0.352	0.584	1.21	2.37	4.11	6.25	7.81	9.35	11.3	12.8
4	0.207	0.297	0.484	0.711	1.06	1.92	3.36	5.39	7.78	9.49	11.1	13.3	14.9
5	0.412	0.554	0.831	1.15	1.61	2.67	4.35	6.63	9.24	11.1	12.8	15.1	16.7
6	0.676	0.872	1.24	1.64	2.20	3.45	5.35	7.84	10.6	12.6	14.4	16.8	18.5
7	0.989	1.24	1.69	2.17	2.83	4.25	6.35	9.04	12.0	14.1	16.0	18.5	20.3
8	1.35	1.65	2.18	2.73	3.49	5.07	7.34	10.2	13.4	15.5	17.5	20.1	22.0
9	1.73	2.09	2.70	3.33	4.17	5.90	8.34	11.4	14.7	16.9	19.0	21.7	23.6
10	2.16	2.56	3.25	3.94	4.87	6.74	9.34	12.5	16.0	18.3	20.5	23.2	25.2
11	2.60	3.05	3.82	4.57	5.58	7.58	10.3	13.7	17.3	19.7	21.9	24.7	26.8
12	3.07	3.57	4.40	5.23	6.30	8.44	11.3	14.8	18.5	21.0	23.3	26.2	28.3
13	3.57	4.11	5.01	5.89	7.04	9.30	12.3	16.0	19.8	22.4	24.7	27.7	29.8
14	4.07	4.66	5.63	6.57	7.79	10.2	13.3	17.1	21.1	23.7	26.1	29.1	31.3
15	4.60	5.23	6.26	7.26	8.55	11.0	14.3	18.2	22.3	25.0	27.5	30.6	32.8
16	5.14	5.81	6.91	7.96	9.31	11.9	15.3	19.4	23.5	26.3	28.8	32.0	34.3
17	5.70	6.41	7.56	8.67	10.1	12.8	16.3	20.5	24.8	27.6	30.2	33.4	35.7
18	6.26	7.01	8.23	9.39	10.9	13.7	17.3	21.6	26.0	28.9	31.5	34.8	37.2
19	6.84	7.63	8.91	10.1	11.7	14.6	18.3	22.7	27.2	30.1	32.9	36.2	38.6
20	7.43	8.26	9.59	10.9	12.4	15.5	19.3	23.8	28.4	31.4	34.2	37.6	40.0
21	8.03	8.90	10.3	11.6	13.2	16.3	20.3	24.9	29.6	32.7	35.5	38.9	41.4
22	8.64	9.54	11.0	12.3	14.0	17.2	21.3	26.0	30.8	33.9	36.8	40.3	42.8
23	9.26	10.2	11.7	13.1	14.8	18.1	22.3	27.1	32.0	35.2	38.1	41.6	44.2
24	9.89	10.9	12.4	13.8	15.7	19.0	23.3	28.3	33.2	36.4	39.4	43.0	45.6
25	10.5	11.5	13.1	14.6	16.5	19.9	24.3	29.3	34.4	37.7	40.6	44.3	46.9
26	11.2	12.2	13.8	15.4	17.3	20.8	25.3	30.4	35.6	38.9	41.9	45.6	48.3
27	11.8	12.9	14.6	16.2	18.1	21.7	26.3	31.5	36.7	40.1	43.2	47.0	49.6
28	12.5	13.6	15.3	16.9	18.9	22.7	27.3	32.6	37.9	41.3	44.5	48.3	51.0
29	13.1	14.3	16.0	17.7	19.8	23.6	28.3	33.7	39.1	42.6	45.7	49.6	52.3
30	13.8	15.0	16.8	18.5	20.6	24.5	29.3	34.8	40.3	43.8	47.0	50.9	53.7

† From C. M. Thompson: *Biometrika,* vol. 32, 1941, as abridged by A. M. Mood and F. A. Graybill, "Introduction to the Theory of Statistics," 2d ed., McGraw-Hill Book Company, New York, 1963.

distribution. We interpret the test by calculating the number of degrees of freedom and χ^2 from the experimental data. Then, consulting Table 3-5, we obtain the probability P that this value of χ^2 or higher values could occur by chance. If $\chi^2 = 0$, then the assumed or expected distribution and measured distribution match exactly. The larger the value of χ^2, the larger is the disagreement between the assumed distribution and the observed values, or the smaller the probability that the observed distribution matches the expected distribution. The reader should consult Refs. [2] and [4] for more specific information on the chi-square test and the derivation of the probabilities associated with it.

One may note that the heading of this section includes the term "goodness of fit." We see that the chi-square test may be used to determine how well a set of experimental observations fits an assumed distribution. In connection with this test we may remark that data may sometimes be "too good" or "too consistent." For example, we would be quite surprised if in the conduct of an experimental test, the results were found to check with theory *exactly* or to follow some well-defined relationship exactly. We might find, for instance, that a temperature controller maintained a set point temperature *exactly*, with no measurable deviation whatsoever. Experienced laboratory people know that controllers usually do not operate this way and would immediately suspect that the temperature recorder might be stuck or otherwise defective. The point of this brief remark is that one must be suspicious of high values of P as well as low values. A good rule of thumb is that if P lies between 0.1 and 0.9, the observed distribution may be considered to follow the assumed distribution. If P is either less than 0.02 or greater than 0.98, the assumed distribution may be considered unlikely.

Let us return for a moment to the tossing of a coin. Suppose a coin is tossed twice, resulting in one head and one tail. This observation certainly matches exactly with what would be expected for an unweighted coin; however, our common sense tells us not to believe the coin is unweighted on the basis of only two tosses. In other words, we must have a certain minimum number of samples for statistics to apply. For the chi-square test the generally accepted minimum number of expected values for each ith cell is 5. If some frequencies fall below 5 it is recommended that the cells or groups be redefined to alleviate the problem.

For example, a plastics company produces two types of styrofoam cups (call them A and B) which can experience eight kinds of defects. One hundred defective samples of each cup are collected and the number of each type of defect determined. The following table results:

Type defect	Cup A	Cup B
1	1	5
2	2	3
3	3	3
4	25	23
5	10	12
6	15	16
7	38	30
8	6	8
Total	100	100

We would like to know if the two cups have the same pattern of defects. To do this we could compute chi squared for cup B assuming cup A has the expected distribution. But we encounter a problem. Defects 1, 2, and 3 do not meet our criterion of a minimum of 5 expected values in each cell. So we must reconstruct the cells by combining 1, 2, and 3 to obtain:

Type defect	Cup A	Cup B
1, 2, 3	6	11
4	25	23
5	10	12
6	15	16
7	38	30
8	6	8
Total	100	100

For the former case we had eight cells or groups and one imposed condition (total observations = 100), so $F = 8 - 1 = 7$. After grouping defects 1, 2, and 3 we have $F = 6 - 1 = 5$.

Example 3-12 Two dice are rolled 300 times and the following results are noted:

Number	Number of occurrences
2	6
3	9
4	27
5	36
6	39
7	57
8	45
9	39
10	24
11	12
12	6

Calculate the probability that the dice are unloaded.

SOLUTION Eleven cells have been observed with only one restriction: the number of rolls of the dice is fixed. Thus, $F = 11 - 1 = 10$. If the dice are unloaded, a short listing of the combinations of the dice will give the probability of occurrence for each number. The expected value of each number is then the probability multiplied by 300, the total number of throws. The values of interest are tabulated below.

Number	Observed	Probability	Expected
2	6	1/36	8.333
3	9	1/18	16.667
4	27	1/12	25.0
5	36	1/9	33.333
6	39	5/36	41.667
7	57	1/6	50.0
8	45	5/36	41.667
9	39	1/9	33.333
10	24	1/12	25.0
11	12	1/18	16.667
12	6	1/36	8.333

From these data the value of chi squared is calculated as 8.034. If Table 3-5 is consulted the probability is given as $P = 0.626$.

Example 3-13 A coin is tossed 20 times, resulting in 6 heads and 14 tails. Using the chi-square test, estimate the probability that the coin is unweighted. Suppose another set of tosses of the same coin is made and 8 heads and 12 tails are obtained. What is the probability of having an unweighted coin based on the information from both sets of data?

SOLUTION For each set of data we may make only two observations: the number of heads and the number of tails. Thus, $n = 2$. Furthermore, we impose one restriction on the data: the number of tosses is fixed. Thus, $k = 1$ and the number of degrees of freedom is

$$F = n - k = 2 - 1 = 1$$

The values of interest are

	Observed	Expected
Heads	6	10
Tails	14	10

For these values χ^2 is calculated as

$$\chi^2 = \frac{(6 - 10)^2}{10} + \frac{(14 - 10)^2}{10} = 3.20$$

Consulting Table 3-5 we find $P = 0.078$; that is, there is an 8 percent chance that this distribution is just the result of random fluctuations and that the coin may be unweighted.

Now consider the additional information we gain about the coin from the second set of observations. We now have four observations: the number of heads and tails in each set. There are only two restrictions on the data: the total number of tosses is fixed in each set. Thus, the number of degrees of freedom is

$$F = n - k = 4 - 2 = 2$$

For the second set of data the values of interest are

	Observed	Expected
Heads	8	10
Tails	12	10

Chi squared is now calculated on the basis of all four observations.

$$\chi^2 = \frac{(6 - 10)^2}{10} + \frac{(14 - 10)^2}{10} + \frac{(8 - 10)^2}{10} + \frac{(12 - 10)^2}{10} = 4.0$$

Consulting Table 3-5 again, we find $P = 0.15$. So, with the additional information we find a stronger likelihood that the tosses are following a random variation and that the coin is unweighted.

Example 3-14 A test is conducted to determine the effect of cigarette smoke on the eating habits and weight of mice. One group is fed a certain diet while being exposed to a controlled atmosphere containing cigarette smoke. A control group is fed the same diet but in the presence of clean air. The observations are given below. Does the presence of smoke cause a loss in weight?

	Gained weight	Lost weight	Total
Exposed to smoke	61	89	150
Exposed to clean air	65	77	142
Total	126	166	292

SOLUTION Clearly, there are four observations in this experiment, but we are faced with the problem of deciding on the expected values. We cannot just take the "clean-air" data as the expected values because some of the behavior might be a result of the special diet that is fed to both groups of mice. Consequently, about the best estimate we can make is one based on the total sample of mice. Thus, the expected frequencies would be

$$\text{Expected fraction to gain weight} = \tfrac{126}{292}$$

$$\text{Expected fraction to lose weight} = \tfrac{166}{292}$$

The expected values for the groups would thus be

	Gained weight	Lost weight
Exposed to smoke	$\frac{126}{292}150 = 64.7$	$\frac{166}{292}150 = 85.3$
Exposed to clean air	$\frac{126}{292}142 = 61.3$	$\frac{166}{292}142 = 80.7$

We observe that there are three restrictions on the data: (1) the number exposed to smoke, (2) the number exposed to clean air, and (3) the additional restriction involved in the calculation of the expected fractions which gain and lose weight. The number of degrees of freedom is thus

$$F = 4 - 3 = 1$$

The value of chi squared is calculated from

$$\chi^2 = \frac{(61 - 64.7)^2}{64.7} + \frac{(89 - 85.3)^2}{85.3} + \frac{(65 - 61.3)^2}{61.3} + \frac{(77 - 80.7)^2}{80.7} = 0.767$$

From Table 3-5 we find $P = 0.41$, or there is a 41 percent chance that the difference in the observations for the two groups is just the result of random fluctuations. One may not conclude from this information that the presence of cigarette smoke causes a loss in weight for the mice.

3-11 METHOD OF LEAST SQUARES

Suppose we have a set of observations $x_1, x_2, \ldots, x_n$. The sum of the squares of their deviations from some mean value is

$$S = \sum_{i=1}^{n} (x_i - x_m)^2 \tag{3-21}$$

Now suppose we wish to minimize S with respect to the mean value x_m. We set

$$\frac{\partial S}{\partial x_m} = 0 = \sum_{i=1}^{n} -2(x_i - x_m) = -2 \left(\sum_{i=1}^{n} x_i - nx_m \right) \tag{3-22}$$

where n is the number of observations. We find that

$$x_m = \frac{1}{n} \sum_{i=1}^{n} x_i \tag{3-23}$$

or the mean value which minimizes the sum of the squares of the deviations is the arithmetic mean. This example might be called the simplest application of the method of least squares. We shall be able to give only two other applications of the method, but it is of great utility in analyzing experimental data.

Suppose that the two variables x and y are measured over a range of values. Suppose further that we wish to obtain a simple analytical expression for y as a function of x. The simplest type of function is a linear one; hence we might try to establish y as a linear function of x. (Both x and y may be complicated functions of other parameters so arranged that x and y vary approximately in a linear manner. This matter will be discussed later.) The problem is one of finding the *best* linear function, for the data may scatter a considerable amount. We could solve the problem rather quickly by plotting the data points on graph paper and drawing a straight line through them by eye. Indeed this is common practice, but the method of least squares gives a more reliable way to obtain a better functional relationship than the guesswork of plotting. We seek an equation of the form

$$y = ax + b \qquad (3\text{-}24)$$

We therefore wish to minimize the quantity

$$S = \sum_{i=1}^{n} [y_i - (ax_i + b)]^2 \qquad (3\text{-}25)$$

This is accomplished by setting the derivatives with respect to a and b equal to zero. Performing these operations, there results

$$nb + a \sum x_i = \sum y_i \qquad (3\text{-}26)$$

$$b \sum x_i + a \sum x_i^2 = \sum x_i y_i \qquad (3\text{-}27)$$

Solving Eqs. (3-26) and (3.27) simultaneously gives

$$a = \frac{n \sum x_i y_i - (\sum x_i)(\sum y_i)}{n \sum x_i^2 - (\sum x_i)^2} \qquad (3\text{-}28)$$

$$b = \frac{(\sum y_i)(\sum x_i^2) - (\sum x_i y_i)(\sum x_i)}{n \sum x_i^2 - (\sum x_i)^2} \qquad (3\text{-}29)$$

Designating the computed value of y as $\hat{y}$, we have

$$\hat{y} = ax + b$$

and the standard error of estimate of y for the data is

$$\text{Standard error} = \left[\frac{\sum (y_i - \hat{y}_i)^2}{n - 2} \right]^{1/2} \qquad (3\text{-}30)$$

$$= \left[\frac{\sum (y_i - ax_i - b)^2}{n - 2} \right]^{1/2} \qquad (3\text{-}31)$$

The method of least squares may also be used for determining higher-order polynomials for fitting data. One only needs to perform additional differentiations to determine additional constants. For example, if it were desired to obtain a least-squares fit according to the quadratic function

$$y = ax^2 + bx + c$$

the quantity

$$S = \sum_{i=1}^{n} [y_i - (ax_i^2 + bx_i + c)]^2$$

would be minimized by setting the following derivatives equal to zero:

$$\frac{\partial S}{\partial a} = 0 = \sum 2[y_i - (ax_i^2 + bx_i + c)](-x_i^2)$$

$$\frac{\partial S}{\partial b} = 0 = \sum 2[y_i - (ax_i^2 + bx_i + c)](-x_i)$$

$$\frac{\partial S}{\partial c} = 0 = \sum 2[y_i - (ax_i^2 + bx_i + c)](-1)$$

Expanding and collecting terms,

$$a \sum x_i^4 + b \sum x_i^3 + c \sum x_i^2 = \sum x_i^2 y_i \qquad (3\text{-}32)$$

$$a \sum x_i^3 + b \sum x_i^2 + c \sum x_i = \sum x_i y_i \qquad (3\text{-}33)$$

$$a \sum x_i^2 + b \sum x_i + cn = \sum y_i \qquad (3\text{-}34)$$

These equations may then be solved for the constants a, b, and c.

In the above discussion of the method of least squares no mention has been made of the influence of experimental uncertainty on the calculation. We are considering the method primarily for its utility in fitting an algebraic relationship to a set of data points. Clearly, the various x_i and y_i could have different experimental uncertainties. To take all these into account requires a rather tedious calculation procedure which we shall not present here; however, we may state the following rules:

1. If the values of x_i and y_i are taken as the data value in y and the value of x *on the fitted curve for the same value of y*, then there is a presumption that the uncertainty in x is large compared with that in y.
2. If the values of x_i and y_i are taken as the data value in y and the value *on the fitted curve for the same value of x*, the presumption is that the uncertainty in y dominates.
3. If the uncertainties in x_i and y_i are believed to be of approximately equal magnitude, a special averaging technique must be used.

In rule 1 we say we are taking a *regression* of x on y, and in rule 2 there is a regression of y on x. In the second case we are minimizing the sum of the squares of the deviations of the actual points from the assumed curve and also assuming *that x does not vary appreciably at each point.* If we obtained

$$y = a + bx$$

and then solved to get

$$x = \frac{1}{b} y - \frac{a}{b}$$

this second relation would *not* necessarily give a good calculation for x, since the minimization was carried out in the y direction, and not the x direction. In Example 3-15, rule 2 is assumed to apply.

Example 3-15 From the following data obtain y as a linear function of x using the method of least squares:

y_i	x_i
1.2	1.0
2.0	1.6
2.4	3.4
3.5	4.0
3.5	5.2
$\sum y_i = 12.6$	$\sum x_i = 15.2$

SOLUTION We seek an equation of the form

$$y = ax + b$$

We first calculate the quantities indicated in the following table:

$x_i y_i$	x_i^2
1.2	1.0
3.2	2.56
8.16	11.56
14.0	16.0
18.2	27.04
$\sum x_i y_i = 44.76$	$\sum x_i^2 = 58.16$

We calculate the value of a and b, using Eqs. (3-28) and (3-29) with $n = 5$:

$$a = \frac{(5)(44.76) - (15.2)(12.6)}{(5)(58.16) - (15.2)^2} = 0.540$$

$$b = \frac{(12.6)(58.16) - (44.76)(15.2)}{(5)(58.16) - (15.2)^2} = 0.879$$

Thus, the desired relation is

$$y = 0.540x + 0.879$$

A plot of this relation and the data points from which it was derived is shown in the accompanying figure.

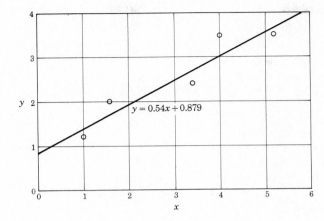

3-12 THE CORRELATION COEFFICIENT

Let us assume that a suitable correlation between y and x has been obtained, either by least-squares analysis or graphical curve fitting. We want to know how good this fit is, and the parameter which conveys this information is the *correlation coefficient r* defined by

$$r = \left[1 - \frac{\sigma_{y,x}^2}{\sigma_y^2} \right]^{1/2} \tag{3-35}$$

where σ_y is the standard deviation of y given as

$$\sigma_y = \left[\frac{\sum\limits_{i=1}^{n} (y_i - y_m)^2}{n-1} \right]^{1/2} \tag{3-36}$$

and

$$\sigma_{y,x} = \left[\frac{\sum\limits_{i=1}^{n} (y_i - y_{ic})^2}{n-2} \right]^{1/2} \tag{3-37}$$

The y_i are the actual values of y, and the y_{ic} are the values computed from the correlation equation for the same value of x. It may be noted that many calculators have built-in routines which calculate the correlation coefficient as well as other statistical functions.

Example 3-16 Calculate the correlation coefficient for the least-square correlation of Example 3-15.

SOLUTION From Example 3-15

$$y_m = \frac{\sum y_i}{n} = \frac{12.6}{5} = 2.52$$

and from the correlating equation $y_{ic} = 0.540x + 0.879$

i	y_i	y_{ic}	$(y_i - y_{ic})^2$
1	1.2	1.419	0.048
2	2.0	1.743	0.066
3	2.4	2.715	0.0992
4	3.5	3.039	0.0212
5	3.5	3.687	0.035
			$\sum = 0.2694$

so that

$$\sigma_{y,\,x} = \left(\frac{0.2694}{3}\right)^{1/2} = 0.2997$$

Also, $\sigma_y = 0.987$, so that the correlation coefficient is

$$r = \left[1 - \left(\frac{0.2997}{0.987}\right)^2\right]^{1/2} = 0.9528$$

3-13 STANDARD DEVIATION OF THE MEAN

We have taken the arithmetic mean value as the best estimate of the true value of a set of experimental measurements. Considerable discussion has been devoted to the gaussian normal error distribution and to an examination of the various types of errors and deviations that may occur in an experimental measurement. But one very important question has not yet been answered; i.e., how *good* (or precise) is this arithmetic mean value which is taken as the best estimate of the true value of a set of readings? To obtain an experimental answer to this question it would be necessary to repeat the set of measurements and find a new arithmetic mean. In general, we would find that this new arithmetic mean would differ from the previous value, and thus we would not be able to resolve the problem until a large number of *sets of data* were collected. We would then know how well the mean of a single set approximated the mean which would be obtained with a large number of sets. The mean value of a large number of sets is presumably the true value. Consequently, we wish to know the standard deviation of the mean of a single set of data from this true value.

It turns out that the problem may be resolved with a statistical analysis, which we shall not present here. The result is

$$\sigma_m = \frac{\sigma}{\sqrt{n}} \tag{3-38}$$

where σ_m = standard deviation of the mean value
   σ = standard deviation of the set of measurements
   n = number of measurements in the set

The following example illustrates the use of Eq. (3-38).

Example 3-17 For the data of Example 3-6, estimate the uncertainty in the calculated mean value of the readings.

SOLUTION We shall make this estimate for the original data and for the reduced data of Example 3-10. For the original data the standard deviation of the mean is

$$\sigma_m = \frac{\sigma}{\sqrt{n}} = \frac{0.627}{\sqrt{10}} = 0.198 \text{ cm}$$

The arithmetic mean value calculated in Example 3-6 was $x_m = 5.613$ cm. We could now specify the uncertainty of this value by using the odds of Table 3-3:

$$x_m = 5.613 \pm 0.198 \text{ cm} \qquad (2.15 \text{ to } 1)$$

$$= 5.613 \pm 0.396 \text{ cm} \qquad (21 \text{ to } 1)$$

$$= 5.613 \pm 0.594 \text{ cm} \qquad (369 \text{ to } 1)$$

Using the data of Example 3-10, where one point has been eliminated by Chauvenet's criterion, we may make a better estimate of the mean value with less uncertainty. The standard deviation of the mean is calculated as

$$\sigma_m = \frac{\sigma}{\sqrt{n}} = \frac{0.465}{\sqrt{9}} = 0.155 \text{ cm}$$

for the mean value of 5.756 cm. Thus, we would estimate the uncertainty as

$$x_m = 5.756 \pm 0.155 \text{ cm} \qquad (2.15 \text{ to } 1)$$

$$= 5.756 \pm 0.310 \text{ cm} \qquad (21 \text{ to } 1)$$

$$= 5.756 \pm 0.465 \text{ cm} \qquad (369 \text{ to } 1)$$

We should note that the calculation of statistical parameters like standard deviation and least-square fits to data is easily performed with standard computer programs which are available on even small hand calculators. Rapid expansion of the availability of compact programs for data analysis is to be expected.

3-14 GRAPHICAL ANALYSIS AND CURVE FITTING

Engineers are well known for their ability to plot many curves of experimental data and to extract all sorts of significant facts from these curves. The better one understands the physical phenomena involved in a certain experiment, the better is one able to extract a wide variety of information from graphical displays of experimental data. Because these physical phenomena may encompass all engineering science, we cannot discuss them here except to emphasize that the person who is usually most successful in analyzing experimental data is the one who understands the physical processes behind the data. Blind curve-plotting and cross-plotting usually generate an excess of displays, which are confusing not only to the management or supervisory personnel who must pass on the experiments, but sometimes even to the experimenter. To be blunt, the engineer should give considerable thought to the kind of information being looked for before even taking the graph paper out of the package.

Assuming that the engineer knows what is to be examined with graphical presentations, the plots may be carefully prepared and checked against appropriate theories. Frequently, a *correlation* of the experimental data is desired in terms of an analytical expression between variables that were measured in the experiment. When the data may be approximated by a straight line, the analytical relation is easy to obtain; but when almost any other functional variation is present, difficulties are usually encountered. This fact is easy to understand since a straight line is easily recognizable on graph paper, whereas the functional form of a curve is rather doubtful. The curve could be a polynomial, exponential, or complicated logarithmic function and still present roughly the same appearance to the eye. It is most convenient, then, to try to plot the data in such a form that a straight line will be obtained for certain types of functional relationships. If the experimenter has a good idea of the type of function that will represent the data, then the type of plot is easily selected. It is frequently possible to estimate the functional form that the data will take on the basis of theoretical considerations and the results of previous experiments of a similar nature.

Table 3-6 summarizes several different types of functions and plotting methods that may be used to produce straight lines on graph paper. The graphical measurements, which may be made to determine the various constants, are also shown. It may be remarked that the method of least squares may be applied to all these relations to obtain the best straight line to fit the experimental data.

Please note that when using logarithmic or semilog graph paper it is unnecessary to make log calculations; the scaling of the paper automatically accomplishes this.

3-15 GENERAL CONSIDERATIONS IN DATA ANALYSIS

Our discussions in this chapter have considered a variety of topics: statistical analysis, uncertainty analysis, curve plotting, least squares, etc. With these tools the reader is equipped to handle a variety of circumstances that may occur in

Table 3-6 Methods of plotting various functions to obtain straight lines

Functional relationship	Method of plot	Graphical determination of parameters
$y = ax + b$	y versus x on linear paper	
$y = ax^b$	$\log y$ versus $\log x$ on log-log paper	
$y = ae^{bx}$	$\log y$ versus x on semilog paper	
$y = \dfrac{x}{a + bx}$	$\dfrac{1}{y}$ versus $\dfrac{1}{x}$ on linear paper	
$y = a + bx + cx^2$	$\dfrac{y - y_1}{x - x_1}$ versus x on linear peper	

(Continued)

Table 3-6 Methods of plotting various functions to obtain straight lines (*Continued*)

Functional relationship	Method of plot	Graphical determination of parameters
$y = \dfrac{x}{a + bx} + c$	$\dfrac{x - x_1}{y - y_1}$ versus x on linear paper	Slope $= b + \dfrac{b^2}{a} x_1$; $a + bx_1$
$y = ae^{bx + cx^2}$	$\log\left[\left(\dfrac{y}{y_1}\right)^{1/(x - x_1)}\right]$ versus x on semilog paper	Slope $= c \log e$; $b + cx_1 \log e$
$y = 1 - e^{-bx}$	$\log\left(\dfrac{1}{1 - y}\right)$ versus x on semilog paper	Slope $= b$
$y = a + \dfrac{b}{x}$	y versus $\dfrac{1}{x}$ on linear paper	Slope $= b$; a
$y = a + b\sqrt{x}$	y versus $\sqrt{x}$ on linear paper	Slope $= b$; a

experimental investigations. As a summary to this chapter let us now give an approximate outline of the manner in which one would go about analyzing a set of experimental data.

1. *Examine the data for consistency.* No matter how hard one tries, there will always be some data points that appear to be grossly in error. If we add heat to a container of water, the temperature must rise, and so if a particular data point indicates a *drop* in temperature for a heat *input*, that point might be eliminated. In other words, the data should follow commonsense consistency, and points that do not appear proper should be eliminated. If very many data points fall in the category of "inconsistent," perhaps the entire experimental procedure should be investigated for gross mistakes or miscalculation.

2. *Perform a statistical analysis of data where appropriate.* A statistical analysis is only appropriate when measurements are repeated several times. If this is the case, make estimates of such parameters as standard deviation, etc.

3. *Estimate the uncertainties in the results.* We have discussed uncertainties at length. Hopefully, these calculations will have been performed in advance, and the investigator will already know the influence of different variables by the time the final results are obtained.

4. *Anticipate the results from theory.* Before trying to obtain correlations of the experimental data, the investigator should carefully review the theory appropriate to the subject and try to glean some information that will indicate the trends the results may take. Important dimensionless groups, pertinent functional relations, and other information may lead to a fruitful interpretation of the data.

5. *Correlate the data.* The word "correlate" is subject to misinterpretation. In the context here we mean that the experimental investigator should make sense of the data in terms of physical theories or on the basis of previous experimental work in the field. Certainly, the results of the experiments should be analyzed to show how they conform to or differ from previous investigations or standards that may be employed for such measurements.

3-16 SUMMARY

By now the reader will have sensed the central theme of this chapter as that of uncertainty analysis and the use of this analysis to influence experiment design, instrument selection, and evaluation of the results of experiments. At this point we must reiterate statements we have made before. We still must recognize that uncertainty is *not* the same as error, even though some people interchange the terms. As we saw in Chap. 2, the determination of "error" is eventually related to a comparison with a standard. Even then, there is still "uncertainty" in the error because the "standard" has its own uncertainty.

In the chapters which follow we shall examine a large number of instruments and measurement devices and will see how the concepts of error, uncertainty, and calibration apply to each.

REVIEW QUESTIONS

3-1 How does an error differ from an uncertainty?

3-2 What is a fixed error; random error?

3-3 Define standard deviation and variance.

3-4 In the normal error distribution, what does $P(x)$ represent?

3-5 What is meant by measure of precision?

3-6 What is Chauvenet's criterion and how is it applied?

3-7 What are some purposes of uncertainty analyses?

3-8 Why is an uncertainty analysis important in the preliminary stages of experiment planning?

3-9 How can an uncertainty analysis help to reduce overall experimental uncertainty?

3-10 What is meant by standard deviation of the mean?

3-11 What is a least-squares analysis?

3-12 What is a least-squares analysis?

PROBLEMS

3-1 The resistance of a resistor is measured 10 times, and the values determined are 100.0, 100.9, 99.3, 99.9, 100.1, 100.2, 99.9, 100.1, 100.0, and 100.5. Calculate the uncertainty in the resistance.

3-2 A certain resistor draws 110.2 V and 5.3 A. The uncertainties in the measurements are ± 0.2 V and ± 0.06 A, respectively. Calculate the power dissipated in the resistor and the uncertainty in the power.

3-3 A small plot of land has measured dimensions of 50.0 by 150.0 ft. The uncertainty in the 50-ft dimension is ± 0.01 ft. Calculate the uncertainty with which the 150-ft dimension must be measured to ensure that the total uncertainty in the area is not greater than 150 percent of that value it would have if the 150-ft dimension were exact.

3-4 Two resistors R_1 and R_2 are connected in series and parallel. The values of the resistances are

$$R_1 = 100.0 \pm 0.1 \ \Omega$$

$$R_2 = 50.0 \pm 0.03 \ \Omega$$

Calculate the uncertainty in the combined resistance for both the series and the parallel arrangements.

3-5 A resistance arrangement of 50 Ω is desired. Two resistances of 100.0 ± 0.1 Ω and two resistances of 25.0 ± 0.02 Ω are available. Which should be used, a series arrangement with the 25-Ω resistors or a parallel arrangement with the 100-Ω resistors? Calculate the uncertainty for each arrangement.

3-6 The following data are taken from a certain heat-transfer test. The expected correlation equation is $y = ax^b$. Plot the data in an appropriate manner, and use the method of least squares to obtain the best correlation.

x	2040	2580	2980	3220	3870	1690	2130	2420	2900	3310	1020	1240	1360	1710	2070
y	33.2	32.0	42.7	57.8	126.0	17.4	21.4	27.8	52.1	43.1	18.8	19.2	15.1	12.9	78.5

Calculate the mean deviation of these data from the best correlation.

3-7 A horseshoes player stands 30 ft from the target. The results of the tosses are

Toss	Deviation from target, ft	Toss	Deviation from target, ft
1	0	6	+2.4
2	+3	7	−2.6
3	−4.2	8	+3.5
4	0	9	+2.7
5	+1.5	10	0

On the basis of these data would you say that this is a good player or a poor player? What advice would you give this player in regard to improving at the game?

3-8 Calculate the probability of drawing a full house (three of a kind and two of a kind) in the first 5 cards from a 52-card deck.

3-9 Calculate the probability of filling an inside straight with one draw from the remaining 48 cards of a 52-card deck.

3-10 A voltmeter is used to measure a known voltage of 100 V. Forty percent of the readings are within 0.5 V of the true value. Estimate the standard deviation for the meter. What is the probability of an error of 0.75 V?

3-11 In a certain mathematics course the instructor informs the class that grades will be distributed according to the following scale provided that the average class score is 75:

Grade	A	B	C	D	F
Score	90–100	80–90	70–80	60–70	Below 60

Estimate the percentage distribution of grades for 5, 10, and 15 percent failing. Assume that there are just as many A's as F's.

3-12 For the following data points y is expected to be a quadratic function of x. Obtain this quadratic function by means of a graphical plot and also by the method of least squares.

x	1	2	3	4	5
y	1.9	9.3	21.5	42.0	115.7

3-13 It is suspected that the rejection rate for a plastic-cup-molding machine is dependent on the temperature at which the cups are molded. A series of short tests is conducted to examine this hypothesis with the following results:

Temperature	Total production	Number rejected
T_1	150	12
T_2	75	8
T_3	120	10
T_4	200	13

On the basis of these data do you agree with the hypothesis?

3-14 A capacitor discharges through a resistor according to the relation $E/E_0 = e^{-t/RC}$ where E_0 = voltage at time zero; R = resistance; C = capacitance. The value of the capacitance is to be measured by recording the time necessary for the voltage to drop to a value E_1. Assuming that the resistance is known accurately, derive an expression for the percent uncertainty in the capacitance as a function of the uncertainty in the measurements of E_1 and t.

3-15 In heat-exchanger applications, a log mean temperature difference is defined by

$$\Delta T_m = \frac{(T_{h_1} - T_{c_1}) - (T_{h_2} - T_{c_2})}{\ln \left[(T_{h_1} - T_{c_1})/(T_{h_2} - T_{c_2}) \right]}$$

where the four temperatures are measured at appropriate inlet and outlet conditions for the heat-exchanger fluids. Assuming that all four temperatures are measured with the same absolute uncertainty w_T, derive an expression for the percentage uncertainty in ΔT_m in terms of the four temperatures and the value of w_T. Recall that the percentage uncertainty is

$$\frac{w_{\Delta T_m}}{\Delta T_m} \times 100$$

3-16 A certain length measurement is made with the following results:

Reading	1	2	3	4	5	6	7	8	9	10
x, in	49.36	50.12	48.98	49.24	49.26	50.56	49.18	49.89	49.33	49.39

Calculate the standard deviation, the mean reading, and the uncertainty. Apply Chauvenet's criterion as needed.

3-17 Devise a method for plotting the gaussian normal error distribution such that a straight line will result. (*Ans.* $(1/\eta) \ln [\sqrt{2\pi} P(\eta)]$ versus η.) Show how such a plot may be labeled so that it can be used to estimate the fraction of points which lie below a certain value of η. Subsequently show that this plot may be used to investigate the normality of a set of data points. Apply this reasoning to the data points of Example 3-6 and Probs. 3-6 and 3-7.

3-18 A citizens' traffic committee decides to conduct its own survey and analysis of the influence of drinking on car accidents. By some judicious estimates the committee determines that in their community 30 percent of the drivers on a Saturday evening between 10 P.M. and 2 A.M. have consumed some alcohol. During this same period there were 50 accidents, varying from minor scratched fenders to fatalities. In these 50 accidents 50 of the drivers had had something to drink (there are 100 drivers for 50 accidents). From these data what conclusions do you draw about the influence of drinking on car accidents? Can you devise a better way to perform this analysis?

3-19 The grades for a certain class fall in the following ranges:

Number	10	30	50	40	10	8
Score	90–100	80–90	70–80	60–70	50–60	Below 50

The arithmetic mean grade is 68. Devise your own grade distribution for this class. Be sure to establish the criteria for the distribution.

3-20 A certain length measurement is performed 100 times. The arithmetic mean reading is 6.823 ft, and the standard deviation is 0.01 ft. How many readings fall within (*a*) ± 0.005 ft, (*b*) ± 0.02 ft, (*c*) ± 0.05 ft, and (*d*) 0.001 ft of the mean value?

3-21 A series of calibration tests is conducted on a pressure gage. At a known pressure of 1000 psia, it is found that 30 percent of the readings are within 1 psia of the true value. At a known pressure of 500 psia, 40 percent of the readings are within 1 psia. At a pressure of 200 psia, 45 percent of the readings are within 1 psia. What conclusions do you draw from these readings? Can you estimate a standard deviation for the pressure gage?

3-22 Two resistors are connected in series and have the following values:

$$R_1 = 10,000\ \Omega \pm 5\% \qquad R_2 = 1\ M\Omega \pm 10\%$$

Calculate the percent uncertainty for the series total resistance.

3-23 Apply Chauvenet's criterion to the data of Example 3-11 and then replot the data on probability paper, omitting any excluded points.

3-24 Plot the data of Example 3-6 on probability paper. Replot the data, taking into account the point eliminated in Example 3-10. Comment on the normality of these two sets of data.

3-25 Two groups of secretaries operate under the same manager. Both groups have the same number of people, use the same equipment, and turn out about the same amount of work. During one maintenance period, group A had 10 service calls on the equipment while group B had only 6 calls. From these data would you conclude that group A was harder on the equipment?

3-26 A laboratory experiment is conducted to measure the viscosity of a certain oil. A series of tests gives the values as 0.040, 0.041, 0.041, 0.042, 0.039, 0.040, 0.043, 0.041, and 0.039 ft^2/s. Calculate the mean reading, the variance, and the standard deviation. Eliminate any data points as necessary.

3-27 The following data are expected to follow a linear relation of the form $y = ax + b$. Obtain the best linear relation in accordance with a least-squares analysis. Calculate the standard deviation of the data from the predicted straight-line relation.

x	0.9	2.3	3.3	4.5	5.7	6.7
y	1.1	1.6	2.6	3.2	4.0	5.0

3-28 The following data points are expected to follow a functional variation of $y = ax^b$. Obtain the values of a and b from a graphical analysis.

x	1.21	1.35	2.40	2.75	4.50	5.1	7.1	8.1
y	1.20	1.82	5.0	8.80	19.5	32.5	55.0	80.0

3-29 The following data points are expected to follow a functional variation of $y = ae^{bx}$. Obtain the values of a and b from a graphical analysis.

x	0	0.43	1.25	1.40	2.60	2.9	4.3
y	9.4	7.1	5.35	4.20	2.60	1.95	1.15

3-30 The following heat-transfer data points are expected to follow a functional form of $N = aR^b$. Obtain the values of a and b from a graphical analysis and also by the method of least squares.

R	12	20	30	40	100	300	400	1000	3000
N	2	2.5	3	3.3	5.3	10	11	17	30

What is the average deviation of the points from the correlating relationship?

3-31 In a student laboratory experiment a measurement is made of a certain resistance by different students. The values obtained were

Reading	1	2	3	4	5	6	7	8	9	10	11	12
Resistance, kΩ	12.0	12.1	12.5	11.8	13.6	11.9	12.2	11.9	12.0	12.3	12.1	11.85

Calculate the standard deviation, the mean reading, and the uncertainty.

3-32 In a certain decade resistance box resistors are arranged so that four resistances may be connected in series to obtain a desired result. The first selector uses 10 resistances of 1000, 2000, ..., 9000, the second uses 10 of 100, 200, ..., 900, the third uses 10 of 10, 20, ..., 90, and the fourth, 1, 2, ..., 9 Ω. Thus, the overall range is 0 to 9999 Ω. If all the resistors have an uncertainty of ± 1.0 percent, calculate the percent uncertainties for total resistances of 9, 56, 148, 1252, and 9999 Ω.

3-33 Calculate the chances and probabilities that data following a normal-distribution curve will fall within 0.2, 1.2, and 2.2 standard deviations of the mean value.

3-34 Suggest improvements in the measurement uncertainties for Example 3–4 which will result in reduction in the overall uncertainty of flow measurement to ± 1.0 percent.

3-35 What uncertainty in the resistance for the first part of Example 3-2 is necessary to produce the same uncertainty in power determination as results from the current and voltage measurements?

3-36 Use the technique of Sec. 3-5 with Example 3-4.

3-37 Use the technique of Sec. 3-5 with Examples 3-3 and 3-2.

3-38 Obtain the correlation coefficient for Prob. 3-27.

3-39 Obtain the correlation coefficient for Prob. 3-28.

3-40 Obtain the correlation coefficient for Probs. 3-29 and 3-30.

3-41 Obtain the correlation coefficient for Probs. 3-6 and 3-12.

3-42 For the heat exchanger of Prob. 3-15 the temperatures are measured as $T_{h_1} = 100°C$, $T_{h_2} = 80°C$, $T_{c_1} = 75°C$, and $T_{c_2} = 55°C$. All temperatures have an uncertainty of $\pm 1°C$. Calculate the uncertainty in ΔT_m using the technique of Sec. 3-5.

3-43 Repeat Prob. 3-41 but with $T_{c_1} = 90°C$ and $T_{c_2} = 70°C$.

3-44 Four resistors having nominal values of 1, 1.5, 3, and 2.5 kΩ are connected in parallel. The uncertainties are ± 10 percent. A voltage of 100 V $\pm$ 1.0 V is impressed on the combination. Calculate the power drawn and its uncertainty. Use Sec. 3-5.

3-45 A radar speed-measurement device for state police is said to have an uncertainty of ± 4 percent when directed straight at an oncoming vehicle. When directed at some angle θ from the straight-on position, the device measures a component of the vehicle speed. The police officer can only obtain a value for the angle $\tilde{\theta}$ through a visual observation having an uncertainty of $\pm 10°$. Calculate the uncertainty of the speed measurement for θ values of 0, 10, 20, 30, and 45°. Use the techniques of both Secs. 3-4 and 3-5.

3-46 An automobile is to be tested for its acceleration performance and fuel economy. Plan this project taking into account the measurements which must be performed and expected uncertainties in these measurements. Assume that three different drivers will be used for the tests. Make plans for the number of runs which will be made for each measurement, taking into account the statistical analysis which will be used to reduce the data. Also prepare a detailed outline with regard to form and content of the report which will be used to present the results.

3-47 A thermocouple is used to measure the temperature of a known standard maintained at 100°C. After converting the electrical signal to temperature the readings are: 101.1, 99.8, 99.9, 100.2, 100.5, 99.6, 100.9, 99.7, 100.1, and 100.3. Using whatever criteria seem appropriate, make some statements about the calibration of the thermocouple.

3-48 Seven students are asked to make a measurement of the thickness of a steel block with a micrometer. The actual thickness of the block is known very accurately as 2.000 cm. The seven measurements are: 2.002, 2.001, 1.999, 1.997, 1.998, 2.003, and 2.003 cm. Comment on these measurements using whatever criteria you think appropriate.

3-49 A collection of 120 rock aggregate samples is taken and the volumes measured for each. The mean volume is 6.8 cm^3 and the standard deviation is 0.7 cm^3. How many rocks would you expect to have volumes ranging from 6.5 to 7.2 cm^3?

3-50 Plot the equation $y = 5e^{1.2x}$ on semilog paper. Arbitrarily assign ficticious data points on both sides of the line so that the line appears by eye as a reasonable representation. Then, using these points, perform a least-squares analysis to obtain the best fit to the points. What do you conclude from this comparison?

REFERENCES

1. Kline, S. J., and F. A. McClintock: Describing Uncertainties in Single-Sample Experiments, *Mech. Eng.*, p. 3, January 1953.
2. Mood, A. M., and F. A. Graybill: "Introduction to the Theory of Statistics," 2d ed., McGraw-Hill Book Company, New York, 1963.
3. Schenck, H.: "Theories of Engineering Experimentation," McGraw-Hill Book Company, New York, 1961.
4. Wilson, E. B.: "An Introduction to Scientific Research," McGraw-Hill Book Company, New York, 1952.
5. Young, H. D.: "Statistical Treatment of Experimental Data," McGraw-Hill Book Company, New York, 1962.
6. Hicks, C. R.: "Fundamental Concepts in the Design of Experiments," Holt, Rinehart, and Winston, Inc., New York, 1964.
7. Moffat, R. J.: Contributions to the Theory of Single-Sample Uncertainty Analysis, *J. Fluids Engr.*, vol. 104, p. 250, June 1982.

CHAPTER

FOUR

BASIC ELECTRICAL MEASUREMENTS AND SENSING DEVICES

4-1 INTRODUCTION

Many measuring devices depend upon some basic electrical principle for their operation, and nearly all data gathering, transmission, and analysis systems depend upon electronic devices. For example, the remote measurement and recording of temperature is ordinarily accomplished in the following way. A transducer is installed at the location of interest, and this device converts the temperature at any given time to an equivalent electric voltage. This voltage is then transmitted to a receiving station where it is displayed in an appropriate fashion. Electrical devices are involved at every stage of this process.

As a consequence of the pervasive nature of electronics in modern engineering, it is desirable to discuss some electrical devices currently employed and to emphasize their uses in the measurement process. We shall first consider the measurement of the basic electrical quantities of current and voltage. Next, we shall examine some simple circuits that may be used for modification and measurement of input signals. Here we shall be concerned specifically with signal amplification and the techniques used to minimize the effects of unwanted but ever-present noise. Finally, the physical principles and operating characteristics of important electrical transducers will be studied and their applications surveyed.

In all the following discussions, due consideration must be given to impedance matching. The principles of matching were discussed in Sec. 2-10, and the reader should keep this material in mind while reading the present chapter.

4-2 FORCES OF ELECTROMAGNETIC ORIGIN

The operation of all electrical devices depends upon two facts: charge exists, and charged entities interact. In terms of basic physics, there are two kinds of charge, positive and negative; like charges repel and unlike charges attract. Such a crude statement suffices to explain simple experiments but is inadequate to allow an understanding of the operation of sophisticated devices. To provide such under-standing, we must carefully model the interaction of charges. Once this is done, we will be able to understand basic electrical phenomena.

Consider a point charge of q coulombs (C). If such a charge were alone in the universe, it would move in a straight line determined by its initial velocity. However, if this charge is moving in a space in which other charges exist, it will "feel" the presence of these other charges. In particular, we define an electric field intensity **E** which exists as a result of the presence of charges other than the point charge q under consideration. Also, we can define a magnetic flux density **B** which exists in the space as a result of the motion of these other charges.

A careful specification of the interaction between the point charge of q C and the other charges is expressed by Lorentz' law. If the point charge is moving with a velocity **v**, the resultant force on the charge is given by the equation

$$\mathbf{F} = q(\mathbf{E} + \mathbf{v} \times \mathbf{B}) \qquad \text{newtons} \qquad (4\text{-}1)$$

where **E** is in volts per meter, **v** is in meters per second, and **B** is in webers per square meter.

In order to apply this equation to a useful case, consider a current-carrying conductor placed in a magnetic field as shown in Fig. 4-1. No electric field is present in this case. The electric current i in the conductor is defined as the ratio of the charge dq passing a cross section of the conductor to the transit time dt.

$$i = \frac{dq}{dt} \qquad (4\text{-}2)$$

We write

$$i \, ds = \frac{dq}{dt} \, ds = v \, dq \qquad (4\text{-}3)$$

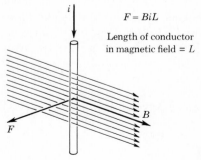

$F = BiL$

Length of conductor
in magnetic field $= L$

Figure 4-1 Current-carrying conductor in a magnetic field.

Figure 4-2 The primitive ammeter.

where ds is an element of length along the conductor covered by the moving charge in a time dt. The force exerted on the charge dq is given by Eq. (4-1) as

$$d\mathbf{F} = dq(\mathbf{v} \times \mathbf{B}) = i(d\mathbf{s} \times \mathbf{B}) \tag{4-4}$$

Remember that the electric field $\mathbf{E}$ is 0. This last equation gives the incremental force $d\mathbf{F}$ exerted on a length ds of the conductor because of the interaction between the charge moving through the conductor and the magnetic flux density in which the conductor is immersed. Carefully observe that this force is perpendicular to the plane defined by the conductor axis and the magnetic field $\mathbf{B}$. To find the total force $\mathbf{F}$ acting on a length L of the conductor, we integrate:

$$\mathbf{F} = \int_0^L i(d\mathbf{s} \times \mathbf{B}) \tag{4-5}$$

In order to simplify our results, consider only the situation in which the conductor axis is perpendicular to $\mathbf{B}$. Then

$$d\mathbf{s} \times \mathbf{B} = dsB \sin 90° = dsB$$

and the total force $\bar{F}$ is given by the expression

$$F = BiL \tag{4-6}$$

To integrate Eq. (4-5) and obtain Eq. (4-6), it was assumed that the magnetic field $\mathbf{B}$ is constant along the length L of the conductor.

Equation (4-6) is very important because it provides a bridge between a basic electrical quantity, i, and a mechanical property, F. In fact, this equation reduces the problem of measuring a current to the more familiar problem of measuring a force. Consider the apparatus shown in Fig. 4-2. With no current flowing through the conductor, the spring will be at its unstretched length. As current flows through the conductor, the spring will stretch and develop the force required to balance the electromagnetic force. The total distance x moved by the spring is found by equating the two forces.

$$Kx = BiL$$

where K is the spring constant. This equation can be solved for x,

$$x = \frac{BL}{K} i$$

and rearranged in the following way:

$$i = \left(\frac{K}{BL}\right)x \tag{4-7}$$

As long as the quantities B, L, and K are known, this last equation offers a direct method of determining the current i flowing in a conductor. We need only measure the total deflection x and then calculate the current. As will be shown, actual current meters are somewhat more complex geometrically, but they do use this basic principle; the current i generates a BiL force which is balanced by a restraining force caused by the stretching or compressing of a spring. The distance the spring moves from equilibrium is a direct measure of the current i.

We will return to careful consideration of electrical measuring devices in Secs. 4-4 and 4-5. Before this, a few pages will be devoted to discussing the nature of electrical signals and the various quantities which are commonly used to characterize such signals. This discussion will allow a better appreciation of measuring instruments.

Consider the primitive current-measuring device shown in Fig. 4-2. If a time-varying current i is applied to the conductor, the position x of the conductor will also change with time. As long as the current changes slowly (so that the mechanical system is in equilibrium at each instant of time), Eq. (4-7) remains valid, and a plot of the current-time function could be obtained by recording the position at different times (by means of a motion picture camera, for example) and applying Eq. (4-7). Figure 4-3 is a possible result of this process. Such a plot is an analog representation of the current i as a function of time t. In this context, the phrase

Figure 4-3 Analog waveform.

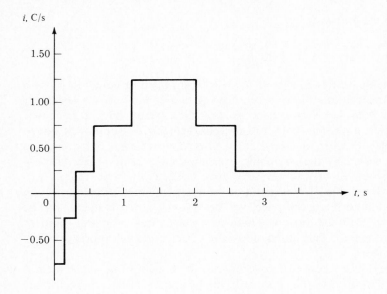

Figure 4-4 Digital waveform.

"analog representation" means that the current is treated as a continuous variable. That is, we assume that we can measure and plot the magnitude of the current with an arbitrarily high degree of precision. In terms of the primitive meter, this implies that quite sophisticated techniques must be used to measure the spring stretching distance x. A laser interferometer could be used, for example. The key idea is that the current i can be treated as a continuous variable. It can take any numerical value. This is the crux of the phrase "analog representation."

Such a representation is the most natural approach to physical quantities. We tend to think of them as continuous in their magnitudes. On the other hand, there are many situations, as we shall see, in which it is helpful to consider *digital representations* of physical quantities. A digital representation allows *only discrete* values. Figure 4-4 shows a digital representation of the analog waveform sketched in Fig. 4-3. In this example, the current is considered in discrete "chunks" of 0.5 C/s.

Several features of digital representation are important. First, as the discrete "chunks" are made smaller, the fit between the digital representation and the analog representation becomes closer. In some situations, the digital representation is a more accurate rendering of the physical reality. For example, if we were forced to measure the deflection of the spring in the primitive instrument of Fig. 4-2 with a meter stick, we could only measure position to, perhaps, the nearest millimeter, and the current i could be found only in discrete pieces.

The concept of a digital representation may appear unfamiliar, but both this chapter and Chap. 14 contain several instances in which the representation and processing of physical signals in digital form is of significant value. Now, we will briefly examine commonly used measures of time-varying quantities.

4-3 WAVEFORM MEASURES

If a scalar physical quantity of interest is constant with respect to time, the specification of this quantity involves only one number. For example, Fig. 4-5 shows a constant current of 7 C/s; to specify this current, one need only state that a current of 7 A is flowing through the conductor.[†] There is no need, in such a situation, to display a plot of current versus time. Engineers describe such a physical situation by simply stating that a 7-A direct current is flowing. The symbol dc is shorthand for the two words *direct current*.

In the case of time-varying physical quantities, the situation is more complex. Consider the current shown in Fig. 4-6 and given by the following equation:

$$i(t) = 7 \sin (377t) \qquad A$$

Is it possible to describe such a waveform meaningfully by a single number? The answer to this question is yes, but in order to appreciate the correct single number, let us examine some incorrect choices. The first possible method is to characterize a time-varying periodic waveform by its average.

$$i_{av} = \frac{1}{T} \int_0^T i(t) \, dt$$

where T is the period of the wave. For the sample wave under consideration, the average current which flows is zero, and this certainly does not constitute a meaningful measure of current $i(t)$. Another measure is the peak-to-peak current; that is, the difference between the maximum and minimum values of the current attained by the waveform:

$$i_{p-p} = 7 - (-7) = 14 \text{ A}$$

† One ampere equals one coulomb per second.

Figure 4-5 A 7-A direct current.

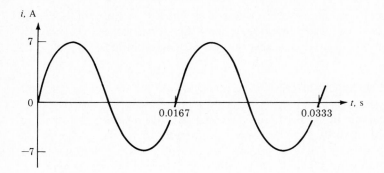

Figure 4-6 A sinusoidal current.

This is a meaningful measure, but many other waveforms also display the same peak-to-peak current. An example is given in Fig. 4-7. Clearly, the current shown in Fig. 4-7 will be significantly different in its physical effects from the current shown in Fig. 4-6.

Suppose that the current $i(t)$ of interest is flowing through a resistor. The power dissipated in the resistor at each instant of time is given by the equation

$$p(t) = Ri^2(t)$$

and the average power dissipated in the resistor is given by the expression

$$P_{av} = \frac{1}{T} \int_0^T p(t)\, dt = R \left\{ \frac{1}{T} \int_0^T i^2(t)\, dt \right\} \qquad (4\text{-}8)$$

Note that we can write

$$P_{av} = RI_{eff}^2$$

as long as we define I_{eff}^2 by the equation

$$I_{eff}^2 = \frac{1}{T} \int_0^T i^2(t)\, dt$$

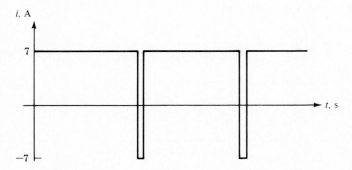

Figure 4-7 Another 14-A peak-to-peak current.

The right-hand side of the last equation defines the root mean square value of the periodic waveform $i(t)$. We write

$$I_{\text{rms}} = \left\{ \frac{1}{T} \int_0^T i^2(t) \, dt \right\}^{1/2} \tag{4-9}$$

and this single number is a meaningful representation of the time-varying current $i(t)$ in that the average power dissipated in a resistor when this current flows is given by the expression

$$P_{\text{av}} = R I_{\text{rms}}^2$$

The rms (root mean square) value associated with a periodic, time-varying waveform is a single number which meaningfully characterizes such a function. It will be very useful in the following sections.

4-4 BASIC ANALOG METERS

Equation (4-6) provided the basic foundation upon which we constructed the primitive current-measuring meter in Fig. 4-2. In order to extend this concept to a more realistic model of actual meters, we construct a coil, as shown in Fig. 4-8, place it in a magnetic field, and measure the force exerted on the coil as a result of the electric current flowing in the coil. If the coil has N turns and the length of each turn in the magnetic field is L, the force on the coil is

$$F = NBiL \tag{4-10}$$

The force is measured by observing the deflection of a spring. The above principles form the basis of the construction of the mirror galvanometer shown in Fig. 4-9. A permanent magnet is used to produce the magnetic field, while the telescope arrangement and expanded scale improve the readability of the instrument. The meter shown in Fig. 4-9 is designated as the *D'Arsonval moving-coil type*. The metal ribbon furnishes the torsional-spring restraining force in this case, while a filamentary suspension would be used for a more sensitive instrument.

Figure 4-8 Current-carrying coil in a magnetic field.

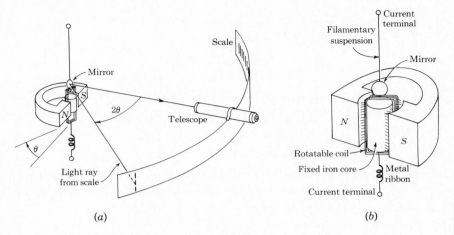

(a)   (b)

Figure 4-9 A typical galvanometer. (a) Optical system; (b) D'Arsonval movement.

Instead of the mirror and light-beam arrangement shown in Fig. 4-9, the D'Arsonval movement could be used as a pointer-type instrument, as shown in Fig. 4-10; however, such an instrument has a lower sensitivity than the mirror galvanometer because of the additional mass of the pointer and decreased readability resulting from the relatively shorter scale length. The sensitivities of various commercial galvanometers as compiled by Sweeney [5] are given in Table 4-1. In this table the period is the undamped natural period of oscillation of the moving coil. The sensitivity values are those obtained with an external circuit resistance necessary to give critical damping.

It is clear that the D'Arsonval movement, in one form or another, may be used for the measurement of direct current. When this movement is connected to an alternating current, the meter will either vibrate or, if the frequency is sufficiently high, indicate zero. In either event, the D'Arsonval movement is not directly applicable to the measurement of alternating current. A detailed analysis of the response of galvanometer movement to various ac waveforms is given by Frank [2].

Figure 4-10 D'Arsonval movement used as a pointer-type instrument.

Table 4-1 Typical galvanometer characteristics[†]

Construction type	Sensitivity per mm or scale division	Period, s	Resistance, Ω	
			External CDR	Coil
Slack suspension, 1-m light beam	0.00001 μA	40	100,000	800
	0.00004 μA	20	70,000	800
	0.0005 μA	6	10,000	650
	0.008 μA	1.5	2500	500
	0.05 μV	7	10	16
	0.2 μV	5	40	16
	0.5 μV	1.5	40	21
Taut suspension, self-contained optical system	0.0005 μA	3	25,000	550
	0.005 μA	3	6000	1000
	0.05 μA	2	2500	350
	0.5 μV	3	50	17
	2.5 μV	2	12	13
	2.5 μV	2	50	13
Taut suspension, pointer type	0.030 μA	5.3	42,000	1900
	0.125 μA	3.5	10,000	1000
	1.0 μA	3	950	250
	46 μV	4.5	30	16
	280 μV	3	50	20

† According to Sweeney [5].

The two most common types of movements used for alternating-current measurement are the iron-vane, or moving-iron, and electrodynamometer arrangements. In the iron-vane instrument, as shown in Fig. 4-11, the current is applied to a *fixed coil*. The iron vane is movable and connected to a restraining spring as shown. The displacement of the vane is then proportional to the inductive force exerted by the coil. The meter is subject to eddy-current losses in the iron vane and various hysteresis effects which limit its accuracy.

The features of the electrodynamometer movement are shown in Fig. 4-12. This movement is similar to the D'Arsonval movement except that the permanent magnet is replaced by an electromagnet, which may be actuated by an alternating current. Consequently, the field in the electromagnet may be made to operate in synchronization with an alternating current in the moving coil. In order to use the electrodynamometer movement for ac measurements, it is necessary to connect the electromagnet and moving coil in series as shown in Fig. 4-13.

Both the iron-vane and the electrodynamometer movements are normally used for low-frequency applications with frequencies from 25 to 125 Hz. Special designs of the electrodynamometer movement may be used to extend its range to

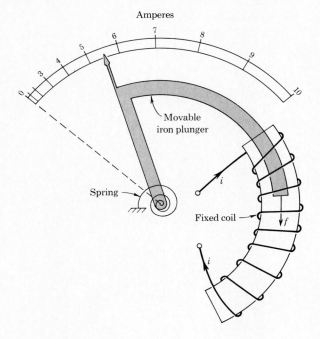

Figure 4-11 Principle of operation of the iron-vane or moving-iron instrument.

Figure 4-12 Basic features of electrodynamometer movement.

$i = i_c = i_p$

Figure 4-13 Electrodynamometer movement used as an ammeter.

about 2000 Hz. Various theoretical considerations for these movements are discussed by Frank [2].

Both the iron-vane and the electrodynamometer instruments indicate the rms value of the alternating current, and the meter deflection varies with I^2_{rms}. The scale of the instrument is not necessarily based on a square law because the proportionality constant between the rms current and the meter deflection changes somewhat with the current.

An important feature of the electrodynamometer instrument is that it may be calibrated with direct current and that the calibration will hold for ac applications within the frequency range of the instrument. The iron-vane instrument is not as versatile because of the residual magnetism in the iron when direct current is used.

A rectifier arrangement may also be used for ac measurements. In this device an ac waveform is modified by some type of rectifier such that current is obtained with a steady dc component. A D'Arsonval movement may then be used to measure the dc component and thereby indicate the value of the alternating current which was applied to the rectifier.

For measurements of high-frequency alternating currents, a thermocouple meter is usually used. This type of meter is indicated in Fig. 4-14. The alternating

Figure 4-14 Schematic of a thermocouple meter.

current is passed through a heater element, and the temperature of the element is indicated by a thermocouple connected to a D'Arsonval meter. The thermocouple indicates the rms value of the current because the average power dissipated in the heater is equal to $I_{rms}^2 R$. The instrument reading is independent of waveform because of this relationship between the thermal emf generated in the thermocouple and the power dissipated in the heater. The thermal emf generated in the thermocouple varies approximately with the square of current, although slight deviations from the square law may be obtained because of change in heater resistance with temperature and other side effects. Alternating currents with frequencies up to 100 MHz may be measured with thermocouple meters.

A dc voltmeter may be constructed very easily by modifying a D'Arsonval movement, as shown in Fig. 4-15. In this arrangement, a large resistor is placed in series with the movement; thus, when the instrument is connected to a voltage source, the current in the instrument is an indication of the voltage. The range of the voltmeter may be altered by changing the internal series resistor. The voltmeter is usually rated in terms of the input voltage for full-scale deflection or in terms of the ratio of internal resistance to the voltage for full-scale deflection. A series-resistor arrangement may also be used with the iron-vane and electrodynamometer instruments for measurements of rms values of ac voltages. For low frequencies up to about 125 Hz the electrodynamometer meter may be calibrated with dc voltage and the calibration used for ac measurements.

Electrostatic forces may also be used to indicate electric potential difference. For this purpose two plates are arranged as shown in Fig. 4-16. One plate is fixed, and the other is mounted in jeweled bearings so that it may move freely. A spiral spring provides a restraining force on the movable plate. The electric potential difference to be measured is impressed on the two plates. If two complete disks were used instead of the sectoral plate arrangement, the net torque would be zero; however, with the arrangement shown, the fringe effects of the electric field produce a net force in the indicated direction which is proportional to the square of the rms voltage. As the movable plate changes position, the capacitance changes, and hence the proportionality between the stored energy and the voltage varies with the impressed voltage.

The electrostatic voltmeter may be used for either ac or dc voltage measurements, but potentials above 100 V are required in order to produce a sufficiently

Figure 4-15 D'Arsonval meter used as a voltmeter.

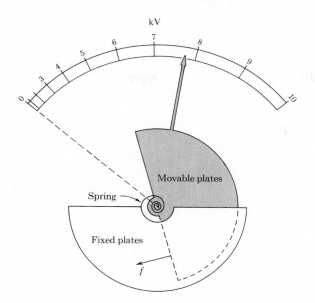

Figure 4-16 Electrostatic-voltmeter movement.

strong torque in the system. The meter may be calibrated with direct current and then used for measurement of rms values of ac voltages, regardless of the waveform. Electrostatic voltmeters are generally applicable up to frequencies of 50 MHz. It may be noted that the electrostatic voltmeter has an extremely high input impedance for dc applications but a much lower ac impedance as a result of the capacitance reactance. The capacitance may be about 20 pF for a 5000-V meter.

   In the preceding paragraphs we have discussed some of the more important analog devices that are used for the measurement of electric current and voltage. Next, we examine digital devices.

4-5 BASIC DIGITAL METERS

We have already discussed analog and digital representations of time-varying signals. Analog meters are reasonably simple in that they depend upon balancing a spring force with a force generated by the interaction between a current and a magnetic field. However, the use of such analog meters necessarily is limited by the accuracy with which the position of the indicator with respect to the scale can be read. Parallax is also a problem, and many times two individuals can read the same meter and arrive at different values of the quantity being measured.

   With a digital meter, the value of the measured quantity is shown directly as a series of digits. The displays of electronic calculators are probably the most familiar example of such outputs. This means that parallax is no longer a problem and different individuals will read the same values. In addition, for

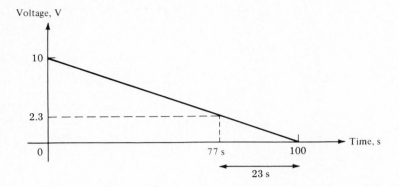

Figure 4-17 Digital voltage measurement.

fundamental reasons which are beyond the scope of this text, the inherent accuracy of digital meters is much greater than that of analog meters. For these reasons, we will briefly discuss the digital voltmeter and related instruments.

The heart of digital meters is an oscillator or "clock." Such a device is often a quartz crystal to which electrical contacts are made. When such a crystal is connected to the proper external electrical components, it establishes an output voltage which is almost sinusoidal with a fixed frequency. This frequency is controlled by the dimensions of the quartz crystal. A familiar example of an application of such devices is the digital watch which has become popular in recent years. All digital meters have such a device, although not all have a quartz oscillator. In some, the timing function is carried out by means of integrated circuits. The clock output is usually "shaped" electronically into a series of pulses, one pulse for each cycle of clock oscillation.

The cornerstone of digital instrumentation is the digital voltmeter. Typical operation is characterized by the following: (1) The capability to generate an internal reference voltage and decrease it linearly from 10 V to 0 at a rate of 0.1 V/s; (2) the ability to compare the reference voltage to the voltage being measured and to generate a signal when the two are equal; and (3) the ability to generate another signal when the reference voltage has reached 0 V.

For simplicity, assume that the clock generates one pulse each second. Let us see how voltage can be determined using this set of capabilities. All we need do is start counting the clock pulses when the reference voltage equals the voltage being measured and stop the counting when the reference voltage reaches zero. If, for example, a total count of 23 is obtained in a given measurement, this means that 23 s elapsed between the "start count" and the "stop count" times. Since the reference voltage was decreasing at 0.1 V/s, the measured voltage is 2.3 V. Figure 4-17 illustrates these concepts.

A digital voltmeter can be converted into an ammeter by including a precision resistor within the instrument and measuring the voltage drop across this resistor caused by an unknown current. In a similar fashion, unknown resistances can be measured by incorporating an accurate current passing through the unknown resistor. In both of these cases, the instrument design is such that the digital readout is the desired current or resistance value.

The simple example given above to illustrate digital meter operation serves its purpose but is somewhat misleading in that we have specified a very slow clock. In fact, the clocks used in available instrumentation emit pulses at much higher rates than those specified above. The digital meter is, of course, naturally wedded to the digital display. Since pulses are available in the system, these are used to drive the display circuits directly. In this way, a set of light-emitting diodes (for example) can be used with little additional circuit complexity.

It is worth mentioning one significant additional difference between analog and digital instruments. The analog meters basically respond to, and measure, currents. The amount of torque produced on a meter movement is directly related to the amount of current flowing through the meter. The situation is complementary in the case of digital instruments. These devices respond to and measure voltages directly. This is a result of the physical behavior of the integrated circuits of which digital instruments are composed.

4-6 BASIC INPUT CIRCUITS

A block diagram of the general method of gathering, processing, and displaying physical information is shown in Fig. 4-18. The first block represents the transducer, which serves to convert the value of the physical property of interest into an electrical signal in some "faithful" fashion. A later section of this chapter concerns itself with such transducers, and we will mention a few examples here. A microphone converts pressure into an equivalent electric voltage. A thermocouple converts temperature into an equivalent electric voltage. A solar cell converts incident light into an equivalent electric current, etc.

The electrical output of the transducer must be connected to some additional circuitry if we are to use it for anything. Care must be taken to ensure that this input circuit does not change the value of the transducer output. We will shortly discuss these issues in detail.

The third block in Fig. 4-18 is labeled "signal conditioning"; this refers to various techniques which are available to reduce the effects of noise. This term is used in a general sense to mean any electrical signals which may be present in addition to the output of the transducer. The next block is the transmission

Figure 4-18 Block diagram of measurement and display.

segment, which simply indicates that we may, in some instances, need to make a measurement at one physical location and transmit the signal to another place for further processing, display, and/or storage.

The next to last step points out the need to process the signal after transmission. We will not go into detail here, but, for example, if you were to listen to the output of an AM radio transmitter, you would hear nothing. That is because the frequency of such transmissions is much higher than the response capabilities of the human ear. A radio receiver provides the processing necessary to convert the signal to an audible frequency.

The display block in Fig. 4-18 indicates the need to ultimately display the electrical signals in some comprehensible form. Examples include cathode-ray screens, paper tape, magnetic tape, lines of printer output, xy plotters, etc. In this section, we will discuss a few representative types of input circuits.

Consider a gas sensor, the resistance of which changes as a function of the gas concentration surrounding the sensor. A simple type of input circuit uses the current flow through the sensor's resistance as an indication of the value of the resistance. Let the sensor be in series with a battery, and represent the battery as a series combination of an ideal voltage source E_i and an internal resistance R_i, as shown in Fig. 4-19. A change in the gas concentration results in a change in the resistance, as indicated by the movable contact. The current is given by

$$ i = \frac{E_i}{R + R_i} \tag{4-11} $$

The maximum resistance of the transducer is R_m, and the current may be written in dimensionless form as

$$ \frac{i}{E_i/R_i} = \frac{1}{(R/R_m)(R_m/R_i) + 1} \tag{4-12} $$

It would be desirable to have the current output vary linearly with the resistance of the transducer. Unfortunately, this is not the case, as the last equation shows, although the output may be approximately linear for some ranges of operation.

The circuit shown above may be modified by using a voltmeter, as shown in Fig. 4-20. Let us assume that the internal impedance of the voltmeter is very large compared with the resistance in the circuit, so that we may neglect the current

Figure 4-19 Current-sensitive input circuit.

Figure 4-20 Voltage-sensitive input circuit. Change in resistance R is indicated through change in voltage indication.

drawn by the meter. That is, the meter looks like an open circuit. The current flow is still given by

$$i = \frac{E_i}{R + R_i} \tag{4-13}$$

Let E be the voltage across the transducer, as indicated in Fig. 4-20. Then

$$\frac{E}{E_i} = \frac{iR}{i(R + R_i)} = \frac{(R/R_m)(R_m/R_i)}{1 + (R/R_m)(R_m/R_i)} \tag{4-14}$$

Now we have obtained a voltage indication as a measurement of the resistance R, but a nonlinear output is still obtained. The advantage of the circuit in Fig. 4-20 over the one in Fig. 4-19 is that a voltage measurement is frequently easier to perform than a current measurement. The voltage-sensitive circuit is called a *ballast circuit*.

We can define the sensitivity of the ballast circuit as the rate of change of the transducer voltage with respect to the transducer resistance R. Thus,

$$S = \frac{dE}{dR} = \frac{E_i R_i}{(R_i + R)^2} \tag{4-15}$$

We would like to design the circuit so that the sensitivity S is a maximum. The circuit-design variable which is at our disposal is the fixed resistance R_i, so that we wish to maximize the sensitivity with respect to this variable. The maximizing condition

$$\frac{dS}{dR_i} = 0 = \frac{E_i(R - R_i)}{(R_i + R)^3} \tag{4-16}$$

is applied. Thus, for maximum sensitivity we should take $R_i = R$. But, since R is a variable, we may select the value of R_i only for the range of R where the sensitivity is to be a maximum.

In both of the above circuits, a current measurement has been used as an indicator of the value of the variable resistance of the transducer. In some instances, it is more convenient to use a voltage-divider circuit as sketched in Fig. 4-21. In this arrangement, a fixed voltage E_0 is impressed across the total transducer resistance R_m while the variable contact is connected to a voltmeter with internal resistance R_i. If the impedance of the meter is sufficiently high, the

Figure 4-21 Simple voltage-divider circuit.

indicated voltage E will be directly proportional to the variable resistance R; that is,

$$\frac{E}{E_0} = \frac{R}{R_m} \qquad \text{for } R_i \gg R \tag{4-17}$$

With a finite meter resistance a current is drawn which affects the voltage measurement. Considering the internal resistance of the meter, the current drawn from the voltage source is

$$i = \frac{E_0}{R_m - R + R_i R/(R + R_i)} \tag{4-18}$$

The indicated voltage is therefore

$$E = E_0 - i(R_m - R)$$

or

$$\frac{E}{E_0} = \frac{R/R_m}{(R/R_m)(1 - R/R_m)(R_m/R_i) + 1} \tag{4-19}$$

As a result of the loading action of the meter, the voltage does not vary in a linear manner with the resistance R. If Eq. (4-19) is taken as the true relationship between voltage and resistance, then an expression for the loading error may be written as

$$\text{Loading error} = \frac{(E/E_0)_{\text{true}} - (E/E_0)_{\text{ind}}}{(E/E_0)_{\text{true}}}$$

We wish to know how much the voltage ratio of Eq. (4-19) differs from the simple linear relation of Eq. (4-17), which would be observed under open circuit conditions or for very large meter impedances. We thus use Eq. (4-19) for the true

Figure 4-22 Simple voltage-balancing potentiometer circuit.

value and Eq. (4-17) for an assumed linear indication and calculate the loading error as

$$\text{Fractional loading error} = -(R/R_m)(1 - R/R_m)(R_m/R_i) \qquad (4\text{-}20a)$$

We may note that the behavior of Eq. (4-20a) is such that the loading error becomes zero at each end of the resistance scale, i.e., for $R/R_m = 0$ and 1.0. Equation (4-20a) expresses the loading error as a fraction of the voltage reading obtained in Eq. (4-19). The raw error or deviation from a linear indication is

$$\text{Deviation from linear} = \frac{-(R/R_m)^2(1 - R/R_m)}{(R/R_m)(1 - R/R_m) + R_i/R_m} \qquad (4\text{-}20b)$$

The voltage-divider circuit shown in Fig. 4-21 has the disadvantage that the indicated voltage is affected by the loading of the meter. This difficulty may be alleviated by utilizing a voltage-balancing potentiometer circuit, as shown in Fig. 4-22. In this arrangement a known voltage E_0 is impressed on the resistor R_m while the unknown voltage is impressed on the same resistor through the galvanometer with internal resistance R_i and the movable contact on the resistor R_m. At some position of the movable contact the galvanometer will indicate zero current, and the unknown voltage may be calculated from

$$\frac{E}{E_0} = \frac{R}{R_m} \qquad (4\text{-}21)$$

Notice that the internal resistance of the galvanometer does not affect the reading in this case; however, it does influence the sensitivity of the circuit. The voltage-balancing potentiometer circuit is widely used for precise measurements of small electric potentials, particularly those generated by thermocouples. In order to accurately determine the unknown voltage E, the supply voltage E_0 must be accurately known. A battery is ordinarily used for the supply voltage, but this represents an unreliable source because of aging characteristics. The aging problem is solved by using a standard-cell arrangement to standardize periodically the battery voltage, as shown schematically in Fig. 4-23. When switch S_2 is

Figure 4-23 Potentiometer circuit incorporating features for standardization of battery voltage.

in position A and switch S_1 is closed, the circuit is the same as Fig. 4-22. Then, when switch S_1 is opened and switch S_2 is placed in position B, the standard cell E_s is connected in the circuit. The variable compensator resistance R_c is adjusted until the galvanometer indicates balance conditions. The protective resistor R_1 may then be bypassed by closing switch S_1 and a fine adjustment of the compensator resistor effected. The protective resistor R_1 is placed in the circuit to avoid excessive current drain on the standard cell and also to protect the galvanometer. Note that the standard-cell voltage is impressed on a fixed portion of the resistance R. Once the battery has been standardized, switch S_2 may be placed in the A position and the unknown voltage E measured. The battery may be standardized as often as necessary for the particular application.

Example 4-1 The output of a transducer with a total resistance of 150 Ω is to be measured with a voltage-sensitive circuit like that shown in Fig. 4-20. The sensitivity is to be a maximum at the midpoint of the transducer. Calculate the sensitivity at the 25 and 75 percent positions, assuming a voltage source E_i of 100 V.

SOLUTION For maximum sensitivity at the midpoint of the range, we take

$$R_i = R = \tfrac{1}{2}R_m = 75 \ \Omega$$

At the 25 percent position, $R = (0.25)(150) = 37.5 \ \Omega$, and the sensitivity is calculated from Eq. (4-15):

$$S = \frac{dE}{dR} = \frac{E_i R_i}{(R_i + R)^2} = \frac{(100)(75)}{(75 + 37.5)^2} = 0.592 \ \text{V}/\Omega$$

At the 75 percent position, the corresponding sensitivity is

$$S = \frac{(100)(75)}{(75 + 112.5)^2} = 0.213 \ \text{V}/\Omega$$

Example 4-2 The voltage-divider circuit is used to measure the output of the transducer in Example 4-1. A 100-V source is used ($E_0 = 100$ V), and the internal resistance of the meter R_i is 10,000 Ω. Calculate the loading error at the 25 and 75 percent positions on the transducer and the actual voltage readings which will be observed at these points.

SOLUTION We have

$$R_m/R_i = 150/10{,}000 = 1.5 \times 10^{-2}$$

At the 25 percent position $R/R_m = 0.25$ and the fractional loading error is calculated from Eq. (4-20a) as

$$\text{Fractional loading error} = -(0.25)(1 - 0.25)$$

$$= 0.281\%$$

The deviation from linear is obtained from Eq. (4-20b)

$$\text{Deviation from linear} = \frac{-(0.25)^2(1 - 0.25)}{(0.25)(1 - 0.25) + 1/0.015}$$

$$= -7.01 \times 10^{-4}$$

The indicated voltage is, therefore,

$$E = (100)(0.25 - 7.01 \times 10^{-4}) = 24.93 \text{ V}$$

At the 75 percent position $R/R_m = 0.75$ and

$$\text{Fractional loading error} = -(0.75)(1 - 0.75)(0.015)$$

$$= 2.81 \times 10^{-3} = 0.281\%$$

The deviation from linear is

$$\text{Deviation from linear} = \frac{-(0.75)^2(1 - 0.75)}{(0.75)(1 - 0.75) + 1/0.015}$$

$$= -2.103 \times 10^{-3}$$

and the indicated voltage is

$$E = (100)(0.75 - 2.103 \times 10^{-3}) = 74.79 \text{ V}$$

Improvement in measurement and accuracy is provided by so-called bridge circuits, which are employed in a variety of applications for the measurement of resistance, inductance, and capacitance under both steady-state and transient conditions. The equivalent electric circuit of some transducers can be represented as an impedance, and, as a consequence, the capability of measuring such electrical quantities accurately is important. Many bridge circuits have been developed, but we shall limit our discussion to some of the more prominent types and their applications to various measurement and control problems.

The Wheatstone bridge is normally used for the comparison and measurement of resistances in the range of 1 Ω to 1 MΩ. A schematic of the bridge is given in Fig. 4-24. The cornerstone of the bridge consists of the four resistances (R_1, R_2, R_3, R_x), which are arranged in a diamond shape. R_2 and R_3 are normally known resistors, R_1 is a variable resistance, and R_x is the unknown resistance value associated with the transducer output. The voltage E is applied to the bridge by closing the switch S_1, and by adjusting the variable resistance R_1, the bridge may be balanced. This means simply that the potential difference between points B and D is zero. This balanced condition may be sensed by closing the switch S_2 and adjusting the value of R_1 until the galvanometer indicates no current flow. When this occurs, the voltage drop across R_2 must equal the voltage drop across R_1, since this implies that the voltage difference between B and D must be zero. Using the currents defined in Fig. 4-24, we obtain

$$i_2 R_2 = i_1 R_1$$

Figure 4-24 Schematic of basic Wheatstone bridge.

Further,

$$i_2 = i_3 = \frac{E}{R_2 + R_3} \qquad \text{if balanced}$$

and

$$i_1 = i_x = \frac{E}{R_1 + R_x} \qquad \text{if balanced}$$

If the currents are eliminated from these relations, the result is

$$\frac{R_2}{R_3} = \frac{R_1}{R_x} \qquad (4\text{-}22)$$

or

$$R_x = \frac{R_1 R_3}{R_2} \qquad (4\text{-}23)$$

This last equation allows the calculation of the value of R_x in terms of the other three resistors in the bridge.

If accurate measurements are to be made with a bridge circuit, the values of the resistors (R_1, R_2, R_3) must be precisely known, and the galvanometer must be sufficiently sensitive to detect small degrees of imbalance. When the unknown resistance R_x is connected to the circuit, care must be taken to use connecting leads which have a resistance that is small in comparison with the unknown.

The term "ratio arms" is frequently used in describing two known adjacent arms in a Wheatstone bridge. The galvanometer is usually connected to the junction of these two known resistors. In Fig. 4-24, R_2 and R_3 would normally be called the ratio arms.

Example 4-3 For the basic Wheatstone bridge in Fig. 4-24, determine the uncertainty in the measured resistance R_x as a result of an uncertainty of 1 percent in the known resistances. Repeat for 0.05 percent.

SOLUTION We use Eq. (3-2) to estimate the uncertainty. We have

$$\frac{\partial R_x}{\partial R_1} = \frac{R_3}{R_2} \qquad \frac{\partial R_x}{\partial R_2} = -\frac{R_1 R_3}{R_2^2} \qquad \frac{\partial R_x}{\partial R_3} = \frac{R_1}{R_2}$$

For a 1 percent uncertainty in the known resistances, this gives

$$\frac{w_{R_x}}{R_x} = (0.01^2 + 0.01^2 + 0.01^2)^{1/2} = 0.01732$$

$$= 1.732\%$$

For a 0.05 percent uncertainty in the known resistances, the corresponding uncertainty in R_x is 0.0866 percent.

The basic Wheatstone bridge circuit may also be employed for the measurement of ac impedances. The main problem is that two balance operations must be made to obtain a null, the first for the real part of the waveforms and the second for the imaginary part. There are some types of ac bridges that may be balanced with two independent adjustments. Several of these types are shown in Table 4-2, along with the balance conditions which may be used to determine the values of the unknown quantities.

Since an alternating current is involved, the null condition may not be sensed by a galvanometer as in the case of the dc Wheatstone bridge; some type of ac instrument must be used. This could be a rectifier meter or an oscilloscope. A pair of headphones can also be used when audio-frequency signals are involved.

It may be noted from Table 4-2 that for some types of ac bridge circuits the balanced condition is independent of frequency (types a, c, and e), while for others the frequency must be known in order to apply the balanced conditions (types b, d, and f). Thus, if the capacitance and resistance values are known for a Wien bridge, the bridge could be used as a frequency-measuring device.

As mentioned, bridge circuits are very useful for experimental measurements. The Wheatstone bridge is widely used for measuring the output resistance of various transducers, such as resistance thermometers, strain gages, and other devices that register the change in the physical variable as a change in output resistance. The ac bridges are used for inductance and capacitance measurements. Many transducers produce a change in either of these quantities upon a change in the physical quantity of interest. Bridge circuits are also useful in an unbalanced condition because a small change in the impedance in one of the bridge arms can produce a relatively large change in the detector signal, which can be used to control other circuits.

Bridge circuits may operate on either a *null* or a *deflection principle*. The null condition has been described above as where the galvanometer or sensing device reads zero at balance conditions. At any other condition the galvanometer reading will be *deflected* from the null condition by a certain amount, which depends on the degree of unbalance. Thus, the signal at the galvanometer or detector may be used as an indication of the unbalance of the bridge and may

Table 4-2 Summary of bridge circuits

Circuit	Balance relations	Name of bridge and remarks
(a)	$$C_x = \frac{C_3 R_2}{R_1}$$ $$R_x = \frac{R_3 R_1}{R_2}$$	Basic Wheatstone bridge. Greatest sensitivity when bridge arms are equal.
(b)	$$\frac{C_x}{C_3} = \frac{R_2}{R_1} - \frac{R_3}{R_x}$$ $$C_x C_3 = \frac{1}{\omega^2 R_3 R_x}$$ If $C_3 = C_x$ and $R_3 = R_x$ $$f = \frac{1}{2\pi R_3 C_3}$$	Wien bridge. May be used for frequency measurement with indicated relations.
(c)	$$L_x = C_1 R_1 R_2$$ $$R_x = \frac{C_1 R_1}{C_2} - R_3$$	Owen bridge.

Circuit	Balance relations	Name of bridge and remarks
(d) 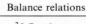	$\omega^2 LC = 1$ $$R_x = \frac{R_3 R_1}{R_2}$$	Resonance bridge. At balance conditions may be used for frequency measurement with $$f = \frac{1}{2\pi\sqrt{LC}}$$
(e)	$L_x = R_1 R_3 C$ $$R_x = \frac{R_1 R_3}{R_2}$$	Maxwell bridge.
(f) 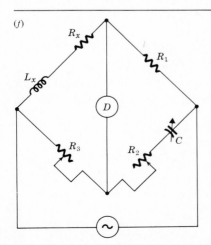	$$L_x = \frac{R_1 R_3 C}{1 + \omega^2 C^2 R_2^2}$$ $$R_x = \frac{\omega^2 C^2 R_1 R_2 R_3}{1 + \omega^2 C^2 R_2^2}$$	Hay bridge.

Figure 4-25 Schematic for analysis of unbalanced bridge.

indicate the deviation of one of the arms from some specified balance condition. The use of the deflection bridge is particularly important for the measurements of dynamic signals where insufficient time is available for achieving balance conditions.

Consider the bridge circuit shown in Fig. 4-25. R_1, R_2, R_3, and R_4 are the four arms of the bridge; R_g is the galvanometer resistance; and i_b and i_g are the battery and galvanometer currents, respectively. R_b represents the resistance of the battery or power-supply circuit. When the bridge is only *slightly* unbalanced, it can be shown that the value of R_b does not appreciably influence the effective resistance of the bridge circuit as presented to the galvanometer (see Ref. [2]). As a result, the following relation for the galvanometer current may be derived:

$$i_g = \frac{E_g}{R + R_g} \tag{4-24}$$

where R is the effective resistance of the bridge circuit presented to the galvanometer and is given by

$$R = \frac{R_1 R_4}{R_1 + R_4} + \frac{R_2 R_3}{R_2 + R_3} \tag{4-25}$$

This effective resistance R is indicated in Fig. 4-26. The voltage presented at the terminals of the galvanometer E_g is

$$E_g = E\left(\frac{R_1}{R_1 + R_4} - \frac{R_2}{R_2 + R_3}\right) \tag{4-26}$$

The voltage impressed on the bridge E depends on the battery or external circuit resistance R_b and the resistance of the total bridge circuit as presented to the

Figure 4-26 Equivalent circuit of bridge as presented to the galvanometer.

battery circuit, which we shall designate as R_0. It can be shown that for small unbalances the resistance R_0 may be calculated by assuming that the galvanometer is not connected in the circuit. Thus,

$$R_0 = \frac{(R_1 + R_4)(R_2 + R_3)}{R_1 + R_4 + R_2 + R_3} \tag{4-27}$$

The voltage impressed on the bridge is then

$$E = E_b \frac{R_0}{R_0 + R_b} \tag{4-28}$$

Because the deflection described above and indicated by Eq. (4-24) is based on the determination of the galvanometer current, the circuit is said to be current-sensitive. If the deflection measurement were made with an electronic voltmeter, oscilloscope, or other high-impedance device, the current flow in the detector circuit would be essentially zero, since as $R_g \to \infty$, $i_g \to 0$ in Eq. (4-24). Even so, there is still an unbalanced condition at the galvanometer or detector terminals of the bridge. This condition is represented by the voltage between terminals A and C for the case of a high-impedance detector. Such an arrangement is called a *voltage-sensitive deflection-bridge circuit*. The voltage indication for such a bridge may be determined from Eq. (4-26).

For relatively large unbalances in a bridge circuit it may become necessary to shunt the galvanometer to prevent damage. Such an arrangement is shown in Fig. 4-27a. In this case the voltage across the galvanometer may still be given to a sufficiently close approximation by Eq. (4-26), but the current through the galvanometer must be computed by considering the shunt resistance R_p. The total resistance of the bridge and shunt combination *as seen by the galvanometer circuit* is given by

$$R_e = \frac{R_p R}{R_p + R} \tag{4-29}$$

and the galvanometer current is now given by

$$i_g = \frac{E_g}{R_e + R_g} \tag{4-30}$$

(a)   (b)

Figure 4-27 (a) Galvanometer shunt arrangement for use with bridge of Fig. 4-26; (b) galvanometer with series resistance for use with bridge in Fig. 4-26.

The galvanometer current may also be reduced by using a series resistor, as shown in Fig. 4-27b. For this case the galvanometer current becomes

$$i_g = \frac{E_g}{R + R_g + R_s} \tag{4-31}$$

where R_s is the value of the series resistor.

Example 4-4 The Wheatstone bridge circuit of Fig. 4-24 has ratio arms (R_2 and R_3) of 6000 and 600 Ω. A galvanometer with a resistance of 70 Ω and a sensitivity of 0.04 μA/mm is connected between B and D, and the adjustable resistance R_1 reads 340 Ω. The galvanometer deflection is 39 mm, and the battery voltage is 4 V. Assuming no internal battery resistance, calculate the value of the unknown resistance R_x. Repeat for R_2 and R_3 having values of 600 and 60 Ω.

SOLUTION In this instance the bridge is operated on the deflection principle. For purposes of analyzing the circuit we use Figs. 4-25 and 4-26. The galvanometer current is calculated from the deflection and sensitivity as

$$i_g = (39)(0.04 \times 10^{-6}) = 1.56 \; \mu A$$

In the circuit of Fig. 4-25 the resistances are:

$R_b = 0 \qquad R_1 = 340 \qquad R_2 = 6000 \qquad R_3 = 600 \qquad R_4 = R_x \qquad R_g = 70$

We also have $E = 4.0$ V.

Combining Eqs. (4-24) to (4-26), we obtain

$$i_g = \frac{E[R_1/(R_1 + R_4) - R_2/(R_2 + R_3)]}{R_g + [R_1 R_4/(R_1 + R_4) + R_2 R_3/(R_2 + R_3)]}$$

Solving for R_4, we have

$$R_4 = \frac{ER_1 R_3 - i_g[R_g R_1(R_2 + R_3) + R_2 R_3 R_1]}{i_g(1 + R_1 + R_g)(R_2 + R_3) + ER_2}$$

Using numerical values for the various quantities, we obtain

$$R_4 = 33.93 \; \Omega$$

Taking $R_2 = 600$ and $R_3 = 60$, we obtain

$$R_4 = 33.98 \; \Omega$$

4-7 AMPLIFIERS

Experimental measurements occur in many forms: the voltage output of a bridge circuit, the frequency signal of a counting circuit, voltage signals representative of a change in capacitance, etc. In many cases, the signals are comparatively weak and must be amplified before they can be used to "drive" an output device. In other instances, there is a serious mismatch between the impedance of the measurement transducer and that of the output circuits, so that some interface must be provided to allow effective impedance matching. An example, in a familiar setting, of the former class of problems involves connecting a phonograph pickup cartridge directly to a loudspeaker. The cartridge is a transducer which converts the pressure produced by mechanical motions in the groove of a record directly into an electric voltage. This voltage could be applied directly to a loudspeaker, but it is so small that no audible sound will result. All sound systems provide a significant amount of amplification to increase this voltage before it is applied to the speakers. As a parenthetical comment, the buyer of such systems generally pays more when he or she insists that the voltage have low distortion as it undergoes this amplification process.

At this point some attention must be paid to the impact of integrated circuits upon electronics. There was a time when engineers of all disciplines had to study electronics in order to cope with amplifier application and design. This is no longer the case. In fact, the efficiencies of integrated circuit manufacturing have made available, at very low prices, amplifiers of extraordinary power.

Most amplifiers today are based on transitor or integrated circuits. Vacuum tubes are employed in a few specialized applications, but we shall be mainly concerned with the overall operating characteristics of amplifiers and the ways they can be used to accomplish a specific objective. We have already mentioned the effects of frequency and amplitude distortion in Sec. 2-9. Such effects can certainly be present in amplifiers if they are driven beyond the limits of their linear operation. We shall examine some techniques to minimize such distortion.

Effect of Feedback

Consider the amplifier schematic, shown in Fig. 4-28, having an input voltage E and an output voltage E_o with a gain of A. Without the feedback-attenuator loop we would have:

$$A = \frac{E_o}{E_i} \tag{4-32}$$

Now, we add the feedback loop which attenuates the output voltage by a factor k and feeds it back into the input. The feedback voltage is $E_f = kE_o$ which is

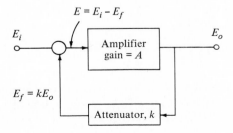

Figure 4-28 Amplifier with feedback loop.

negative and subtracted from the input voltage resulting in an input to the amplifier of $E = E_i = E_f$. Now, the gain of the amplifier and feedback system is

$$A_f = \frac{E_o}{E_i} = \frac{E_o}{E_o/A + kE_o} = \frac{A}{1 + kA} \tag{4-33}$$

If the amplifier gain is very large, i.e., $kA \gg 1$, then the gain is very nearly

$$A_f \approx \frac{A}{kA} = \frac{1}{k} \tag{4-34}$$

Note now that the gain of the overall system becomes constant so long as kA is sufficiently large. We may define the input and output impedances for the feedback system as

$$Z_{i,f} = \frac{E_i}{I_i} \quad \text{and} \quad Z_{o,f} = E_o/I_o$$

respectively. The input impedance for the amplifier alone is

$$Z_i = \frac{E}{I_i} = \frac{E_o}{AZ_i}$$

From Eq. (4-33)

$$E_i = E_o\left(k + \frac{1}{A}\right)$$

Combining these relations we obtain:

$$Z_{i,f} = \frac{E_o(k + 1/A)}{E_o/AZ_i} = Z_i(1 + kA) \tag{4-35}$$

Thus, we find that by the use of feedback we can increase the input impedance by a factor of $1 + kA$, or by a very large factor when the amplifier gain is large. The voltage appearing to drive the output impedance is AE and the output voltage is:

$$E_o = AE - I_o Z_o$$

For $E_i = 0$ the input voltage to the amplifier will be $E = -kE_o$ and

$$E_o = -AkE_o - I_o Z_o$$

$$E = \frac{I_o Z_o}{kA + 1} \tag{4-36}$$

The effective output impedance for the system is then

$$Z_{o,f} = \frac{E_o}{I_o} = \frac{Z_o}{1 + kA} \tag{4-37}$$

So the effect of feedback is to lower the output impedance by a factor of $1 + kA$. We have already mentioned that an amplifier will only be able to accept a limited range of input voltages before it can be driven into nonlinear behavior. The effect

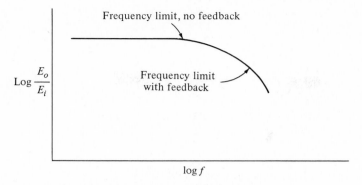

Frequency limit, no feedback

$\text{Log } \dfrac{E_o}{E_i}$

Frequency limit
with feedback

log f

Figure 4-29 Typical amplifier frequency response curve.

of feedback is to greatly extend this range because the feedback voltage lowers the voltage which is presented to the input of the amplifier.

Now consider a typical frequency response curve for an amplifier as shown in Fig. 4-29. The curve is flat up to a certain frequency and then drops off, i.e., the gain which can be delivered by the amplifier is reduced at higher frequencies. However, when negative feedback is employed lower gain is experienced (by a factor of k), but this has the effect of extending the frequency response.

Negative feedback also has the effect of reducing amplifier noise which is present in the output signal. Consider the two cases shown in Fig. 4-30. In (a) we have the open-loop amplifier and the noise N is assumed to appear at the output terminal. In (b) we have the negative feedback configuration. As before, we can write $E = E_i - kE_o$ and the output voltage is

$$E_o = AE + N = A(E_i - kE_o) + N \tag{4-38}$$

so that

$$E_o = \frac{AE_i + N}{1 + kA} \tag{4-39}$$

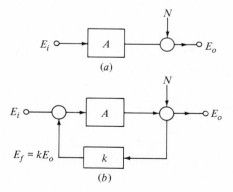

Figure 4-30 Effect of feedback on noise.

For $kA \gg 1$ this reduces to

$$E_o = \frac{E_i}{k} \pm \frac{N}{kA} \tag{4-40}$$

where the $\pm$ indicates that the noise is random.

Example 4-5 An amplifier has a gain of 1000 and a signal-to-noise ratio of 1000 in the open-loop configuration. Calculate the S/N ratio when negative feedback with $k = 0.1$ is used.

SOLUTION For $E_i = 1$ the open-loop output would be $AE_i = E_o$.

$$E_o = 1000 \pm 1$$

i.e., the noise is $1/1000$ of the output signal.

To produce this same output signal with the feedback arrangement we would need an input signal of $E_i = 100$ because the gain is $A_f = 1/k = 10$. Then, from Eq. (4-40) the output would be (N is still 1.0)

$$E_o = \frac{100}{0.1} \pm \frac{1}{(0.1)(1000)}$$

$$= 1000 \pm 0.01$$

Now the S/N ratio is

$$\text{S/N} = \frac{1000}{0.01} = 100,000$$

So by the use of feedback we have increased the S/N ratio by a factor of 100.

4-8 DIFFERENTIAL AMPLIFIERS

A *differential* or *balanced* amplifier is a device which provides for two inputs and an output proportional to the *difference* in the two input voltages. For good operation it is necessary to have well-matched internal components and provisions for trimming the system so that a zero differential voltage input will produce a zero output. The differential amplifier is particularly useful for amplification and measurement of small signals subjected to stray electric fields (typically line voltage at 60 Hz and 115 V). Such stray fields can induce substantial voltages in the input lines unless stringent efforts are made to provide shielding. However, the application of equal voltages to the inputs or equal ac voltages of the same phase will cancel out. This feature is called *common mode rejection*. The descriptive property of the amplifier is called the *common mode rejection ratio*

(CMRR), defined as the common mode signal at input divided by the common mode signal at output.

$$CMRR = \frac{CM_i}{CM_o} \tag{4-41}$$

Values greater than 10^6 (120 dB) are not uncommon.

4-9 OPERATIONAL AMPLIFIERS

An operational amplifier (op-amp) is a dc differential amplifier incorporating many solid-state elements in a compact package and shown schematically in Fig. 4-31. It operates from dc to some upper frequency limit of the order of 1.0 MHz. The (+) input is called the *noninverting* input because the output from this source is in phase with the input. The *inverting* input (−) has the opposite behavior, i.e., the output resulting from that source is 180° out of phase with the input. Op-amps using npn transistors will have input impedances of greater than 1 MΩ, while those constructed with field-effect transistors may have input imped-ances of the order of 10^{12} Ω. Because of these high impedances the input current for op-amps is essentially zero.

The output impedance of op-amps is very low, typically less than 1.0 Ω, and the open-loop gain is very high, of the order of 10^6 (120 dB).

In drawing the schematic diagrams for op-amps it is common practice to leave off the power supply and null adjustment inputs. Of course, one recognizes that these inputs must always be provided. With this brief description of the general characteristics of operational amplifiers we now examine some specific amplifier-feedback configurations which are useful in experimental measure-ments. Fig. 4-32 gives the schematics for the examples to be discussed.

The open-loop configurations in (*a*) and (*b*) are not often used because even a very small input will drive the output to full capacity of the power supply. Because the input impedance is high, the current presented to the input of the

Figure 4-31 Schematic for operational amplifier.

Figure 4-32 Operational amplifier configurations. (*a*) Open-loop, differential input; (*b*) open loop, single input; (*c*) inverting, differential input; (*d*) noninverting; (*e*) noninverting voltage follower; (*f*) voltage follower with gain greater than unity; (*g*) voltage follower with resistance in feedback.

$$E_o = -R_f \left(\frac{E_1}{R_1} + \frac{E_2}{R_2} + \frac{E_3}{R_3} \right)$$

(h)

$$E_o = E_1 + E_2$$

(i)

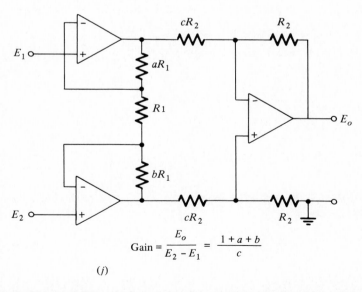

$$\text{Gain} = \frac{E_o}{E_2 - E_1} = \frac{1 + a + b}{c}$$

(j)

Figure 4-32 *Continued* (h) Inverting summer or adder; (i) noninverting adder; (j) circuit for high input impedance and common mode rejection.

$$E_o = -\frac{1}{R_i C} \int E_i \, dt \quad \text{for } E_i = 0 \text{ at } t = 0$$

(k)

$$E_o = R_f C \frac{dE_i}{dt}$$

(l)

$$E_o = R_f I_i$$

(m)

$$E_o = \frac{-Q(t)}{C_f} \quad \text{for } C_f \gg C_i$$

(n)

Figure 4-32 *Continued* (k) Integrator; (l) differentiator; (m) current-to-voltage converter; (n) charge amplifier.

inverting terminal in (c) will be nearly zero. Designating the voltage present at the amplifier input as E_s this requires that

$$\frac{E_i - E_s}{R_1} + \frac{E_o - E_s}{R_f} = 0$$

But $E_o = -AE_s$, or $E_s = -E_o/A$ so that

$$\frac{E_o}{R_f} = -\frac{E_i}{R_1} - \frac{E_o}{A}\left(\frac{1}{R_f} + \frac{1}{R_1}\right) \tag{4-42}$$

Because the gain is very large the last term drops out and the gain of the negative feedback arrangement becomes

$$\frac{E_o}{E_i} = -\frac{R_f}{R_1} \tag{4-43}$$

The very high gain has the effect of causing E_s to be very nearly zero. A similar deviation could be used to obtain the gain for the noninverting configuration in Fig. 4-32d. The voltage follower in (e) has a gain of unity and can serve as a power amplifier to convert a high impedance input into a low impedance output. The potentiometer arrangement in (f) can be used to produce a gain greater than unity through the split between nR and R. Addition of an arbitrary resistor as in (g) still produces unity gain with the voltage follower.

Still further use of the zero-input current condition could be used to derive the relations shown for the inverting and noninverting adder configurations shown in (h) and (i). A configuration for amplifying a high impedance input with excellent common mode rejection is shown in (j).

In many experimental applications a transducer will be employed which gives a signal proportional to a certain rate, such as a fluid-flow meter which indicates the flow rate of the fluid through the device. In addition to the rate of flow the experimentalist may also wish to know the total quantity of fluid flowing over a given length of time. In other words, we need to integrate the input signal over a given time. The configuration in (k) accomplishes this objective. For very high gain we again have nearly zero input voltage and sum the input currents to the amplifier as

$$\frac{E_i}{R_i} + C\frac{dE_o}{dt} = 0$$

Thus

$$E_o = -\frac{1}{R_i C} \int E_i \, dt \tag{4-44}$$

Normally, a shorting switch is installed across the capacitor to make the voltage zero at the time the integration is to begin.

In a similar way the op-amp can be used as a differentiator as shown in Fig. 4-32l. If one has a constant *current* input as in (m) we would have

$$I_i - \frac{E_o}{R_f} = 0 \tag{4-45}$$

and we have a current-to-voltage convertor with $E_o = R_f I_i$.

With some transducers the input signal is electric charge. Examples are piezoelectric transducers which convert an impressed-force input to charge and condenser microphones used to measure acoustic phenomena. The configuration in Fig. 4-32n is used for such measurements.

The particular amplifier configuration to be employed depends on the application, and one must carefully observe the operating characteristics of the sensors or transducers connected to the amplifier inputs. Impedance matching is quite important. If the source impedance is large the amplifier circuit might cause sufficient loading so that most of the signal power is lost. In this case the input noise level in relation to the reduced signal may become excessive and reduce the overall accuracy of measurement. One way to avoid this problem is to first employ a voltage follower to reduce the impedance (the output impedance of an op-amp is low) and then follow with whatever other amplification and signal processing are required.

4-10 TRANSFORMERS

Transformers are also used to match impedances in many experimental situations. In addition, they are still commonly used in audio systems to match the impedance of the loudspeakers to the output impedance of the last amplifier stage. An ideal n-turn transformer is shown in Fig. 4-33 along with the associated four terminal variables, two voltages and two currents. The equations relating the voltages and currents are

$$v_2 = nv_1$$

$$i_2 = \frac{1}{n} i_1 \tag{4-46}$$

Dividing the first by the second results in the following:

$$\frac{v_2}{i_2} = n^2 \frac{v_1}{i_1}$$

and as long as the voltages and the currents are single-frequency sinusoids, we may state that the output impedance Z_2 is released to the input impedance Z_1 by the equation

$$Z_2 = n^2 Z_1 \tag{4-47}$$

This equation states that the output impedance of an n-turn ideal transformer is n^2 greater than the input impedance. As a concrete example, consider an experiment in which we need to apply the 1000-Hz output voltage of an operational amplifier to a loudspeaker to produce an audible sound. The op-amp has an output impedance of 50 Ω at this frequency, and the loudspeaker has an equivalent impedance of 3 Ω. For efficient power transfer, we wish the loudspeaker impedance to "look" as much as possible like 50 Ω. This can be accomplished by using a transformer as shown in Fig. 4-34. Here,

$$Z_1 = 3 \ \Omega$$

and, by the definition of an ideal transformer,

$$Z_2 = n^2 Z_1$$

or

$$50 = n^2(3)$$

and taking the closest integer value for n, we obtain

$$n = 4$$

Figure 4-33 Ideal transformer.

$n:1$

$50\,\Omega$

v_s

$3\,\Omega$

Figure 4-34 Impedance matching via transformer.

The insertion of a four-turn transformer between the output of the operational amplifier and the loudspeaker will ensure a close-to-optimum transfer of energy from the amplifier to the speaker.

This completes our discussion of the basics of operational amplifiers, their uses in measurement procedures, and impedance matching. We next turn to a condensed examination of power supplies, since they are needed if one is to use operational amplifiers and other integrated circuits. Parenthetically, they are also essential if one is working with transistors or vacuum tubes.

4-11 POWER SUPPLIES

There are two commonly used sources of dc energy: batteries and electronic modules which convert 60-Hz electric energy (available from power lines) to dc. These modules are called *power supplies.*

Batteries are constantly being improved in terms of both their cost per unit energy and their weight and volume per unit energy. Despite this progress, they are still relatively expensive and bulky; they are preferred only in situations in which access to electric utility lines is very difficult. In such remote locations, it is becoming more common to connect the batteries to a charging circuit which is powered by either sunlight or wind. The battery supplies the electronic circuits, and the charger restores the battery whenever the sun is bright enough or the wind is blowing hard enough. In general, such sources of dc energy account for a very small percentage of the total presently in use.

The specific details underlying the operation of dc power supplies are of minor interest here, but some general features will be pointed out. In almost all cases, combinations of solid-state diodes are used to convert the pure 60-Hz sine

Figure 4-35 Rectification.

Figure 4-36 Power-supply voltage regulation.

wave available at an electric outlet to the positive, "rectified" wave shown in Fig. 4-35. This voltage contains a dc component as well as a series of sinusoids with frequencies which are integer multiples of 60 Hz. The "ripple" caused by these harmonics is undesirable and is reduced by further filtering, as will be discussed in the next section. The output of a properly designed power supply is similar to that shown in Fig. 4-35 and is nearly constant. There is, to be sure, a nonzero amount of residual ripple, and this is normally specified by the manufacturer in percentage terms.

It is also important that the output voltage of a dc power supply remain constant as the current supplied to the circuits it is driving increases. If we represent the power supply as an ideal constant voltage source in series with some characteristic resistance, the goal of the power-supply designer is to reduce the value of this resistance as much as possible. Figure 4-36 illustrates various possible voltage-vs.-current characteristics for power supplies. The curve labeled "perfect" indicates a practically unattainable ideal in that the voltage supplied is independent of the current drawn. The degree to which the output voltage is independent of the current supplied is termed *voltage regulation*. There are available integrated circuits which, using operational amplifiers, do an excellent job of maintaining a constant output voltage. For example, one such circuit will maintain an output voltage to within 0.05 V of 5 V for output currents ranging from 5 to 500 mA. Such regulation is more than adequate for most electronic circuits.

We next consider those elements of proper experimental procedure which are necessary to minimize the effects of noise upon measurement accuracy.

4-12 SIGNAL CONDITIONING

Noise is present in all physical situations in which measurements are attempted or information is conveyed, and noise is, in fact, a rather subtle ramification of the second law of thermodynamics. Noise is fundamentally another manifestation of the attempts of the physical universe to attain a state of randomness. Proper experimental design and procedures can greatly reduce its effects in many situations, as we shall now discuss.

A significant help to the experimenter is any a priori knowledge which may be available concerning the frequency ranges in which the desired signal exists. For example, if atmospheric temperatures are being measured, it is reasonable to expect that such temperatures will not change significantly over times shorter than several minutes. As a consequence, the meaningful output voltage of a thermocouple includes only the low-frequency part. Higher-frequency components are noise due to random fluctuations and can be eliminated without loss of experimental information, if, indeed, such elimination is possible. Another example would be the investigation by an electric utility company of the audible noise produced by an ultra-high-voltage transmission line. Because the human ear responds to acoustic signals only in the range of frequencies between 20 Hz and 18 kHz, the processing of sensor outputs can be safely limited to this frequency range.

This discussion naturally leads to an examination of the possibility of filtering electrical signals. This term means that we desire circuits which transmit or "pass" only certain bands or ranges of frequencies of an input signal. The unwanted parts of the signal can be characterized as noise, but, in addition, there is also noise present in the frequency band of interest. Filtering will not solve all the problems, but it does provide a significant degree of experimental improvement. If a relatively narrow range of frequencies is important, it is usually possible to operate with much simpler electronic circuits for amplification and processing purposes than when a broad frequency range must be examined.

Various arrangements may be used for filter circuits, but they all fall into three categories: (1) lowpass, (2) highpass, and (3) bandpass circuits. Some filters are composed only of passive elements, while others involve amplification to eliminate losses and are termed *active filters* (discussed in Chap. 14). A lowpass filter permits the transmission of signals with frequencies below a certain cutoff value with little or no attenuation, while a highpass filter allows the transmission

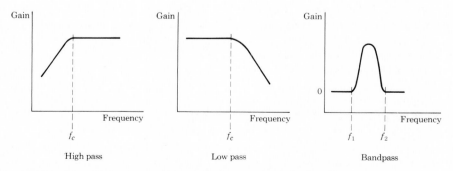

Figure 4-37 Approximate performance curves for three types of filters.

of signals with frequencies above a cutoff value. The bandpass filter permits the transmission of signals with frequencies in a certain range or band while attenuating signals with frequencies both above and below the limits of this band. The approximate performance for the three types of filters is shown in Fig. 4-37. The cutoff frequency is designated by f_c for both the highpass and lowpass filters, while the limits of frequency transmission for the bandpass filter are given by the symbols f_1 and f_2. Note carefully that the filters do not provide discontinuities in attenuation at the transition frequencies; i.e., there is transmission of signals with frequencies above or below a given cutoff value, although the signal attenuation becomes more pronounced as the frequency moves away from the cutoff point.

A summary of several types of passive filters is given in Fig. 4-38 as recommended in Ref. [8]. The various filter sections may be used separately or in combination to produce sharper cutoff performance. In the m-derived sections, the quantity m is defined by

$$m = \sqrt{1 - \left(\frac{f_c}{f_\infty}\right)^2} \tag{4-48}$$

for a lowpass filter and

$$m = \sqrt{1 - \left(\frac{f_\infty}{f_c}\right)^2} \tag{4-49}$$

for a highpass filter, where f_c is the desired cutoff frequency and f_∞ is a frequency having high attenuation. When only one m-derived section is used, a value of $m = 0.6$ is recommended [8]. When a filter is designed for a particular application, it is usually important that the impedance of the filter circuit be matched to the connecting circuit. Consideration of the matching of impedances in such circuits is beyond the scope of our discussions, and the interested reader should consult Ref. [9] for more information.

Table 4-3 presents some basic RL and RC filter sections. In a surprisingly large number of experimental situations, these simple two-element filters offer significant improvement in noise reduction. The filters in this table do, of course, attenuate signal levels because they contain resistances. After all, any current flowing through a resistor results in a conversion of electric energy into heat. The designer often has to compensate for this resistive attenuation by including some additional amplifying circuits.

The natural next topic is a discussion of *active filters*, but this area has been significantly influenced by progress in converting signals from analog to digital form before performing the filtering operations. In order to allow a unified discussion of active filters, we will defer considering them until Chap. 14, where analog-to-digital and digital-to-analog conversion will be discussed, since this topic is a prerequisite for a unified discussion of active filtering. Numerous references are available on filter design and the interested reader may wish to consult Refs. [16] to [19] for more information.

Figure 4-38 Basic filter sections and design formulas according to Ref. [8]. R is in ohms, C is in farads, L is in henrys, and f is in hertz.

Table 4-3 Some simple RL and RC filter sections according to Ref. [9]

Diagram	Type	Time constant	Formula and asymtopic approximation
	A lowpass RC	$T = RC$	$\dfrac{E_o}{E_i} = \dfrac{1}{\sqrt{1 + \omega^2 T^2}} \approx \dfrac{1}{\omega T}$ $\phi_A = -\tan^{-1}(R\omega C)$
	B highpass RC	$T = RC$	$\dfrac{E_o}{E_i} = \dfrac{1}{\sqrt{1 + \dfrac{1}{\omega^2 T^2}}} \approx \omega T$ $\phi_B = \tan^{-1}\dfrac{1}{R\omega C}$
	C lowpass RL	$T = \dfrac{L}{R}$	$\dfrac{E_o}{E_i} = \dfrac{1}{\sqrt{1 + \omega^2 T^2}} \approx \dfrac{1}{\omega T}$ $\phi_C = -\tan^{-1}\dfrac{\omega L}{R}$
	D highpass RL	$T = \dfrac{L}{R}$	$\dfrac{E_o}{E_i} = \dfrac{1}{\sqrt{1 + \dfrac{1}{\omega^2 T^2}}} \approx \omega T$ $\phi_D = \tan^{-1}\dfrac{R}{\omega L}$

A measurement of the degree of amplification or attenuation provided by a circuit is given by its gain or amplification ratio. Gain is defined as

$$\frac{\text{Output}}{\text{Input}}$$

The output and input quantities may be voltage, current, or power, depending on the application. Gain is a dimensionless quantity, but engineers speak of decibels of gain or loss in terms of the following definition involving a logarithmic ratio:

$$\text{Decibels} = 10 \log \frac{P_2}{P_1} \qquad (4\text{-}50)$$

where P_1 and P_2 are the input and output powers, respectively. The voltage or current gain may be defined in a similar manner.

Since power may be expressed as E^2/R or I^2R, for a resistive load we would have

$$\frac{P_2}{P_1} = \frac{E_2^2}{E_1^2} = \frac{I_2^2}{I_1^2}$$

Expressed in terms of voltages or current, the decibel notation would then be

$$\text{Decibels} = 20 \log \frac{E_2}{E_1} \tag{4-51}$$

$$\text{Decibels} = 20 \log \frac{I_2}{I_1} \tag{4-52}$$

For passive filters, the ratio of output power to input power is less than 1, and, as a consequence, the gain in decibels is negative. In order to characterize such circuits with a positive quantity, engineers define an *insertion loss* by the following equation:

$$10 \log \frac{P_1}{P_2} \tag{4-53}$$

You should readily note that the number of decibels of gain is simply the negative of the decibels of insertion loss.

The filter circuits shown in Fig. 4-38 are the most commonly employed and are generally the most efficient passive-element types. It is possible to construct filters in RC and RL arrangements instead of the LC designs given in Fig. 4-38; however, the circuits which employ resistive elements are generally not as efficient because they remove energy from the system. Some simple RL and RC filter sections are shown in Table 4-3, along with the appropriate equation describing their input-output voltage characteristics. The phase-lag angle ϕ is also given.

Example 4-6 A 1.0-mV signal is applied to an amplifier such that an output of 1.0 V is produced. The 1.0-V signal is then applied to a second amplification stage to produce an output of 25 V. For the three voltage points indicated, calculate the voltage in decibel notation referenced to (a) 1.0 mV and (b) 1.0 V. Also calculate the overall voltage amplification in decibels.

SOLUTION The schematic for the amplifiers is shown in Fig. Example 4-6.
(a) For $E_0 = 1.0$ mV, we have

$$20 \log \frac{E_1}{E} = \text{dB}_1$$

$$20 \log \frac{1.0 \times 10^{-3}}{1.0 \times 10^{-3}} = \text{dB}_1 \qquad \text{dB}_1 = 0$$

$$20 \log \frac{1.0}{1.0 \times 10^{-3}} = \text{dB}_2 \qquad \text{dB}_2 = 60$$

$$20 \log \frac{25}{1.0 \times 10^{-3}} = \text{dB}_3 \qquad \text{dB}_3 = 87.92$$

Figure Example 4-6

(b) For $E_0 = 1.0$ V, we have

$$20 \log \frac{1.0 \times 10^{-3}}{1.0} = dB_1 \qquad dB_1 = -60$$

$$20 \log \frac{1.0}{1.0} = dB_2 \qquad dB_2 = 0$$

$$20 \log \frac{.25}{1.0} = dB_3 \qquad dB_3 = 27.92$$

The overall amplification is calculated from

$$20 \log \frac{E_3}{E_1} = 20 \log \frac{25}{1.0 \times 10^{-3}} = 87.92 \text{ dB}$$

Notice that this result could have been obtained from the decibel calculations in (a) and (b) for either reference level. In other words,

$$\text{Gain}_{13}(\text{dB}) = dB_3 - dB_1$$

when all are expressed on the same reference basis.

Example 4-7 A simple RC circuit is to be used as a lowpass filter. It is desired that the output voltage be attenuated 3 dB at 100 Hz. Calculate the required value of the time constant $T = RC$.

SOLUTION Network A of Table 4-3 is the desired arrangement. We wish to have

$$\frac{E_o}{E_i} = -3 \text{ dB}$$

so that

$$-3 = 20 \log \frac{E_o}{E_i}$$

or

$$\frac{E_o}{E_i} = 0.708$$

Thus,

$$\frac{E_o}{E_i} = \frac{1}{(1 + \omega^2 T^2)^{1/2}} = 0.708$$

where $\omega^2 = (2\pi f)^2 = [2\pi(100)]^2$.

We find that $T = 1.59 \times 10^{-3}$ s. The circuit might be constructed using a 1.0-μF capacitor and a 1.59-kΩ resistor.

Passive filters, as discussed and characterized above, offer one possible strategy for combating the effects of noise in experimental measurements. Remember that we must have some knowledge of where we expect our signals to appear in the frequency spectrum if filtering is to be of any value. We conclude this section by discussing a few other methods which are useful for minimizing noise.

For these techniques to be effective, it is necessary that the signals of interest be repetitive. This may seem to be an exceptionally stringent requirement, but clever experimental procedures can result in its satisfaction in many situations. For example, to measure the vibration of a wall which results from an impulse of force, the proper experimental procedure is to make the force repetitive. In this way, the signal of interest (the wall's vibration) will also be repetitive. In addition, any dc signal can always be converted into a repetitive form by "chopping" it by electrical or mechanical means. As long as such steps have been taken to ensure that the signal to be measured is repetitive, either of the following techniques may be used to reduce the effects of noise.

The first is based on the observation that the signal will be repetitive, but the associated noise will not be. Consider taking the output of the transducer for two successive signals, adding them, dividing the resultant sum by 2, and then displaying the output. What will have happened? The signal of interest will still be present, but two different noise components will have been added together and then divided by 2. Since the noise will be different in each of the two successive times, it follows, as long as the noise is uncorrelated, that the amount of noise present in the resultant signal will decrease. This can be more clearly seen by considering the following two sums which represent the first and second "pieces" of output from the transducer.

$$v_s(t) + n_1(t)$$

$$v_s(t) + n_2(t)$$

After these are added together and divided by 2, the result is

$$v_s(t) + \frac{n_1(t) + n_2(t)}{2}$$

and this sum is the desired signal v_s as well as an average of the noise present in the two "pieces." After N samples of the desired signal plus noise are added and averaged, the result is

$$v_s(t) + \frac{n_1(t) + n_2(t) + \cdots + n_N(t)}{N} \tag{4-54}$$

This shows the output of such a noise reduction system to be composed of the desired signal $v_s(t)$ as well as an average of various samples of the noise. One may wonder what assurances can be offered that the average of noise samples will be smaller than a given segment $n(t)$. A detailed answer to that question involves a careful study of stochastic processes and is not necessary for our purposes. In physical terms, we can be sure that this averaging process will reduce the effective

Figure 4-39 CAT filtering.

level of noise as long as the origins of the noise are truly unrelated to the processes giving rise to the signal of interest v_s. If such is the case, the noise and signal are said to be uncorrelated, and averaging will improve the ratio of signal to noise in the system and, as a consequence, improve the quality and accuracy of our measurements.

This averaging process clearly requires some physical method of storing "pieces" of the transducer output. Some form of memory is necessary. There are available instruments which go by the shorthand CAT (for Computer Analysis of Transients). Such devices provide the necessary memory and analysis capability to allow the implementation of the storage and averaging procedure we have just described. Figure 4-39 is a sketch of the improvement that such instruments offer.

We conclude this section with a discussion of the lock-in amplifier. This instrument uses a principle different from that used by a CAT to reduce noise. An electronic circuit known as a *phase-sensitive detector* is the key to understanding the lock-in amplifier. Such a detector has two input terminals, and, assuming that the two input signals are sinusoids with the same frequency, it produces an output which depends upon the cosine of the phase angle between the two signals. For the detector sketched in Fig. 4-40, the input voltages are given by

$$v_1(t) = V_1 \sin(\omega t)$$

$$v_2(t) = V_2 \sin(\omega t + \phi)$$

and the output voltage is

$$v_0 = K V_1 V_2 \cos \phi \tag{4-55}$$

where K is a constant.

With an understanding of this element, the basic behavior of a lock-in amplifier can be described quite simply. Recall that we have a repetitive signal of

Figure 4-40 Phase-sensitive detector.

Figure 4-41 Lock-in amplifier.

interest accompanied by some inevitable noise. This entire transducer output is passed through a narrow bandpass filter which has its pass frequency located at the frequency of repetition of the input signal. The output of this filter constitutes one of the inputs to the phase-sensitive detector. The other necessary input is a sinusoid with the same frequency and a phase which is maintained constant with respect to the chopper used to modulate the transducer's output. This means that the desired signal (or, at least, that part of it which survived the filtering operation) and the reference signal have a fixed phase (ϕ_f) maintained between them. Any noise which is present in the filter output will not maintain such a fixed phase, and the output of the phase-sensitive detector can be written as

$$v_0(t) = K V_s V_R \cos \phi_f + K V_N V_R \cos \phi(t) \qquad (4\text{-}56)$$

The system is sketched in Fig. 4-41.

The phase between the reference voltage and the noise will change as time progresses, and the cosine of this angle will take on both positive and negative values. If the noise is truly uncorrelated, the time average of the second term in the last equation will be zero. The last part of a lock-in amplifier is a suitable integrator which serves to time-average. The output of the system depends only upon the amplitude and phase of the signal of interest v_s.

$$\overline{v_0(t)} = \frac{1}{T} \int_0^T v_0(t)\, dt = K V_s V_R \cos \phi_f \qquad (4\text{-}57)$$

Well-designed lock-in amplifiers result in significant signal-to-noise ratio improvements. In fact, the degree of improvement made possible by their careful design and use is nearly incredible. When we are faced with what appears to be a nearly hopeless situation in terms of the desired signal being buried in noise, this is a method which should be considered. Manufacturers of these instruments provide very helpful documentation and application assistance.

4-13 THE ELECTRONIC VOLTMETER (EVM)

The electronic voltmeter employs either vacuum-tube or solid-state elements for operation and is particularly useful because of its high impedance. Most such meters now employ transistors and other solid-state devices for amplification and signal-modification stages.

The EVM is one of the most useful laboratory devices for the measurement of voltage. It may be used for both ac and dc measurements and is particularly valuable because of its high input-impedance characteristics, which make it applicable to the measurement of voltages in electronic circuits.

A block-diagram schematic of a simple EVM is shown in Fig. 4-42. The input voltage is connected through appropriate terminals to the function switch. If a dc voltage is to be measured, the signal is fed directly to the range selector switch operating as a voltage-divider circuit where the signal is reduced to a suitable range for the succeeding amplifier circuit. The voltage signal from the voltage divider is then applied to the amplifier stage, whose voltage output is used to drive a conventional D'Arsonval movement for readout purposes. The output from the amplifier stage could also be connected to a recorder, or oscilloscope, or for the measurement of transient voltages. In solid-state or DVM devices, the signal from the voltage divider is applied to the gate of a field-effect transistor which also displays a large input impedance.

For an ac voltage measurement the signal is fed through a voltage divider and then to a rectifier circuit, which produces a dc voltage proportional to the input ac signal. The dc voltage is then impressed on the voltage-divider network of the range-selector switch. The remainder of the circuit is the same as in the case of the dc measurement.

Various modifications of the above arrangement may be used, depending on the range of voltages to be measured. For very low voltage ac signals an amplification stage may be added to the input instead of the voltage-divider arrangement. For very low dc signals a chopper device might be used at the input to

Figure 4-42 Block-diagram schematic of an electronic voltmeter.

Figure 4-43 Photograph of an EVM.

produce an ac signal, which is more easily amplified to the voltage levels that may be handled with conventional circuitry.

The important point is that the input impedance of the EVM is very high, usually greater than 50 MΩ, so that the measured circuit is not loaded appreciably and the indicated voltage more closely represents the true voltage to be measured. Example 4-8 illustrates the influence that the meter can exert on a circuit and the reduction in error of measurement that can result when an EVM is used. It may be noted that a correction may be made for the meter impedance when the impedance of the measured circuit is known.

A photograph of a typical commercial EVM is shown in Fig. 4-43.

4-14 DIGITAL VOLTMETERS

As mentioned earlier in this chapter, a wide variety of voltmeters are now available that provide a digital output instead of the conventional pointer-scale arrangement. With suitable field-effect transistors used in the input stages, their input impedance can be as large as that of vacuum tube or other solid-state instruments. A wide choice of input-signal modifiers, ac-dc converters, resistance dc converters, amplifiers, etc., is available, so that, in a real sense, the digital voltmeter now enables the experimentalist to make precision measurements over a broad range of variables. Not surprisingly, the cost of digital voltmeters is directly related to their accuracy and versatility.

Example 4-8 The voltage at points *A* and *B* is to be measured. A constant 100 V is impressed on the circuit as shown. Two meters are available for the measurement: a small volt-ohmmeter with an internal impedance of 100,000 Ω and a range of 100 V, and an EVM with an input impedance of 17 MΩ.

Compare the error in measurement with each of these devices.

SOLUTION The true voltage, by inspection, is 50 V. With the volt-ohmmeter connected in the circuit there results

and the voltage at *A* and *B* is

$$E_{AB} = 100 \frac{1/[(1/0.5) + (1/0.1)]}{1/[(1/0.5) + (1/0.1)] + 0.5}$$

$$= 100 \left(\frac{0.0833}{0.0833 + 0.5} \right) = 14.3 \text{ V}$$

or an error of −71 percent.

With the EVM connected in the circuit there results

and the voltage at A and B is

$$E_{AB} = 100 \frac{1/[(1/0.5) + (1/17)]}{1/[(1/0.5 + 1/17)] + 0.5}$$

$$= 100\left(\frac{0.4857}{0.4857 + 0.5}\right) = 49.27 \text{ V}$$

or an error of -1.46 percent.

4-15 THE OSCILLOSCOPE

We have seen that the EVM offers the advantage that it can be used to measure voltage without a substantial "loading" of the transducer or other source supplying the signal. The cathode-ray oscilloscope (CRO) is similar to the EVM in that it also has a high input impedance and is a voltage-measuring device. In addition, the CRO is capable of displaying voltage signals as functions of time.

The heart of any oscilloscope is the cathode-ray tube (CRT), which is shown schematically in Fig. 4-44. Electrons are released from the hot cathode and accelerated toward the screen by the use of a positively charged anode. An appropriate grid arrangement then governs the focus of the electron beam on the screen. The exact position of the spot on the screen is controlled by the use of the horizontal and vertical deflection plates. A voltage applied on one set of plates produces the x deflection, while a voltage on the other set produces the y deflection. Thus, with appropriate voltages on the two sets of plates, the electron beam may be made to fall on any particular spot on the screen of the tube. The screen is coated with a phosphorescent material, which emits light when struck by the electron beam. If the deflection of the beam against a known voltage input is calibrated, the oscilloscope may serve as a voltmeter. Since voltages of the order of several hundred volts are usually required to produce beam deflections across the entire diameter of the screen, the cathode-ray tube is not directly applicable for many low-level voltage measurements, and amplification must be provided to bring the input signal up to the operating conditions for the CRT.

A schematic of the CRO is shown in Fig. 4-45. The main features are the CRT, as described above, the horizontal and vertical amplifiers, and the sweep

Figure 4-44 Schematic diagram of a cathode-ray tube (CRT).

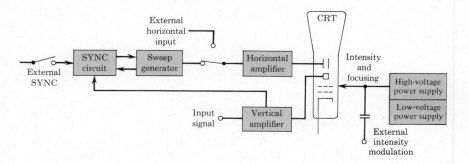

Figure 4-45 Block-diagram schematic of an oscilloscope.

Figure 4-46 Photograph of an oscilloscope.

and synchronization circuits. The sweep generator produces a sawtooth wave which may be used to provide a periodic horizontal deflection of the electron beam, in accordance with some desired frequency. This sweep then provides a time base for transient voltage measurements by means of the vertical deflection. Oscilloscopes provide internal circuits to vary the horizontal sweep frequency over a rather wide range as well as external connections for introducing other sweep frequencies. Internal switching is also provided, which enables the operator of the scope to "lock" the sweep frequency onto the frequency impressed on the vertical input. Provisions are also made for external modulation of the intensity of the electron beam. This is sometimes called the *z-axis input*. This modulation may be used to cause the trace to appear on the screen during certain portions of a waveform and disappear during other portions. It may also be used to produce traces of a specified time duration on the screen of the CRT so that a time base is obtained along with the waveform under study. A photograph of a commercial single-beam oscilloscope is given in Fig. 4-46.

A *dual-beam oscilloscope* provides for amplification and display of two signals at the same time, thereby permitting direct comparison of the signals on the CRT screen. A single-beam oscilloscope may be given a dual-beam function through the use of an electronic switch. This device switches between two inputs and alternately impresses the two signals at the input terminals of the single-beam oscilloscope. The switching rate may be varied to accommodate different frequency input signals. A schematic of the system is given in Fig. 4-47. Notice the chopped nature of the signal displayed on the CRT screen; this is a result of the switching process. The width of the gaps in the displays is determined by the rate with which the electronic switches are able to respond.

The CRO may be used to measure phase shift in an electronic circuit, as shown in Fig. 4-48. An oscillator is connected to the input of the circuit under test. The output of the circuit is connected to the CRO vertical input, whereas the oscillator signal is connected directly to the horizontal input. The phase-shift angle ϕ may be determined from the relation

$$\phi = \sin^{-1}\frac{B}{A} \tag{4-58}$$

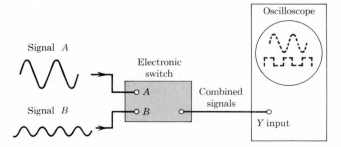

Figure 4-47 Use of an electronic switch to achieve a dual-beam function with a single-beam oscilloscope.

Oscillator

$B \; A \quad \sin \phi = B/A$

Vertical input

Oscilloscope

Horizontal input

Network under test

Figure 4-48 Schematic illustrating use of oscilloscope for phase measurements.

where B and A are measured as shown in Fig. 4-48. For zero phase shift the ellipse will become a straight line with a slope of 45° to the right; for 90° phase shift it will become a circle; and for 180° phase shift it will become a straight line with a slope of 45° to the left.

The CRO offers a convenient means of comparing signal frequencies through the use of Lissajous diagrams. Two frequencies are impressed on the CRO inputs, one on the horizontal input and one on the vertical input. One of these frequencies may be a known frequency as obtained from a variable frequency oscillator or signal generator. If the two input frequencies are the same, an ellipse will be displayed on the CRT screen. If the two input frequencies are not the same, the patterns that are displayed on the CRT screen are called Lissajous diagrams. There is a distinct relationship that governs the shape of these diagrams in accordance with the input frequencies:

$$\frac{\text{Vertical input frequency}}{\text{Horizontal input frequency}}$$

$$= \frac{\text{number of vertical maxima on Lissajous diagram}}{\text{number of horizontal maxima on Lissajous diagram}} \qquad (4\text{-}59)$$

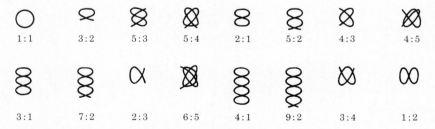

| 1:1 | 3:2 | 5:3 | 5:4 | 2:1 | 5:2 | 4:3 | 4:5 |

| 3:1 | 7:2 | 2:3 | 6:5 | 4:1 | 9:2 | 3:4 | 1:2 |

Figure 4-49 Lissajous diagrams for various frequency ratios as indicated.

Some typical shapes for the Lissajous diagrams are shown in Fig. 4-49. It may be noted that these shapes can vary somewhat depending on the phase relation between the input signals.

Oscilloscope traces may be recorded by various photographic methods, and several cameras are manufactured especially for oscilloscope applications.

Example 4-9 An oscilloscope is used for a phase-shift measurement in accordance with the arrangement in Fig. 4-48. The measurements are

$$A = 2.5 \pm 0.05 \text{ cm}$$

$$B = 1.5 \pm 0.05 \text{ cm}$$

Calculate the nominal value of the phase angle, and estimate the uncertainty in the measurement.

SOLUTION We calculate the nominal value with Eq. (4-58).

$$\phi = \sin^{-1} \frac{B}{A} = \sin^{-1} \frac{1.5}{2.5} = 36.9°$$

We may estimate the uncertainty with the use of Eq. (3-2). Observing that

$$\frac{\partial[\sin^{-1}(B/A)]}{\partial(B/A)} = \frac{1}{[1-(B/A)^2]^{1/2}}$$

we have

$$\frac{\partial \phi}{\partial B} = \frac{1}{[1-(B/A)^2]^{1/2}} \frac{1}{A} = \frac{1}{(1-0.36)^{1/2}} \frac{1}{1.5} = 0.833 \text{ rad/cm}$$

$$\frac{\partial \phi}{\partial A} = -\frac{B}{[1-(B/A)^2]^{1/2}} = \frac{-1.5}{(1-0.36)^{1/2}} \left(\frac{1}{2.5}\right)^2 = -0.3 \text{ rad/cm}$$

Also,

$$w_B = 0.05 \text{ cm}$$

$$w_A = 0.05 \text{ cm}$$

so that the uncertainty in ϕ is

$$w_\phi = [(0.833)^2(0.05)^2 + (0.3)^2(0.05)^2]^{1/2}$$

$$= 0.0443 \text{ rad} = 2.54°$$

or

$$\frac{w_\phi}{\phi} = \frac{2.54}{36.9} = 0.0687 = 6.87 \text{ percent}$$

4-16 SAMPLING OSCILLOSCOPES

Each oscilloscope is characterized by its manufacturer by a set of performance specifications. One of the most important is the maximum response frequency. This number indicates the highest-frequency sinusoid which the instrument will be able to follow and display accurately. In terms of internal components, this maximum frequency is primarily determined by the frequency response of the amplifiers used in the oscilloscope. In an excellent state-of-the-art scope, this maximum frequency can be 200 MHz or more. The attainment of such high-frequency response necessarily involves much clever engineering design, but even this is not sufficient for all signals of interest. There will arise experimental situations in which sinusoids and other signals with frequencies greater than 200 MHz are of interest. Fortunately, there is a technique for examining such signals as long as they are repetitive.

This procedure is known as *sampling* and is the electronic equivalent of the stroboscope. Figure 4-50 illustrates the operation of a sampling oscilloscope. The first waveform shown is repetitive and has a frequency which exceeds the maximum response rate of the oscilloscope amplifiers. In order to "slow" this waveform, a series of electronic "snapshots" are taken and displayed, as shown

Figure 4-50 Waveforms in a sampling oscilloscope.

in the second waveform. This "sampling" occurs at slightly later times in each cycle of the original waveform. As a consequence, the oscilloscope is able to display a digital approximation of the original signal.

4-17 OSCILLOGRAPHS

The oscillograph is basically a recording D'Arsonval movement. Various input amplifiers are available for raising the signal to an acceptable level for driving the movement. Two types of galvanometer movements are used: the direct-writing, or stylus, movement and the light-beam movement. In the stylus movement the galvanometer deflection is recorded directly on recording paper by action of the stylus. Three types of stylus are in common use: ink-writing, pressure-writing, and heated styli. The latter two write on either pressure- or heat-sensitive paper. The light-beam oscillograph uses a light-beam galvanometer, a light source to produce a fine beam, and light-sensitive paper for recording. The movement of a stylus oscillograph is similar to that of the indicating D'Arsonval movement shown in Fig. 4-10 except that the pointer is replaced by a stylus, which writes on the recording paper. The construction of the light-beam oscillograph is somewhat similar to that of the light-beam galvanometer except that a lens system is used to focus the light beam to a small point in order to produce a fine trace on the recording paper. Since input amplification is provided, the galvanometer may be constructed in a more substantial fashion than in the system of Fig. 4-9. Typically, the mirror and coil are placed in an integral unit, which may be inserted in the oscillograph instrument. The ends of the unit allow it to pivot freely and reflect the light beam onto the recording paper. This unit is frequently not much larger than a matchstick. A photograph of a typical mirror or galvanometer for use in an oscillograph is shown in Fig. 4-51a, and the construction details are shown in Fig. 4-51b.

The frequency response of an oscillograph is primarily a function of the mechanical frequency response of the galvanometer, which, in turn, is primarily dependent on its mass moment of inertia and damping characteristics. The smaller the mass of the galvanometer movement, the higher the frequency response. For this reason, the light-beam oscillograph has a much higher frequency response than the stylus type because of the higher mass associated with the stylus movement. In general, stylus-type oscillographs exhibit flat frequency responses from 0 to 100 Hz, while light-beam oscillographs may be used up to 5000 Hz.

A multitude of preamplifiers, chart speeds, and remote operation facilities are available on commercial oscillographs. Commercial models are available with 1 to 50 separate channels. A photograph of a typical oscillograph is shown in Fig. 4-52.

The oscillograph has an accuracy of about ± 2 percent of full scale. For more precise recording of signals a chart recorder is used with an internal servo control, which acts as a self-balancing potentiometer circuit. Such recorders have

Figure 4-51 (*a*) Photograph of a typical oscillograph galvanometer (the galvanometer is approximately the size of a matchstick); (*b*) schematic diagram of matchstick galvanometer and electric connections. (*Courtesy of Honeywell, Inc.*)

Figure 4-52 Photograph of a commercial oscillograph system.

accuracies of the order of 0.2 percent of full scale but are only adaptable to low-frequency signals below about 5 Hz. These recorders are widely used for thermocouple measurements, and standard models are available with a full-scale range as low as 1 mV. Most of these recorders use a zener-diode device to produce the reference voltage rather than a standard cell because such a source is quite stable and does not require frequent standardization.

4-18 COUNTERS—TIME AND FREQUENCY MEASUREMENTS

The engineer is called upon to perform counting-rate and frequency measurements over an extremely broad range of time intervals. A determination of revolutions per minute on a slow-speed diesel engine might involve a simple mechanical revolutions counter and a hand-operated stopwatch. In this case, the accuracy of the measurement would depend on the human-response time in starting and stopping the watch. The same measurement performed on an automobile engine might utilize an electric transducer which generates a pulse for each revolution of the engine. The pulses could be fed to some type of counting device, which would establish the number of pulses produced for a given increment of time. The accuracy of measurement would again depend on the accuracy of the specification of the time interval. In both of these simple cases a circular frequency measurement is being performed through the combination of a counting measurement and time-interval measurement. Aside from the simple frequency measurements discussed with the CRO, most frequency measurements are performed by some type of counting operation.

A large variety of electronic counters are available commercially. These instruments have internal circuitry which enables them to be used for measurement of frequency, period, or time intervals over a very wide range. These instruments usually contain four sets of internal circuits:

1. Input-signal conditioning circuits, which transform the input signal into a series of pulses for counting
2. The time base, which provides precise time increments during which the pulses are counted
3. A signal "gate," which starts and stops the counting device
4. A decade counter and display, which counts the pulses and provides a digital readout

Figure 4-53 Photograph of an electronic counter.

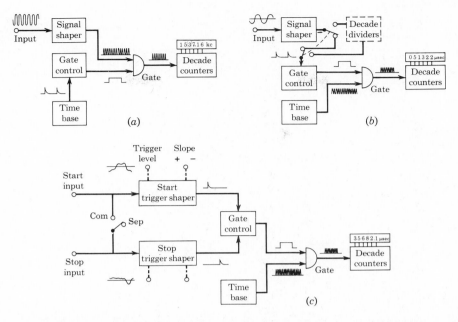

Figure 4-54 Block diagrams indicating functions of electronic counter. (*a*) Frequency measurement: function selector is set to "frequency," and appropriate gate time is selected by time-base switch (combined with function selector on some counters); (*b*) period measurement: function selector is set to "period," and counted frequency is chosen by time-base switch; (*c*) time-interval measurements: start and stop triggers may be derived from two separate sources, as shown, or from different points of one waveform by setting COM-SEP switch to COM. (*Courtesy of Hewlett-Packard Company*)

A photograph of a typical commercial counter for frequency measurement is shown in Fig. 4-53.

The functions of the electronic counter are shown schematically in Fig. 4-54. In Fig. 4-54*a* a frequency measurement is performed. The input signal is fed to the signal shaper, which produces an equivalent pulse train, which in turn is fed to the gate along with a predetermined time-base signal. The counters subsequently register the number of pulses in the given time, which is interpreted in terms of frequency. A period measurement is indicated in Fig. 4-54*b*. In this instance the time base is used to provide a known number of pulses per unit time, while the input signal is used to actuate the gate circuit. The measured count is thus an indication of the period of the input signal. For time-interval measurements the arrangement in Fig. 4-54*c* is used. Again, the time base provides a known number of pulses per unit time. Either external or internal trigger circuits provide the signals to the gate circuits to start and stop the count. These trigger circuits may be made to actuate on certain polarities or peak values of a waveform, depending on the measurement which is desired.

Phase-angle measurements may be made by measuring the time interval between similar points on the two waveforms. Calibration of the time-base cir-

Table 4-4 Summary of transducer characteristics

Type of transducer and principle of operation	Type of input	Input range or level	Input-impedance characteristics	Input sensitivity	Error and noise characteristics
Variable resistance: Movement of contact on slide-wire; also called resistance potentiometers	Linear displacement or angular displacement	Minimum level as low as 0.1% of total resistance	Varies widely, depending on the total resistance characteristics and physical size	Commercial potentiometers can have sensitivity of less than 0.002 in, or 0.2° in an angular measurement	Deviation from nonlinearity of the order of 0.5% of total resistance. Noise is usually negligible of the order of 10 μV at the contact. Noise increases with "chatter" of contact
Differential transformer: See fig. and text discussion; linear variable differential transformer most widely used; converts displacement to voltage	Linear displacement	Total range from ± 0.005 to ± 3 in	Depends on size; forces from 0.1 to 0.3 g usually required	0.5% of total input range	Deviation from linearity about 0.5%; generally accurate to $\pm 1\%$
Capacitive transducer: Variable distance between plates registered as a change in capacitance	Displacement or change in dielectric constant between plates; also change in area of plates	Very broad; from 10^{-8} cm to several meters	The input force requirements are very small, of the order of a few dynes	Highly variable; can obtain sensitivities of the order of 1 pF/0.0001 in for air-gap measurements of displacement	Errors may result from careless mechanical construction, humidity variations, noise, and stray capacitance in cable connections
Piezoelectric effect: Force impressed on crystals with asymmetrical charge distributions produces a potential difference at the surface of the crystal	Force or stress	Varies widely with crystal material; see sensitivity	Input force requirements are relatively large compared with other transducers	Varies with material: Quartz 0.05 V·m/N Rochelle salt 0.15 V·m/N Barium titanate 0.007 V·m/N	Subject to hysteresis and temperature effects
Photoelectric effect: Light striking metal cathode causes liberation of electrons which may be attracted to anode to produce electric current	Light	Wavelength range depends on glass-tube enclosure; photoemissive materials respond between 0.2 and 0.8 μm	Not applicable	Vacuum tubes: 0.002–0.1 μA/μW Gas-filled tubes: 0.01–0.15 μA/μW	Depends on plate voltage but of the order of 10^{-8} A at room temperature
Photoconductive transducer: Light striking a semiconductor material, such as selenium, metallic sulfides, or germanium, produces a decrease in resistance of the material.	Light	Very broad; from thermal radiation through the X-ray region	Not applicable	About 300 μA/μW at maximum sensitivity of the device	Very low noise, usually less than associated circuit
Photovoltaic cell: Light falling upon a semiconducting material in contact with a metal plate produces a potential	Light	Depends on material; selenium 0.2–0.7 μm CuO 0.5–1.4 μm germanium 1.0–1.7 μm	Not applicable	1 mA/lm or 10^{-7} W/cm²-lm	Low noise
Ionization transducer: Displacement converted to voltage through a capacitance change	Displacement, 0.1–10 MHz excitation frequency	Less than 1 mm to several in	Small force required	1–10 V/mm	Can be accurate to microinches
Magnetometer search coil: Changing magnetic field impressed on coil generates an emf proportional to the time rate of change of the field	Changing magnetic field	10^{-3} oersted to highest values obtainable	Not applicable	Depends on coil dimensions but can be of the order of 10^{-5} oersted	Accuracies of 0.05% have been obtained
Hall-effect transducer: Magnetic field impressed on a plate carrying an electric current generates a potential difference in a direction perpendicular to both the current and the magnetic field	Magnetic field	1–20,000 gauss	Not applicable	Depends on plate thickness and current; of the order of -1×10^{-8} V·cm/A·G for bismuth	Can be calibrated within 1%

Frequency response	Temperature effects	Type of output	Output range or level	Output-impedance characteristics	Application	Remarks
Generally not above 3 Hz for commercial potentiometers	0.002–0.15% °C⁻¹ due to a change in resistance; also, some thermo-electric effects depending on types of contacts used	Voltage or current depending on connecting circuit	Wide	Variable	Used for measurement of displacement	Simple, inexpensive, easy to use, many types available commercially
Frequency of applied voltage must be 10 times desired response; mechanical limitations also	Small influence of temperature may be reduced by using a thermistor circuit	Voltage proportional to input displacement	0.4–4.0 mV/ 0.001-in/V input depending on frequency; lower frequency produces lower output	Mainly resistive; low to medium impedance, as low as 20 Ω, depends on size	Used for measurement of displacement	Simple, rugged, inexpensive, high output, requires simple accessory equipment. Care must be taken to eliminate stray magnetic fields
Depends strongly on mechanical construction but may go to 50,000 Hz	Not strong if design allows for effects	Capacitance	Usually between 10^{-3} and 10^3 pF change in capacitance over output range	Usually 10^3– 10^7 Ω	Displacement, area liquid level, pressure, sound-level measurements, and others. Particularly useful where small forces are available for driving the transducer	High output impedance may require careful construction of output circuitry
Depends on external circuitry and mechanical mounting; 20–20 kHz easily obtained; no response to steady-state forces	Wide variation in crystal properties with temperature	Voltage proportional to input force	Wide, depends on crystal size and material; see sensitivity; can have output of several volts	High, of the order of 10^3 MΩ	Measurement of force, pressure, sound level (microphone use)	Simple, inexpensive, rugged
Linear 0–500 Hz; current response drops off 15% at 10,000 Hz	Generally not operable above 75–100°C	Current	Of the order of 2 μA	High, of the order of 10 MΩ	Very useful for counting purposes	Inexpensive, high output
Rise time varies widely with material and incident radiation from 50 μs to min	Response to longer wavelengths increases with reduction in temperature	Current drawn in the external circuit	Depends on incident intensity; see input sensitivity	High, varies from 1–10^3 MΩ; in commercial devices	Widely used for radiant measurements at all wavelengths	Fairly expensive; calls for precise circuitry to utilize its full potential
Rise time of the order of 1 μs, response into the megacycle range	Variations of 10% over 40°C, range depending on external resistance load	Voltage	100–250 mV in normal room light; selenium cells up to 500 V at high illumination	3,000–10,000 Ω capacitance of the order of 0.05 μF/ cm² in selenium cells	Widely used for exposure meters, selenium cells, responds to X-rays	Inexpensive, nonlinear behavior, some aging effects
0–3000 Hz	Small	Voltage	Depends on excitation circuit; see input sensitivity	High, of the order of 1 MΩ	Can be used where accurate measurement of displacement is needed	Relatively insensitive to frequency of excitation circuitry
0 to radio-frequency	Small	Voltage	Depends on coil size; see input sensitivity	Depends on coil size	Measurement of magnetic field	Simple, although accurate, coil dimensions must be maintained for high accuracy
High	Large but can be calibrated	Voltage	Millivolts and microvolts	Low, of the order of 100 Ω for a bismuth detector	Measurement of magnetic field	Each transducer usually must be calibrated because of nonuniformity in semiconductors used for construction

cuits may be achieved by the use of special frequency standards based on crystal oscillators, or with NBS standard broadcast frequencies. A very detailed discussion of frequency and time standards is given in Ref. [6].

4-19 TRANSDUCERS

A large number of devices transform values of physical variables into equivalent electrical signals. Such devices are called *transducers* and have been mentioned frequently in the preceding sections of this chapter. We shall now discuss some of the more widely used transducers and their principles of operation. Table 4-4 presents a compact summary of transducer characteristics, and the reader should consult this table throughout the following discussions to maintain an overall perspective. It may be noted that the subsequent paragraphs will be primarily concerned with principles of operation and will not cover much of the detailed information contained in Table 4-4. Thus, the written discussion and the tabular presentation are complementary and should be used jointly.

The discussions of transducers that follow in this section are concerned primarily with electric effects. Other transducers of a more specialized nature (thermocouples, strain gages, pressure transducers, and nuclear radiation detectors) will be discussed in subsequent chapters.

4-20 THE VARIABLE-RESISTANCE TRANSDUCER

The variable-resistance transducer is a very common device which may be constructed in the form of a moving contact on a slide-wire, a moving contact on a coil of wire, through either linear or angular movement, or a contact that moves through an angular displacement on a solid conductor like a piece of graphite. The device may also be called a *resistance potentiometer* or *rheostat* and is available commercially in many sizes, designs, and ranges. Costs can range from a few cents for a simple potentiometer used as a volume control in a radio circuit to hundreds of dollars for a precision device used for accurate laboratory work.

The variable-resistance transducer fundamentally is a device for converting either linear or angular displacement into an electric signal; however, through mechanical methods it is possible to convert force and pressure to a displacement so that the device may also be useful in force and pressure measurements.

4-21 THE DIFFERENTIAL TRANSFORMER (LVDT)

A schematic diagram of the differential transformer is shown in Fig. 4.55. Three coils are placed in a linear arrangement as shown with a magnetic core which may move freely inside the coils. The construction of the device is indicated in Fig. 4-56. An alternating input voltage is impressed in the center coil, and the

E_o Output voltage

Secondary coils

Core

Primary coils

Displacement

E_i
Input voltage

Figure 4-55 Schematic diagram of a differential transformer.

Protective shield

Insulating form

Core

Cross section

Primary coil

Secondary coils

Figure 4-56 Construction of a commercial linear variable differential transformer (LVDT). (*Courtesy of Schaevitz Engineering Company*)

Voltage out +

A

0

B

Core position

Voltage out, opposite phase −

Core displacement

Core at A

Core at 0
(Null position)

Core at B

Figure 4-57 Output characteristics of an LVDT, according to Ref. [7].

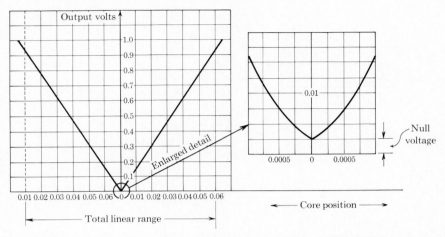

Figure 4-58 V graph for an LVDT showing slight nonlinear behavior in the null region, according to Ref. [7].

output voltage from the two end coils depends on the magnetic coupling between the core and the coils. This coupling is, in turn, dependent on the position of the core. Thus, the output voltage of the device is an indication of the displacement of the core. As long as the core remains near the center of the coil arrangement, the output is very nearly linear, as indicated in Fig. 4-57. The linear range of commercial differential transformers is clearly specified, and the devices are seldom operated outside this range. When operating in the linear range, the device is called a *linear variable differential transformer* (LVDT). Near the null position a slight nonlinearity condition is encountered, as illustrated in Fig. 4-58. It will be noted that Fig. 4-57 considers the phase relationship of the output voltage, while the " V " graph in Fig. 4-58 indicates the absolute magnitude of the output. There is a 180° phase shift from one side of the null position to the other.

The frequency response of LVDTs is primarily limited by the inertia characteristics of the device. In general, the frequency of the applied voltage should be 10 times the desired frequency response.

Commercial LVDTs are available in a broad range of sizes and are widely used for displacement measurements in a variety of applications. Force and pressure measurements may also be made after a mechanical conversion. Table 4-4 indicates the general characteristics of the LVDT. The interested reader should consult Ref. [7] for more detailed information.

4-22 CAPACITIVE TRANSDUCERS

Consider the capacitive transducer shown in Fig. 4-59. The capacitance (in pico-farads) of this arrangement is given by

$$C = 0.225\epsilon \frac{A}{d} \tag{4-60}$$

Dielectric

d

Overlapping area, A

Figure 4-59 Schematic of a capacitive transducer.

where d = distance between the plates, in
A = *overlapping* area, in^2
ϵ = dielectric constant ($\epsilon = 1$ for air; $\epsilon = 3$ for plastics)

The constant is 0.0885 when the area is in square centimeters and the separation distance is in centimeters.

This plate arrangement may be used to measure a change in the distance d through a change in capacitance. A change in capacitance may also be registered through a change in the overlapping area A resulting from a relative movement of the plates in a lateral direction or a change in the dielectric constant of the material between the plates. The capacitance may be measured with bridge circuits. The output impedance of a capacitor is given by

$$Z = \frac{1}{2\pi f C} \qquad (4\text{-}61)$$

where Z = impedance, Ω
f = frequency, Hz
C = capacitance, F

In general, the output impedance of a capacitive transducer is high; this fact may call for careful design of the output circuitry.

The capacitive transducer may be used for displacement measurements through a variation of either the spacing distance d or the plate area. It is commonly used for liquid-level measurements, as indicated in Fig. 4-60. Two electrodes are arranged as shown, and the dielectric constant varies between the electrodes according to the liquid level. Thus, the capacitance between the electrodes is a direct indication of the liquid level.

Liquid level

Figure 4-60 Use of a capacitive transducer for liquid-level measurement.

Example 4-10 A capacitive transducer is constructed of two 1-in^2 plates separated by a 0.01-in distance in air. Calculate the displacement sensitivity of such an arrangement. The dielectric constant for air is 1.0006.

SOLUTION The sensitivity is found by differentiating Eq. (4-60).

$$S = \frac{\partial C}{\partial d} = -\frac{0.255\epsilon A}{d^2}$$

Thus,
$$S = -\frac{(0.255)(1.0006)(1)}{(0.01)^2} = -2.25 \times 10^3 \text{ pF/in}$$

Example 4-11 For the capacitive transducer in Example 4-10 the allowable uncertainty in the spacing measurement is $w_d = \pm 0.0001$ in, while the estimated uncertainty in the plate area is ± 0.005 in^2. Calculate the tolerable uncertainty in the capacitance measurement in order to achieve the allowable uncertainty in the spacing measurement.

SOLUTION Solving Eq. (4-60) for d, we have

$$d = 0.225 \frac{\epsilon A}{C} \tag{a}$$

Making use of Eq. (3-2), we obtain

$$\frac{w_d}{d} = \left[\left(\frac{w_c}{C} \right)^2 + \left(\frac{w_A}{A} \right)^2 \right]^{1/2} \tag{b}$$

We have

$$\frac{w_d}{d} = \frac{0.0001}{0.01} = 0.01 \qquad \frac{w_A}{A} = \frac{0.005}{1.0} = 0.005$$

so that

$$\frac{w_c}{C} = 0.00866 = 0.866\%$$

The nominal value of C is

$$C = \frac{(0.255)(1.0006)(1.0)}{0.01} = 22.513 \text{ pF}$$

so that the tolerable uncertainty in C is

$$w_C = (22.513)(0.00866) = \pm 0.195 \text{ pF}$$

4-23 PIEZOELECTRIC TRANSDUCERS

Consider the arrangement shown in Fig. 4-61. A piezoelectric crystal is placed between two plate electrodes. When a force is applied to the plates, a stress will be produced in the crystal and a corresponding deformation. With certain crystals this deformation will produce a potential difference at the surface of the crystal, and the effect is called the *piezoelectric effect*. The induced charge on the crystal is proportional to the impressed force and is given by

$$Q = dF \tag{4-62}$$

where Q is in coulombs, F is in newtons, and the proportionality constant d is called the *piezoelectric constant*. The output voltage of the crystal is given by

$$E = gtp \tag{4-63}$$

where t is the crystal thickness in meters, p is the impressed pressure in newtons per square meter, and g is called voltage sensitivity and is given by

$$g = \frac{d}{\epsilon} \tag{4-64}$$

Values of the piezoelectric constant and voltage sensitivity for several common piezoelectric materials are given in Table 4-5.

The voltage output depends on the direction in which the crystal slab is cut in respect to the crystal axes. In Table 4-5 an X (or Y) cut means that a perpendicular to the largest face of the cut is in the direction of the x axis (or y axis) of the crystal.

Piezoelectric crystals may also be subjected to various types of shear stresses instead of the simple compression stress shown in Fig. 4-61, but the output voltage is a complicated function of the exact crystal orientation. Piezoelectric crystals are widely used as inexpensive pressure transducers for dynamic measurements and are commonly employed as phonograph pickups. General information on the piezoelectric effect is given by Cady [1].

Crystal

Figure 4-61 The piezoelectric effect.

Table 4-5 Piezoelectric constants†

Material	Orientation	Charge sensitivity d, $\dfrac{C/m^2}{N/m^2}$	Voltage sensitivity g, $\dfrac{V/m}{N/m^2}$
Quartz	X cut; length along Y length longitudinal	2.25×10^{-12}	0.055
	X cut; thickness longitudinal	-2.04	-0.050
	Y cut; thickness shear	4.4	-0.108
Rochelle salt	X cut 45°; length longitudinal	435.0	0.098
	Y cut 45°; length longitudinal	-78.4	-0.29
Ammonium dihydrogen phosphate	Z cut 0°; face shear	48.0	0.354
	Z cut 45°; length longitudinal	24.0	0.177
Commercial barium titanate	To polarization	130–160	0.0106
ceramics	To polarization	-56.0	0.0042–0.0053

† According to Lion [3].

Example 4-12 A quartz piezoelectric crystal having a thickness of 2 mm and a voltage sensitivity of 0.055 V·m/N is subjected to a pressure of 200 psi. Calculate the voltage output.

SOLUTION We calculate the voltage output with Eq. (4-63):

$$p = (200)(6.895 \times 10^3) = 1.38 \times 10^6 \ N/m^2$$

$$t = 2 \times 10^{-3} \ m$$

Thus,
$$E = (0.055)(2 \times 10^{-3})(1.38 \times 10^6) = 151.8 \ V$$

4-24 PHOTOELECTRIC EFFECTS

A photoelectric transducer converts a light beam into a usable electric signal. Consider the circuit shown in Fig. 4-62. Light strikes the photoemissive cathode and releases electrons, which are attracted toward the anode, thereby producing

Figure 4-62 The photoelectric effect.

an electric current in the external circuit. The cathode and anode are enclosed in a glass or quartz envelope, which is either evacuated or filled with an inert gas. The photoelectric sensitivity is defined by

$$I = S\Phi \tag{4-65}$$

where I = photoelectric current
Φ = illumination of the cathode
S = sensitivity

The sensitivity is usually expressed in units of amperes per watt or amperes per lumen.

Photoelectric-tube response to different wavelengths of light is influenced by two factors: (1) the transmission characteristics of the glass-tube envelope and (2) the photoemissive characteristics of the cathode material. Photoemissive materials are available which will respond to light over a range of 0.2 to 0.8 μm. Most glasses transmit light in the upper portion of this range, but many do not transmit below about 0.4 μm. Quartz, however, transmits down to 0.2 μm. Various noise effects are present in photoelectric tubes, and the interested reader should consult the discussion by Lion [3] for more information.

Photoelectric tubes are quite useful for measurement of light intensity. Inexpensive devices can be utilized for counting purposes through periodic interruption of a light source.

4-25 PHOTOCONDUCTIVE TRANSDUCERS

The principle of the photoconductive transducer is shown in Fig. 4-63. A voltage is impressed on the semiconductor material as shown. When light strikes the semiconductor material, there is a decrease in the resistance, thereby producing an increase in the current indicated by the meter. A variety of substances are used for photoconductive materials, and a rather detailed discussion of the pertinent literature on the subject is given in Refs. [3] and [10].

Photoconductive transducers enjoy a wide range of applications and are useful for measurement of radiation at all wavelengths. It must be noted, however, that extreme experimental difficulties may be encountered when operating with long-wavelength radiation.

Figure 4-63 Schematic of a photoconductive transducer.

The responsivity R_v of a detector is defined as

$$R_v = \frac{\text{rms output voltage}}{\text{rms power incident upon the detector}} \qquad (4\text{-}66)$$

The *noise equivalent power* (NEP) is defined as the minimum-radiation input that will produce a signal-to-noise ratio of unity. The *detectivity* D is defined as

$$D = \frac{R_v}{\text{rms noise-voltage output of cell}} \qquad (4\text{-}67)$$

The detectivity is the reciprocal of NEP. A normalized detectivity D^* is defined as

$$D^* = (A \ \Delta f)^{1/2} D \qquad (4\text{-}68)$$

where A is the area of the detector and Δf is a noise-equivalent bandwidth. The units of D^* are usually $\text{cm} \cdot \text{Hz}^{1/2}/\text{W}$, and the term is used in describing the performance of detectors so that the particular surface area and bandwidth will not affect the results. Figures 4-64 and 4-65 illustrate the performance of several photoconductive detectors over a range of wavelengths. In these figures the wavelength is expressed in micrometers, where $1 \ \mu m = 10^{-6}$ m. The symbols represent lead sulfide (PbS), lead selenide (PbSe), lead telluride (PbTe), indium antimonide (InSb), and gold- and antimony-doped germanium (Ge : Au, Sb). For these figures D^* is a monochromatic detectivity for an incident radiation, which is chopped at 900 Hz and a 1-Hz bandwidth.

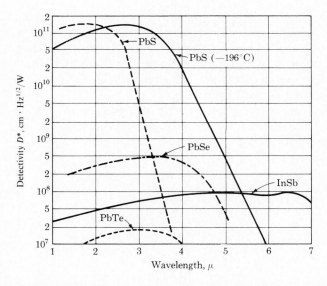

Figure 4-64 Absolute spectral response of typical detectors at room temperature, according to Ref. [11].

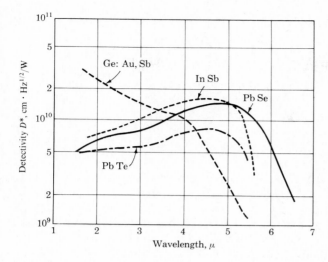

Figure 4-65 Absolute spectral response of typical detector cooled to liquid nitrogen temperature, according to Ref. [11].

The lead-sulfide cell is very widely used for detection of thermal radiation in the wavelength band of 1 to 3 μm. By cooling the detector, more favorable response at higher wavelengths can be achieved up to about 4 or 5 μm. For measurements at longer wavelengths the indium-antimonide detector is preferred, but it has a lower detectivity than lead sulfide. Some of the applications of these cells are discussed in Chap. 12.

Example 4-13 Calculate the incident radiation at 2 μm that is necessary to produce a signal-to-noise (S/N) ratio of 40 dB with a lead-sulfide detector at room temperature, having an area of 1 mm^2.

SOLUTION We first insert Eqs. (4-66) and (4-67) in Eq. (4-68) to express the detectivity in terms of voltages and incident power. Thus,

$$D^* = (A \ \Delta f)^{1/2} \ \frac{E_o}{E_{noise}} \ \frac{1}{P_{incident}} \qquad (a)$$

From Fig. 4-64, $D^* = 1.5 \times 10^{11}$ cm $\cdot$ Hz$^{1/2}$/W for $\Delta f = 1$ Hz. For a S/N ratio of 40 dB, we have

$$40 = 20 \log \frac{E_o}{E_{noise}}$$

so that

$$\frac{E_o}{E_{noise}} = 100$$

Using $A = 10^{-2}$ cm^2, Eq. (a) yields

$$1.5 \times 10^{11} = \frac{(10^{-2})^{1/2}(1)(100)}{P_{\text{incident}}}$$

Thus, $$P_{\text{incident}} = 6.7 \times 10^{-11} \text{ W}$$

4-26 PHOTOVOLTAIC CELLS

The photovoltaic-cell principle is illustrated in Fig. 4-66. The sandwich construction consists of a metal base plate, a semiconductor material, and a thin transparent metallic layer. This transparent layer may be in the form of a sprayed, conducting lacquer. When light strikes the barrier between the transparent metal layer and the semiconductor material, a voltage is generated as shown. The output of the device is strongly dependent on the load resistance R. The open-circuit voltage approximates a logarithmic function, but more linear behavior may be approximated by decreasing the load resistance.

Perhaps the most widely used application of the photovoltaic cell is the light exposure meter in photographic work. The logarithmic behavior of the cell is a decided advantage in such applications because of its sensitivity over a broad range of light intensities.

4-27 IONIZATION TRANSDUCERS

A schematic of the ionization transducer is shown in Fig. 4-67. The tube contains a gas at low pressure while the RF generator impresses a field on this gas. As a result of the RF field, a glow discharge is created in the gas, and the two electrodes 1 and 2 detect a potential difference in the gas plasma. The potential

Figure 4-66 Schematic diagram of a photovoltaic cell.

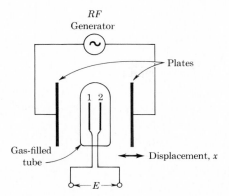

Figure 4-67 Schematic diagram of an ionization-displacement transducer.

difference is dependent on the electrode spacing and the capacitive coupling between the RF plates and the gas. When the tube is located at the central position between the plates, the potentials on the electrodes are the same; but when the tube is displaced from this central position, a dc potential difference will be created. Thus, the ionization transducer is a useful device for measuring displacement. Some of the basic operating characteristics are given in Table 4-4, and a detailed description of the output characteristics is given by Lion [4].

4-28 MAGNETOMETER SEARCH COIL

A schematic of the magnetometer search coil is shown in Fig. 4-68. A flat coil with N turns is placed in the magnetic field as shown. The length of the coil is L, and the cross-sectional area is A. The magnetic field strength H and the magnetic flux density B are in the direction shown, where

$$B = \mu H \tag{4-69}$$

and μ is the magnetic permeability. The voltage output of the coil E is given by

$$E = NA \cos \alpha \, \frac{dB}{dt} \tag{4-70}$$

Figure 4-68 Schematic of a magnetometer search coil.

Figure 4-69 Use of a rotating search coil for measurement of steady magnetic fields.

where α is the angle formed between the direction of the magnetic field and a line drawn perpendicular to the plane of the coil. The total flux through the loop is

$$\phi = A \cos \alpha B \tag{4-71}$$

so that

$$E = N \frac{d\phi}{dt} \tag{4-72}$$

Note that the voltage output of the device is dependent on the rate of change of the magnetic field and that a stationary coil placed in a steady magnetic field will produce a zero-voltage output. The search coil is thus a transducer that transforms a magnetic field signal into a voltage.

In order to perform a measurement of a steady magnetic field it is necessary to provide some movement of the search coil. A typical method is to use a rotating coil as shown in Fig. 4-69. The rms value of the output voltage for such a device is

$$E_{\text{rms}} = \frac{1}{\sqrt{2}} NAB\omega \tag{4-73}$$

where ω is the angular velocity of rotation. Oscillating coils are also used. The accuracy of the search coil device depends on the accuracy with which the dimensions of the coil are known. The coil should be small enough that the magnetic field is constant over its area.

In the above equations the magnetic flux density is expressed in webers per square meter, the area is in square meters, the time is in seconds, the magnetic flux is in webers, the magnetic field strength (magnetic intensity) is in amperes per meter, and the magnetic permeability for free space is $4\pi \times 10^{-7}$ H/m. An alternative set of units uses B in gauss, H in oersteds, A in centimeters squared, and μ in abhenrys per centimeter. The magnetic permeability for free space in this instance is unity.

Example 4-14 A rotating search coil has 10 turns with a cross-sectional area of 5 cm². It rotates at a constant speed of 100 rpm. The output voltage is 40 mV. Calculate the magnetic field strength.

SOLUTION According to Eq. (4-73),

$$B = \frac{\sqrt{2}E_{rms}}{NA\omega} = \frac{\sqrt{2}(0.04)}{(10)(5 \times 10^{-4})[(100)(2\pi)/60]}$$

$$= 1.08 \text{ Wb/m}^2$$

$$H = \frac{B}{\mu} = \frac{1.08}{4\pi \times 10^{-7}}$$

$$= 8.6 \times 10^5 \text{ A/m}$$

4-29 HALL-EFFECT TRANSDUCERS

The principle of the Hall effect is indicated in Fig. 4-70. A semiconductor plate of thickness t is connected as shown so that an external current I passes through the material. When a magnetic field is impressed on the plate in a direction perpendicular to the surface of the plate, there will be a potential E_H generated as shown. This potential is called the Hall voltage and is given by

$$E_H = K_H \frac{IB}{t} \qquad (4.74)$$

where I is in amperes, B is in gauss, and t is in centimeters. The proportionality constant is called the *Hall coefficient* and has the units of volt-centimeters per ampere-gauss. Typical values of K_H for several materials are given in Table 4-6.

Table 4-6 Hall coefficients for different materials†

Material	Field strength, G	Temp., °C	K_H, $\dfrac{V \cdot cm}{A \cdot G}$
As	4,000–8,000	20	4.52×10^{-11}
C	4,000–11,000	Room	-1.73×10^{-10}
Bi	1,130	20	-1×10^{-8}
Cu	8,000–22,000	20	-5.2×10^{-13}
Fe	17,000	22	1.1×10^{-11}
n-Ge	100–8,000	25	-8.0×10^{-5}
Si	20,000	23	4.1×10^{-8}
Sn	4,000	Room	-2.0×10^{-14}
Te	3,000–9,000	20	5.3×10^{-7}

† According to Lion [3].

Figure 4-70 The Hall effect.

Example 4-15 A Hall-effect transducer is used for the measurement of a magnetic field of 5000 G. A 2-mm slab of bismuth is used with a current of 3 A. Calculate the voltage output of the device.

SOLUTION We use Eq. (4-74) and the data of Table 4-6.

$$E_H = K_H \frac{IR}{t}$$

$$= \frac{(-1 \times 10^{-8})(3)(5000)}{(2 \times 10^{-1})}$$

$$= -7.5 \times 10^{-4} \text{ V}$$

4-30 DIGITAL DISPLACEMENT TRANSDUCERS

A digital displacement transducer can be used for both angular and linear measurements. In Fig. 4-71 an angular measurement device is shown. As the wheel rotates, light from the source is alternately transmitted and stopped, thereby submitting a digital signal to the photodetector. The signal is then amplified and

Figure 4-71 Digital transducer for angular displacement.

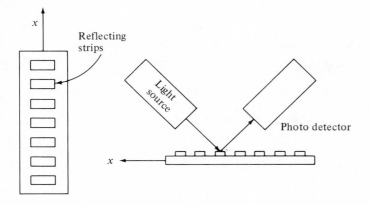

Figure 4-72 Digital transducer for linear displacement.

sent to a counter. The number of counts is proportional to angular displacement. The *frequency* of the signal is proportional to angular velocity. Sensitivity of the device may be improved by increasing the number of cutouts.

A linear transducer which operates on a reflection principle is shown in Fig. 4-72. Small reflecting strips are installed on the motion device. Light from the source is then alternately reflected and absorbed with linear motion, thereby presenting a digital signal to the photodetector. Readout is the same as with the angular instrument. Calibration with a known displacement standard must be performed.

4-31 COMPARISON OF ANALOG AND DIGITAL INSTRUMENTS

As more and more instrument circuits and readout devices employ digital techniques one might automatically assume that everything should be converted to digital. But analog instruments have their place too. For example, when balancing a circuit to obtain a null condition an analog instrument may be easier to use. When a panel of instruments is used to indicate the operating condition of a complicated plant or process, analog instruments might be preferable because a trained operator can visually sense the position of all the indicators more quickly than with several digital readouts.

Most physical measurements, such as those of resistance, voltage, force, and displacement, occur in the form of some analog signal. Certainly, if direct computer processing of the data is to be performed, analog-to-digital conversion must be performed, and a wide variety of commercial equipment is available for such conversion. However, in low-cost simple applications we may expect analog instruments to continue to have their own place for some time to come.

4-32 SUMMARY

This chapter began by considering the basic physical phenomena which underlie electrical instruments and circuits. We then examined the meaningful measures which can be used to characterize time-varying waveforms. The basic instrumentation, both analog and digital, used to measure these waveforms was discussed.

This set the stage for an examination of the general experimental setting in which measurements are made with the assistance of electronic systems. In general, the value of the particular physical property of interest is converted to an electrical signal by some suitable transducer, and the output of this transducer is fed into an input circuit. This circuit is intended to efficiently couple the transducer's signal to the additional equipment which is necessary to reduce the effects of noise upon the signal's distinguishability, transmit the signal, and display its form.

We have discussed the elements of this general procedure. In Chaps. 5 through 13 you will gain familiarity with the measurement procedures to be used for a wide range of physical quantities. Chapter 14 will complete the story because it is there that we will examine, in more complete terms, the problem of data acquisition and display. In particular, Chap. 14 will focus on digital techniques which are now dominant in these applications.

REVIEW QUESTIONS

4-1 What is the "bridge" between electric current and mechanical force?

4-2 State the differences between analog and digital representations of signals.

4-3 Why is the root mean square value of a periodic waveform useful?

4-4 How does an electrodynamometer differ from a D'Arsonval meter? What advantages does it have over the D'Arsonval meter?

4-5 How do measurements of alternating currents differ from measurements of direct currents?

4-6 How are very high-frequency currents measured?

4-7 What are the applications for an electrostatic voltmeter?

4-8 How can an ammeter be converted to a voltmeter?

4-9 What are the advantages of digital instruments?

4-10 How is the current-sensitive input circuit dependent on internal meter impedance?

4-11 What is meant by a ballast circuit? How does it differ from a voltage-divider circuit?

4-12 What are the advantages of a high meter impedance when it is used with a voltage-divider circuit?

4-13 What is meant by loading error?

4-14 Why does the internal resistance of the galvanometer not influence the reading in a potentiometer circuit?

4-15 Differentiate between a bridge operated on the null and on the deflection principle.

4-16 What is meant by a voltage-sensitive deflection-bridge circuit?

4-17 How can an operational amplifier be used to average two inputs?

4-18 What purpose does a voltage follower serve?

4-19 How do transformers aid in matching impedances?

4-20 Why are dc power supplies needed?

4-21 How does the decibel notation for voltage level differ from that for power level?

4-22 What is the function of the phase-sensitive detector in a lock-in amplifier?

4-23 Why is an EVM useful for electrical measurements?

4-24 Why are sampling oscilloscopes used?

PROBLEMS

4-1 Expand Eq. (4-12) in series form and indicate the relation which may be used to obtain a linear approximation for the current as a function of the transducer resistance R. Also, show the error which results from this approximation.

4-2 A charged particle ($q = 0.5$ C) is moving with a velocity of 10 m/s in a magnetic field of 10 Wb/m^2.

    (a) What is the magnitude of the largest force the charge can experience under these conditions?

    (b) What is a physical geometry in which the particle will experience no force?

4-3 The basic meter of Fig. 4-2 is designed so that

$$B = 1 \text{ Wb/m}^2$$

$$L = 0.1 \text{ m}$$

$$K = 1 \text{ N/m}$$

    (a) What deflection x results when a direct current of 4 A flows through the meter?

    (b) Find an expression for $x(t)$ if $i(t) = 2 \cos t$ A.

4-4 Calculate the rms values of the following periodic currents:

    (a) $i(t) = 10 \cos (t)$ A

    (b) $i(t) = 10 \cos (377t)$ A

    (c) $i(t) = t \qquad\quad 0 \le t \le 1$

             $= t - 1 \qquad 1 \le t \le 2$

             $= t - 2 \qquad 2 \le t \le 3 \qquad$ etc.

4-5 Expand Eq. (4-14) in series form and indicate the relation which may be used to obtain a linear approximation for the current as a function of the transducer resistance R. Also, show the error which results from this approximation.

4-6 Derive an expression for the sensitivity S of the circuit of Fig. 4-19 where

$$S = \frac{di}{dR}$$

Find the condition for maximum sensitivity.

4-7 Obtain a linear approximation for the sensitivity of the ballast circuit of Fig. 4-20. Under what conditions would this relation apply? Estimate the error in this approximation.

4-8 Show that the output voltage v_o of the following circuit is proportional to the time derivative of the input voltage v_i

Figure Prob. 4-8

If $C = 10^{-5}$ F, what value of R is required if

$$v_o = -\frac{dv_i}{dt}$$

Assume the operational amplifier is ideal.

4-9 A diode has a current-voltage relation of the form

$$i = \alpha e^{\beta v}$$

where α and β are constants.

Find the relation between input and output voltages for the following circuit.

Figure Prob. 4-9

4-10 The transformer in the following circuit has a turns ratio of 10. A sinusoidal voltage of 10 V rms is applied at terminals $11'$.

Figure Prob. 4-10

What is I_2?

4-11 A Wheatstone bridge circuit has resistance arms of 400, 40, 602, and 6000 Ω taken sequentially around the bridge. The galvanometer has a resistance of 110 Ω and is connected between the junction of the 40- and 602-Ω resistors to the junction of the 400- and 6000-Ω resistors. The battery has an emf of 3 V and negligible internal resistance. Calculate the voltage across the galvanometer and the galvanometer current.

4-12 It is known that a certain resistor has a resistance of approximately 800 Ω. This resistor is placed in a Wheatstone bridge the other three arms of which have resistances of exactly 800 Ω. A 4-V battery with negligible internal resistance is used in the circuit. The galvanometer resistance is 100 Ω, and the indicated galvanometer current is 0.08 μA. Calculate the resistance of the unknown resistor.

4-13 Two galvanometers are available for use with a Wheatstone bridge having equal ratio arms of 100 Ω. One galvanometer has a resistance of 100 Ω and a sensitivity of 0.05 μA/mm, whereas the other has a resistance of 200 Ω and a sensititivy of 0.01 μA/mm. A 4-V battery is used in the circuit, and it has negligible internal resistance. An unknown resistance of approximately 500 Ω is to be measured with the bridge. Calculate the deflection of each galvanometer. State assumptions necessary to make this calculation.

4-14 Two known ratio arms of a Wheatstone bridge are 4000 and 400 Ω. The bridge is to be used to measure a resistance of 100 Ω. Two galvanometers are available: one with a resistance of 50 Ω and a sensitivity of 0.05 μA/mm, and one with a resistance of 500 Ω and a sensitivity of 0.2 μA/mm. Which galvanometer would you prefer to use? Assume that the galvanometer is connected from the junction of the ratio arms to the opposite corner of the bridge.

4-15 A Wheatstone bridge is constructed with ratio arms of 60 and 600 Ω. A 4-V battery with negligible internal resistance is used and is connected from the junction of the ratio arms to the opposite corner. A galvanometer having a resistance of 50 Ω and a sensitivity of 0.05 μA/mm is connected between the other corners. When the adjustable arm reads 200 Ω, the galvanometer deflection is 30 mm. What is the value of the unknown resistance?

4-16 The four arms of a Wheatstone bridge have resistances of 500, 1000, 600, and 290 Ω taken in sequence around the bridge. The battery of 3 V connects between the 1000- and 600-Ω resistors and the 500- and 290-Ω resistors. A galvanometer with a resistance of 50 Ω and a sensitivity of 0.05 μA/mm is connected across the other two terminals. The galvanometer is shunted by a resistance of 30 Ω. Calculate the galvanometer deflection. Repeat the calculation for a series resistance of 30 Ω connected in the galvanometer circuit instead of the shunt arrangement.

4-17 Suppose an *end resistor* is added to the circuit of Fig. 4-21 so that there results

Figure Prob. 4-17

where R_1 is the end resistor. Derive an expression for the voltage output and loading error of such an arrangement. What advantage does it offer over the circuit in Fig. 4-21? Suppose the end resistor were attached to the other end of the variable resistance. What would be the advantage in this circumstance? What would be the advantage if an end resistor were placed on each end of the variable resistor?

4-18 Design a bandpass filter to operate between the limits of 500 and 2000 Hz, with a load resistance of 16 Ω.

4-19 Design a lowpass filter with a cutoff frequency of 500 Hz with a load resistance of 1000 Ω.

4-20 Design a highpass filter with a cutoff frequency of 1000 Hz with a load resistance of 1000 Ω.

4-21 A voltage of 500 V is impressed on a 150-kΩ resistor. The impedance of the voltage source is 10 kΩ. Two meters are used to measure the voltage across the 150-kΩ resistor: a volt-ohmmeter with an internal impedance of 1000 Ω/V and an EVM with an impedance of 11 MΩ. Calculate the voltage indicated by each of these devices.

4-22 Plot the gain of the filter circuit of Example 4-7.

4-23 Five 1-in^2 plates are arranged as shown. The plate spacings are 0.01 in. The arrangement is to be used for a displacement transducer by observing the change in capacitance with the distance x.

x **Figure Prob. 4-23**

Calculate the sensitivity of the device in picofarads per inch. Assume that the plates are separated by air.

4-24 Calculate the voltage-displacement sensitivity for the LVDT whose characteristics are shown in Fig. 4-58.

4-25 The piezoelectric crystal of Example 4-12 is used for a pressure measurement at a nominal 100 psi. The uncertainty in the voltage measurement is ± 0.5 V, and the uncertainty in the crystal thickness is ± 0.0003 in. Estimate the uncertainty in the pressure measurement.

4-26 A rotating search coil like that shown in Fig. 4-69 has a nominal area of 1 cm^2 with 50 turns of small-diameter wire. The rotational speed is nominally 180 rpm. Calculate the voltage output when the coil is placed in a magnetic field of 1 Wb/m^2.

4-27 An amplifier has a stated power bandwidth of $+0$, -1 dB over a frequency range of 10 Hz to 30 kHz. If the 0-dB level is 35 W, what is the power that would occur at the lower limit of this specification?

4-28 An amplifier is stated to produce a 35-W output into an 8-Ω load when driven with an input voltage of 2.2 mV. The hum and noise is stated as -65 dB referenced to 10 mV at the input. What is the voltage output of the noise when operating into the 8-Ω load?

4-29 Suppose the amplifier of Prob. 4-28 requires an input of 180 mV to produce the 35-W output and has a hum and noise level of -75 dB referenced to 0.25 V at the input. What is the voltage output of the noise under these conditions? Calculate the relative decibel level of this noise and the noise calculated in Prob. 4-28.

4-30 An emf of 100 V is connected across a load of 10,000 Ω. In order to measure this voltage, a meter having an internal impedance of 100 kΩ is connected across the load. Calculate the percent error that will result from the loading by the meter.

4-31 In Prob. 4-30 calculate the number of decibels below the true voltage indicated by the meter.

4-32 A certain transducer has an internal resistance of 10 kΩ. The output of the device is to be measured with a voltage-divider circuit using a voltage source of 50 V and a meter with an impedance of 100 kΩ. Calculate the maximum loading error between the ranges of 10 and 90 percent of the measured variables.

4-33 A simple ballast circuit is used to measure the output of a pressure pickup. The circuit is designed so that the internal resistance is six times the total transducer resistance. A source of 100 V is used to energize the circuit. Calculate the voltage output at 25, 50, 60, and 80 percent of full load on the transducer. What error would result if the voltage output were assumed to vary linearly with the load?

4-34 A simple Wheatstone bridge has arms of $R_1 = 121$ Ω, $R_2 = 119$ Ω, and $R_3 = 121$ Ω. What is the value of R_4 for balance? If $R_4 = 122$ Ω and the bridge is driven with a source at 100 V, what is the open circuit output voltage?

4-35 A lead-sulfide detector is used to measure a radiation signal at 3.5 μm. What reduction in incident power, measured in decibels, is afforded when the detector is cooled to $-196°$C, assuming a constant S/N ratio?

4-36 An indium antimonide detector is used to sense radiant energy at 5 μm. What incident flux on a 4-mm^2 detector at room temperature is necessary to produce a S/N ratio of 45 dB?

4-37 A current-sensitivity input circuit is used to measure the output of a transducer having a total resistance of 500 Ω. A 12-V source is used, and the internal meter impedance is 7000 Ω. What is the current indication for 25, 50, 75, and 100 percent of full scale? What error would result at each of these readings if the current output were assumed to vary linearly with the transducer resistance?

4-38 A simple voltage-divider circuit is used to measure the output of a transducer whose total resistance is 1000 Ω $\pm$ 1 percent. The internal resistance of the voltmeter is 10,000 Ω $\pm$ 5 percent, and the meter is calibrated to read within $\frac{1}{2}$ percent of the true voltage at a full scale of 30 V. The source voltage is 24 V $\pm$ 0.05 V. Calculate the percent full-scale reading of the transducer for voltage readings of 6, 12, 18, and 24 V. Estimate the uncertainty for each of these readings.

4-39 Suppose the transducer of Prob. 4-38 is connected to an end resistor of 500 Ω, as shown in Prob. 4-17. Calculate the percent full-scale reading of the transducer for voltage readings of 12, 15, 18, and 24 V. Estimate the uncertainty for each of these readings, assuming the end resistor is accurate within 1 percent.

4-40 An amplifier has a power bandwidth of $+0.5$, -1.0 dB over a frequency range of 20 to 100 kHz. The nominal power output at 0 dB is 0.1 W. Calculate the power levels at the upper and lower limits of the specification for each end of the frequency range.

4-41 A voltage divider circuit measures the output of a transducer having a resistance of 800 $\Omega \pm 2$ percent. The internal resistance of the voltage measuring device is 15,000 $\Omega \pm 10$ percent. Calibration is performed so it reads within 0.5 percent at a full-scale reading. The source voltage is 30 V $\pm$ 0.1 V. Calculate the percent full-scale reading of the transducer and uncertainty for readings of 5, 10, 15 V.

4-42 The noise level of a certain amplifier is stated to be 80 dB below the power output of 100 W, referenced to an input level of 1.0 mV. Calculate the noise level input.

REFERENCES

1. Cady, W. G.: " Piezoelectricity," McGraw-Hill Book Company, New York, 1946.
2. Frank, E.: " Electrical Measurement Analysis," McGraw-Hill Book Company, New York, 1959.
3. Lion, K. S.: " Instrumentation in Scientific Research," McGraw-Hill Book Company, New York, 1959.
4. Lion, K. S.: Mechanic-electric Transducer, *Rev. Sci. Instr.*, col. 27, no. 2, pp. 222–225, 1956.
5. Sweeney, R. J.: " Measurement Techniques in Mechanical Engineering," John Wiley & Sons, Inc., New York, 1953.
6. ———, " Frequency and Time Standards," Application Note 52, Hewlett-Packard Co., Palo Alto, Calif.
7. ———, " Notes on Linear Variable Differential Transformers," Bulletin AA-la, Schaevitz Engineering Co., Camden, N.J.
8. ———, " The Radio Amateur's Handbook," 36th ed., Amateur Radio Relay League, West Hartford, Conn., 1959.
9. ———, " Reference Data for Radio Engineers," International Telephone and Telegraph Corp., New York, 1956.
10. International Conference on Photoconductivity, *J. Phys. Chem. Solids*, vol. 22, pp. 1–409, December 1961.
11. Beyen, W., P. Bratt, W. Engeler, L. Johnson, H. Levinstein, and A. MacRaw: Infrared Detectors Today and Tomorrow, *Proc. IRIS*, vol. 4, no. 1, March 1959.
12. Bartholomew, D.: " Electrical Measurements and Instrumentation," Allyn and Bacon, Inc., Boston, Mass., 1963.
13. Doebelin, E. O.: " Measurement Systems: Analysis and Design," 2d ed., McGraw-Hill Book Company, New York, 1983.
14. Wait, J. V., L. P. Huelsman, and G. A. Korn: " Introduction to Operational Amplifier Theory and Applications," McGraw-Hill Book Company, New York, 1975.
15. Rutkowski, G. B.: " Handbook of Integrated-Circuit Operational Amplifiers," Prentice-Hall, Inc., Englewood Cliffs, N.J., 1975.
16. Williams, A. B., " Electronic Filter Design Handbook," McGraw-Hill Book Company, New York, 1981.
17. Huelsman, L. P., " Active Filters: Lumped, Distributed, Integrated, Digital, and Parametric," McGraw-Hill Book Company, New York, 1970.
18. Antoniou, A., " Digital Filters, Analysis and Design," McGraw-Hill Book Company, New York, 1980.
19. Huelsman, L. P., and P. E. Allen, " Introduction to the Theory and Design of Active Filters," McGraw-Hill Book Company, New York, 1980.

FIVE

DISPLACEMENT AND AREA MEASUREMENTS

5-1 INTRODUCTION

Many of the transducers discussed in Chap. 4 represent excellent devices for measurement of displacement. In this chapter we wish to examine the general subject of dimensional and displacement measurements and indicate some of the techniques and instruments that may be utilized for such purposes, making use, where possible, of the information in the preceding sections.

Dimensional measurements are categorized as determinations of the size of an object, while a displacement measurement implies the measurement of the movement of a point from one position to another. An area measurement on a standard geometric figure is a combination of appropriate dimensional measurements through a correct analytical relationship. The determination of areas of irregular geometric shapes usually involves a mechanical, graphical, or numerical integration.

Displacement measurements may be made under both steady and transient conditions. Transient measurements fall under the general class of subjects discussed in Chap. 11. The present chapter is concerned only with static measurements.

5-2 DIMENSIONAL MEASUREMENTS

The standard units of length were discussed in Chap. 2. All dimensional measurements are eventually related to these standards. Simple dimensional measurements with an accuracy of ± 0.01 in (0.25 mm) may be made with graduated

Fixed jaw

Movable jaw

Vernier

Slider

Fine-adjustment screw

Figure 5-1 Vernier caliper.

metal machinists' scales or wood scales which have accurate engraved markings. For large dimensional measurements metal tapes are used to advantage. The primary errors in such measurement devices, other than readability errors, are usually the result of thermal expansion or contraction of the scale. On long metal tapes used for surveying purposes this can represent a substantial error, especially when used under extreme temperature conditions. It may be noted, however, that thermal expansion effects represent *fixed* errors and may easily be corrected when the measurement temperature is known.

Vernier calipers represent a convenient modification of the metal scale to improve the readability of the device. The caliper construction is shown in Fig. 5-1, and an expanded view of the vernier scale is shown in Fig. 5-2. The caliper is placed on the object to be measured and the fine adjustment rotated until the jaws fit tightly against the workpiece. The increments along the primary scale are 0.025 in. The vernier scale shown in used to read to 0.001 in (0.025 mm), so that it has 25 equal increments (0.001 is $\frac{1}{25}$ of 0.025) and a total length of $\frac{24}{25}$ times the length of the primary scale graduations. Consequently, the vernier scale does not line up exactly with the primary scale, and the ratio of the last coincident number on the vernier to the total vernier length will equal the fraction of a whole primary scale division indicated by the index position. In the example shown in Fig. 5-2 the reading would be $2.350 + (\frac{14}{25})(0.025) = 2.364$ in.

The micrometer calipers shown in Fig. 5-3 represent a more precise measurement device than the vernier calipers. Instead of the vernier scale arrangement, a calibrated screw thread and circumferential scale divisions are used to indicate the fractional part of the primary scale divisions. In order to obtain the maximum effectiveness of the micrometer, care must be exerted to ensure that a consistent pressure is maintained on the workpiece. The spring-loaded ratchet device on the handle enables the operator to maintain such a condition. When properly used, the micrometer can be employed for the measurement of dimensions within 0.0001 in (0.0025 mm).

Index

Vernier scale

Figure 5-2 Expanded view of vernier scale.

Main scale
0.025″ divisions

Frame

Thimble scale
25 divisions

Ratchet

Figure 5-3 Micrometer calipers.

Dial indicators are devices that perform a mechanical amplification of the displacement of a pointer or follower in order to measure displacements within about 0.001 in. The construction of such indicators provides a gear rack, which is connected to a displacement-sensing shaft. This rack engages a pinion which in turn is used to provide a gear-train amplification of the movement. The output reading is made on a circular dial.

Example 5-1 A 30-m (at 15°C) steel tape is used for surveying work in the summer such that the tape temperature in the sun is 45°C. A measurement indicates 24.567 ± 0.001 m. The linear thermal coefficient of expansion is $11.65 \times 10^{-6}/°C$ at 15°C. Calculate the true distance measurement.

SOLUTION The indicated tape length would be the true value if the measurement were taken at 15°C. At the elevated temperature the tape has expanded and consequently reads too small a distance. The actual length of the 30-m tape at 45°C is

$$L(1 + \alpha \, \Delta T) = [1 + (11.65 \times 10^{-6})(45 - 15)](30) = 30.010485 \text{ m}$$

Such a true length would be indicated as 30 m. The true reading for the above situation is thus

$$(24.567)[1 + (11.65 \times 10^{-6})(45 - 15)] = 24.576 \text{ m}$$

5-3 GAGE BLOCKS

Gage blocks represent industrial dimension standards. They are small steel blocks about $\frac{3}{8} \times 1\frac{3}{8}$ in with highly polished parallel surfaces. The thickness of the blocks is specified in accordance with the following tolerances:

Grade of block	Tolerance, μin†
AA	2
A	4
B	8

† Tolerances are for blocks less than 1 in thick; for greater thickness the same tolerances are per inch.

Gage blocks are available in a range of thicknesses that make it possible to stack them in a manner such that with a set of 81 blocks any dimension between 0.100 and 8.000 in can be obtained in increments of 0.0001 in. The blocks are stacked through a process of *wringing*. With surfaces thoroughly clean, the metal surfaces are brought together in a sliding fashion while a steady pressure is exerted. The surfaces are sufficiently flat so that when the wringing process is correctly executed, they will adhere as a result of molecular attraction. The adhesive force may be as great as 30 times atmospheric pressure.

Because of their high accuracy, gage blocks are frequently used for calibration of other dimensional measurement devices. For very precise measurements they may be used for direct dimensional comparison tests with a machined item. A discussion of the methods of producing gage-block standards is given in Ref. [3]. The literature of manufacturers of gage blocks furnishes an excellent source of information on the measurement techniques which are employed in practice.

5-4 OPTICAL METHODS

An optical method for measuring dimensions very accurately is based on the principle of light interference. The instrument based on this principle is called an *interferometer* and is used for the calibration of gage blocks and other dimensional standards. Other optical instruments in wide use are various types of microscopes and telescopes, including the conventional surveyor's transit, which is employed for measurement of large distances.

Consider the two sets of light beams shown in Fig. 5-4. In Fig. 5-4a the two beams are in phase, so that the brightness at point P is augmented when they intersect. In Fig. 5-4b the beams are out of phase by half a wavelength, so that a cancellation is observed, and the light waves are said to *interfere* with each other. This is the essence of the interference principle. The effect of the cancellation is brought about by allowing two light waves from a single source to travel along

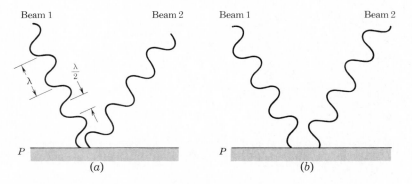

Figure 5-4 Interference principle. (a) Beams in phase; (b) beams out of phase.

Screen, S

Glass
optically flat

Reflecting
surface

Figure 5-5 Application of interference principle.

paths of different lengths. When the difference in the distance is an integral multiple of wavelengths, there will be a reinforcement of the waves, while there will be a cancellation when the difference in the distances is an odd multiple of half-wavelengths.

Now let us apply the interference principle to dimensional measurements. Consider the two parallel plates shown in Fig. 5-5. One plate is a transparent, strain-free glass accurately polished flat within a few microinches. The other plate has a reflecting metal surface. The glass plate is called an *optical flat*. Parallel light beams A and B are projected on the plates from a suitable collimating source. The separation distance between the plates d is assumed to be quite small. The reflected beam A intersects the incoming beam B at the point P. Since the reflected beam has traveled further than beam B by a distance of $2d$, it will create an interference at point P if this incremental distance is an odd multiple of $\lambda/2$. If the distance $2d$ is an even multiple of $\lambda/2$, the reflected beam will augment beam B. Thus, for $2d = \lambda/2$, $3\lambda/2$, etc., the screen S will detect no reflected light. Now consider the same two plates, but let them be tilted slightly so that the distance

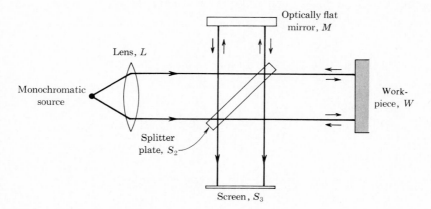

Figure 5-6 Schematic of interferometer.

Table 5-1 Monochromatic Light Sources

Source	Wavelength, μm	Half-wavelength fringe interval, μm
Helium	0.589	0.295
Krypton 86	0.606	0.303
Mercury 198	0.546	0.273
Sodium	0.598	0.299

between the plates is a variable. Now, if one views the reflected light beams, alternate light and dark regions will appear on the screen, indicating the variation in the plate spacing. The dark lines or regions are called *fringes*, and the change in the separation distance between the positions of two fringes corresponds to

$$\Delta(2d) = \frac{\lambda}{2} \tag{5-1}$$

The interference principle offers a convenient means for measuring small surface defects and for calibrating gage blocks. The use of a tilted optical flat as in Fig. 5-5 is an awkward method of utilizing the principle, however. For practical purposes the interferometer, as indicated schematically in Fig. 5-6, is employed. Monochromatic light from the source is collimated by the lens L onto the splitter plate S_2, which is a half-silvered mirror that reflects half of the light toward the optically flat mirror M and allows transmission of the other half toward the workpiece W. Both beams are reflected back and recombined at the splitter plate S_2 and then transmitted to the screen. Fringes may appear on the screen resulting from differences in the optical path lengths of the two beams. If the instrument is properly constructed, these differences will arise from dimensional variations of the workpiece. The interferometer is primarily used for calibration of gage blocks and other applications where extremely precise absolute dimensional measurements are required. For detailed information on experimental techniques used in interferometry, the reader should consult Refs. [3] and [5]. The use of the interferometer for fluid-flow measurements will be discussed in Chap. 7.

As shown in Eq. (5-1) the wavelength of the monochromatic light source will influence the fringe spacing. Table 5-1 lists the wavelengths of some common light sources and the corresponding half-wavelength fringe interval.

Example 5-2 A mercury light source employs a green filter such that the wavelength is 5460 Å. This light is collimated and directed onto two tilted surfaces like those shown in Fig. 5-5. At one end the surfaces are in precise contact. Between the point of contact and a distance of 3.000 in, five interference fringes are observed. Calculate the separation distance between the two surfaces and the tilt angle at this position.

SOLUTION The five fringe lines correspond to $\lambda/2, 3\lambda/2, \ldots, 9\lambda/2$; that is, for the fifth fringe line

$$2d = \frac{9\lambda}{2}$$

We have $\lambda = 5460 \times 10^{-8}$ cm $= 2.15 \times 10^{-5}$ in, so that

$$d = \tfrac{9}{4}(2.15 \times 10^{-5}) = 48.4 \ \mu\text{in}$$

The tilt angle is

$$\phi = \tan^{-1}\frac{48.4 \times 10^{-6}}{3.000} = \frac{48.4 \times 10^{-6}}{3.000} = 16.1 \times 10^{-6} \ \text{rad}$$

Mechanical displacement may also be measured with the aid of the electric transducers discussed in Chap. 4. The LVDT, for example, can be used to sense displacements as small as 1 μin. Resistance transducers are primarily of value for measurement of fairly large displacements because of their poor resolution. Capacitance and piezoelectric transducers, on the other hand, provide high resolution and are suitable for dynamic measurements.

5-5 PNEUMATIC DISPLACEMENT GAGE

Consider the system shown in Fig. 5-7. Air is supplied at a constant pressure p_1. The flow through the orifice and through the outlet of diameter d_2 is governed by the separation distance x between the outlet and the workpiece. The change in flow with x will be indicated by a change in the pressure downstream from the orifice p_2. Thus, a measurement of this pressure may be taken as an indication of the separation distance x. For purposes of analysis we assume incompressible flow. (See Sec. 7-3 for a discussion of the validity of this assumption.) The volumetric flow through an orifice may be represented by

$$Q = CA\sqrt{\Delta p} \tag{5-2}$$

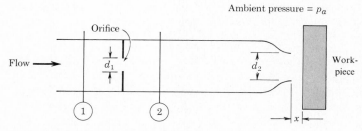

Figure 5-7 Pneumatic displacement device.

where C = discharge coefficient
$\quad\quad A$ = flow area of the orifice
$\quad\quad \Delta p$ = pressure differential across the orifice

There are *two* orifices in the situation depicted in Fig. 5-7, the obvious one and the orifice formed by the flow restriction between the outlet and the work-piece. We shall designate the area of the first orifice A_1 and that of the second A_2. Then, Eq. (5-2) becomes

$$Q = C_1 A_1 \sqrt{p_1 - p_2} = C_2 A_2 \sqrt{p_2 - p_a} \tag{5-3}$$

where p_a is the ambient pressure and is assumed constant. Equation (5-3) may be rearranged to give

$$r = \frac{p_2 - p_a}{p_1 - p_a} = \frac{1}{1 + (A_2/A_1)^2} \tag{5-4}$$

where it is assumed that the discharge coefficients C_1 and C_2 are equal. We may now observe that

$$A_1 = \frac{\pi d_1^2}{4} \tag{5-5}$$

$$A_2 = \pi d_2 x \tag{5-6}$$

Thus we see the relation between the pressure ratio r and the workpiece displacement x. Graneek and Evans [1] have shown experimentally that the relation between r and the area ratio A_2/A_1 is very nearly linear for $0.4 < r < 0.9$ and that

$$r = 1.10 - 0.50 \frac{A_2}{A_1} \tag{5-7}$$

for this range. Introducing Eqs. (5-5) and (5-6), we have

$$r = \frac{p_2 - p_a}{p_1 - p_a} = 1.10 - 2.00 \left(\frac{d_2}{d_1^2}\right) x \qquad \text{for } 0.4 < r < 0.9 \tag{5-8}$$

The pneumatic displacement gage is mainly used for small displacement measurements.

Example 5-3 A pneumatic displacement gage like the one shown in Fig. 5-7 has $d_1 = 0.030$ in and $d_2 = 0.062$ in. The supply pressure is 10.0 psig, and the differential pressure $p_2 - p_a$ is measured with a water manometer which may be read with an uncertainty of 0.05 inH_2O. Calculate the displacement range for which Eq. (5-8) applies and the uncertainty in this measurement, assuming that the supply pressure remains constant.

SOLUTION We have

$$\frac{d_2}{d_1^2} = \frac{0.062}{(0.030)^2} = 68.8$$

When $r = 0.4$, we have, from Eq. (5-8),

$$x = \frac{1.10 - 0.4}{(2.00)(68.8)} = 0.0509 \text{ in } (0.129 \text{ cm})$$

When $r = 0.9$, $x = 0.0145$ in (0.0368 cm)

Utilizing Eq. (3-2) as applied to Eq. (5-8), we have

$$w_r = \left[\left(\frac{\partial r}{\partial x} \right)^2 w_x^2 \right]^{1/2} = \pm \left| \frac{\partial r}{\partial x} \right| w_x$$

Furthermore,

$$w_r = \frac{w_{\Delta p}}{p_1 - p_a} \qquad \frac{\partial r}{\partial x} = -(2.00)(68.8) = -137.6$$

The uncertainty in the measurement of $p_2 - p_a$ is

$$w_{\Delta p} = (0.05)(0.0361) = 1.805 \times 10^{-3} \text{ psig } (12.44 \text{ N/m}^2)$$

Thus, the uncertainty in x is given by

$$w_x = \frac{1.805 \times 10^{-3}}{(137.6)(10.00)} = 1.313 \times 10^{-6} \text{ in } = 1.313 \ \mu\text{in } (0.033 \ \mu\text{m})$$

From this example we see that the pneumatic gage can be quite sensitive, even with modest pressure-measurement facilities at hand.

In addition to its application as a steady-state–displacement-measurement device, the pneumatic gage may be employed as a dynamic sensor in conjunction with properly designed fluidic circuits. By periodically interrupting the discharge, the device may serve as a periodic signal generator for fluidic circuits. Different interruption techniques may be employed to generate square-, triangle-, or sine-wave signals [7]. The device may also be employed for fluidic reading of coded information on plates or cards that pass under the outlet jet [8 and 9]. The rapidity with which such readings may be made depends on the dynamic response of the gage and its associated connecting lines and pressure transducers. Studies [11] have shown that the device can produce good frequency response for signals up to 500 Hz.

5-6 AREA MEASUREMENTS

There are many applications that require a measurement of a plane area. Graphical determinations of the area of survey plots from maps, the integration of a function to determine the area under a curve, and analyses of experimental data plots all may rely on a measurement of a plane area. There are also many applications for the measurement of surface areas, but such measurements are considerably more difficult to perform.

5-7 THE PLANIMETER

The planimeter is a mechanical integrating device that is used for measurement of plane areas. Consider the schematic representation shown in Fig. 5-8. The point O is fixed, while the tracing point T is moved around the periphery of the figure whose area is to be determined. The wheel W is mounted on the arm BT so that it is free to rotate when the arm undergoes an angular displacement. The wheel has engraved graduations and a vernier scale so that its exact number of revolutions may be determined as the tracing point moves around the curve. The planimeter and area are placed on a flat, relatively smooth surface so that the wheel W will only slide when the arm BT undergoes an axial translational movement. Thus the wheel registers zero angular displacement when an axial translational movement of arm BT is experienced. Let the length of the tracing arm BT be L and the distance from point B to the wheel be a. The diameter of the wheel is D. The distance OB is taken as R. Now suppose the arm BT is rotated an angle $d\theta$ and the arm OB through an angle $d\phi$ as a result of movement of the tracing point. The area swept out by the arms BT and OB is

$$dA = \tfrac{1}{2}L^2 \, d\theta + LR \cos \beta \, d\phi + \tfrac{1}{2}R^2 \, d\phi \qquad (5\text{-}9)$$

where β is the angle between the two arms. Similarly, the distance traveled by the rim of the wheel owing to rotation is

$$ds = a \, d\theta + R \cos \beta \, d\phi \qquad (5\text{-}10)$$

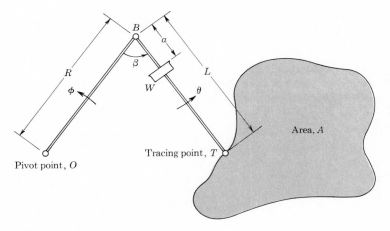

Figure 5-8 Schematic of a polar planimeter.

We may now integrate these expressions and obtain

$$A = \tfrac{1}{2}L^2 \int d\theta + LR \int \cos \beta \, d\phi + \tfrac{1}{2}R^2 \int d\phi \qquad (5\text{-}11)$$

$$s = a \int d\theta + \int R \cos \beta \, d\phi \qquad (5\text{-}12)$$

Thus,

$$\int R \cos \beta \, d\phi = s - a \, d\theta$$

and

$$A = (\tfrac{1}{2}L^2 - aL) \int d\theta + Ls + \tfrac{1}{2}R^2 \int d\phi \qquad (5\text{-}13)$$

If the pole is outside the area as shown in Fig. 5-8 we have both $\int d\theta = 0$ and $\int d\phi = 0$, and the area is obtained as

$$A = Ls \qquad \text{(pole outside area)}$$

When the pole is inside the area both $\int d\phi$ and $\int d\theta$ are equal to 2π because both arms make complete rotations. The area is then

$$A = Ls + (R^2 + L^2 - 2aL)\pi \qquad \text{(pole inside area)} \qquad (5\text{-}14)$$

The last term in Eq. (5-14) represents the area of the *zero circle*, which is the area the tracing point would sweep out when the pivot point is inside the area and the wheel reading is zero.

The instrument described above is called a *polar planimeter*. Typical commercial devices have a wheel circumference of 2.50 in, and models are available with both fixed- and adjustable-length tracing arms. The area of the zero circle is frequently given by the manufacturer but may also be determined as in Example 5-4. The construction of the polar planimeter is indicated in Fig. 5-9.

The polar planimeter is not generally suitable for the measurement of long, narrow areas because the pivot point must be fixed. The *roller planimeter* shown

Figure 5-9 Construction of a polar planimeter.

Figure 5-10 Roller planimeter.

in Fig. 5-10 is suitable for these measurements. The two rollers R are connected through a bevel-gear arrangement to the table A. The wheel W is in contact with the table and rotates in accordance with the number of revolutions of the table. The tracing point T is connected to the mounting for the wheel so that its movement causes the wheel to move inward and outward on the rotating table. The rotation of the wheel depends upon its distance from the center of the table, so that it is, in turn, dependent on the displacement of the tracing point in a direction parallel to the axis of the two rollers. Thus, the total indication of the wheel is a function of the distance moved by the rollers and the perpendicular displacement of the tracing point and is consequently a function of the area traversed. The device may be used for a large class of area measurements since it can move in two directions.

Example 5-4 A planimeter with a fixed tracing arm 4.00 in long has a wheel circumference of 2.50 in. The instruction sheet for the instrument is lost, so that the area of the zero circle is not known. To determine the area of the zero circle, a certain area is measured with the pole both inside and outside the area. With the pole outside the area the wheel records 2.55 revolutions. With the pole inside the area the wheel records -13.01 revolutions. Calculate the area of the zero circle and the reading of the planimeter when it is used to trace out a circle with a 9-in radius.

SOLUTION From Eq. (5-14) we have

$$A = Ls + A_0$$

where A_0 is the area of the zero circle. The true area is the value of Ls when the pole is outside the area. Thus,

$$A = (4.00)(2.55)(2.50) = 25.50 \text{ in}^2 \ (164.5 \text{ cm}^2)$$

and

$$A_0 = 25.50 - (4.00)(-13.01)(2.50) = 155.6 \text{ in}^2 \ (1003.9 \text{ cm}^2)$$

When the planimeter is used to measure a 9-in-radius circle, we have

$$s = \frac{A - A_0}{L} = \frac{\pi(9)^2 - 155.6}{4.00} = 24.717$$

The number of wheel revolutions is thus

$$n = \frac{s}{2.50} = \frac{24.717}{2.50} = 9.887$$

5-8 GRAPHICAL AND NUMERICAL METHODS FOR AREA MEASUREMENT

A very simple method of plane-area measurement is to place the figure on coordinate paper and count the number of squares enclosed by the figure. An appropriate scale factor is then applied to determine the area. Numerical integration is commonly applied to determine the area under an irregular curve. Perhaps the two most common methods are the trapezoidal rule and Simpson's rule. Consider the area shown in Fig. 5-11. The area under the curve is

$$A = \int y \, dx \tag{5-15}$$

If the figure is divided into equal increments Δx along the x axis, the trapezoidal rule gives for the area

$$A = \left(\frac{y_0 + y_n}{2} + \sum_{i=1}^{n-1} y_i \right) \Delta x \tag{5-16}$$

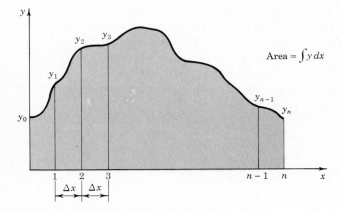

Figure 5-11 Plane-area determination.

When the area is divided into an *even* number of increments, Simpson's rule gives

$$A = \frac{\Delta x}{3} \left\{ y_0 + y_n + \sum_{i=1}^{n-1} y_i [3 + (-1)^{i+1}] \right\} \tag{5-17}$$

The trapezoidal rule is obtained by joining the ordinates of the curve with straight lines, while the result given by Eq. (5-17) is obtained by joining three points at a time with a parabola. Both of the above equations are special cases of a general class of equations called the *Newton-Cotes integration formulas*. A derivation of these formulas is based on an approximation of the actual curve with a polynomial which agrees with it at $n + 1$ equally spaced points. The general form of the integration formulas is derived in Ref. [2] and may be written in the following way:

$$\int_{x_1}^{x_2} f(x) \, dx = \Delta x \sum_{k=0}^{n} C_k f(x_k) \tag{5-18}$$

where

$$\Delta x = \frac{x_2 - x_1}{n} \tag{5-19}$$

$$x_k = x_1 + k \, \Delta x \tag{5-20}$$

and the coefficients C_k are given by

$$C_k = \int_0^n \frac{s(s-1) \cdots (s-k+1)(s-k-1) \cdots (s-n)}{k(k-1) \cdots (k-k+1)(k-k-1) \cdots (k-n)} \, ds \tag{5-21}$$

The variable s is defined according to

$$s = \frac{x - x_1}{n} \tag{5-22}$$

Note that the Newton-Cotes formulas use n increments in the independent variable and that this requires a matching of a polynomial at $n + 1$ points in the region that is to be integrated. A discussion of the errors involved in the above formulas is given by Hildebrand [2]. Because of the ready availability of computing power, the simple trapezoidal or Simpson's rule with a large number of small increments may be preferred over the more complex Newton-Cotes formulas.

5-9 SURFACE AREAS

Consider the general three-dimensional surface shown in Fig. 5-12. The surface is described by the function

$$z = f(x, y)$$

and the surface area is given in Ref. [4] as

$$A = \iint \left[\left(\frac{\partial z}{\partial x} \right)^2 + \left(\frac{\partial z}{\partial y} \right)^2 + 1 \right]^{1/2} dx \, dy \tag{5-23}$$

Figure 5-12 Surface-area determination.

If the function z is known and well behaved, the integral in Eq. (5-23) may be evaluated directly. Let us consider the case where the function is not given but specific values of z are known for incremental changes in x and y. The increments in x and y are denoted by Δx and Δy, while the value of z is denoted by z_n^p, where the subscript n refers to the x increments and the superscript p refers to the y increment. We thus have the approximations

$$\frac{\partial z}{\partial x} \approx \frac{z_{n+1}^p - z_n^p}{\Delta x}$$

$$\frac{\partial z}{\partial y} \approx \frac{z_n^{p+1} - z_n^p}{\Delta y}$$

The integral Eq. (5-23) is now replaced by the double sum

$$A = \sum_p \sum_n \left[\left(\frac{z_{n+1}^p - z_n^p}{\Delta x} \right)^2 + \left(\frac{z_n^{p+1} - z_n^p}{\Delta y} \right)^2 + 1 \right]^{1/2} \Delta x \, \Delta y \qquad (5-24)$$

The surface area may be determined by performing this numerical summation.

Standard computer routines and calculator programs are available for performing the numerical computations indicated above.

PROBLEMS

5-1 A 12-in steel scale is graduated in increments of 0.01 in and is accurate when used at a temperature of 60°F. Calculate the error in an 11-in measurement when the ambient temperature is 100°F. Should the person using the scale be concerned about this error? Why?

5-2 Calculate the temperature error in a 76-ft measurement with a steel surveyor's tape at $-10°F$ when the tape is accurate at 60°F.

5-3 Show that the spacing distance d in Fig. 5-5 can be represented by

$$d = \frac{2n - 1}{4}\lambda$$

where n is the number of fringe lines.

5-4 A pneumatic displacement gage is designed according to the arrangement in Fig. 5-7. An air supply pressure of 20 psig is available, and displacements are to be measured over a range of 0.050 in. The orifice diameter is 0.025 in. Calculate the maximum displacement which may be measured in the linear range of operation and the outlet tube diameter d_2.

5-5 A planimeter is used to measure the area of an irregular plane figure. The wheel circumference is 2.50 in, and the tracing arm is adjustable. With a tracing-arm length of 2.00 in and the pole outside the area, the wheel indicates 9.52 r. With a tracing-arm length of 4.00 in and the pole outside the area, the wheel indicates 4.70 r. Do the two readings correspond to the same area? If not, how do you explain the differences?

5-6 Plot the equation

$$y = 3 + 4x - 6x^2 + 2.8x^3 - 0.13x^4$$

for the range $0 < x < 5$. Determine the area under the curve by counting squares and also by numerical integration using the trapezoidal relation and Simpson's rule. Calculate the error in each of these three cases by comparing the results with that obtained analytically.

5-7 Use the trapezoidal rule and Simpson's method to perform the integration

$$A = \int_0^{\pi} \sin x \, dx$$

Use 4, 8, and 12 increments of x and calculate the error for each case.

5-8 Consider the sphere given by

$$x^2 + y^2 + z^2 = 25$$

Using the summation of Eq. (5-24), calculate the surface area bounded by $x = \pm 1$ and $y = \pm 1$. Use $\Delta x = \Delta y = 0.5$. Determine the error in the calculation by comparing it with the true value as calculated from Eq. (5-23).

5-9 By suitable numerical integration determine the surface area and volume of a right circular cone having a height of 12.5 cm and a base diameter of 15 cm. Compare the result with that obtained by an exact calculation.

5-10 A steel tape is used to measure a distance of 20 m at 10°C. Assuming that the tape will measure the true distance when the temperature is 20°C, what would be the indicated distance at 10°C? What would be the indicated distance at 40°C?

5-11 A mercury light source and green filter (5460 Å) are used with an interferometer to determine the distance between an optically flat glass plate and a metal plate, as shown in Fig. 5-5. For this arrangement nine interference fringes are observed. Calculate the separation distance between the two surfaces.

5-12 A pneumatic displacement gage is to be used to measure displacements between 0.1 and 0.2 mm. The output of the device is a hypodermic needle having a diameter of 0.4 mm. The measurement is to be performed with an uncertainty of 25 nm. Specify values of the orifice diameter, upstream pressure, and allowable uncertainty in the pressure measurement to accomplish this objective.

5-13 The planimeter of Example 5-4 is used to measure the area of an irregular-shaped figure. With the pole outside the area, the wheel records 5.23 r. What is the area of the figure? What would be the reading if the pole were placed inside the figure?

5-14 The pneumatic displacement gage of Example 5-3 is operated under the same conditions as given, but it is discovered that the supply pressure has a random fluctuation of ± 0.07 psia during the measurements. Calculate the uncertainty in the dimensional measurement under these new conditions, assuming that the uncertainty in $p_2 - p_a$ remains at 0.05 in H_2O.

5-15 The function y is given as

$$y = 1 + 3 + 4x^2$$

Perform the integration $\int y \, dx$ between $x = 0$ and $x = 4$ using the trapezoidal rule and Simpson's rule. Use four increments in x and calculate the error for each case. How many more increments would have to be taken for the trapezoidal rule to equal the accuracy of Simpson's rule?

5-16 Draw some kind of irregular figure on graph paper having small subdivisions. Determine the area by (a) counting squares, (b) the trapezoidal rule, (c) Simpson's rule, and (d) a planimeter. Repeat by drawing a 10-cm-diameter circle using a sharp pencil and accurately measured radius from a metal scale. Discuss the "uncertainties" in these two experiments.

5-17 Plot the function $y = 3xe^{0.4x}$ on graph paper as accurately as you can. Determine the area under the curve between $x = 1$ and $x = 3$ using several methods. Estimate the uncertainties.

5-18 Assemble a group of at least five people. Have each person measure two objects using a vernier caliper and a micrometer, with each person taking at least five readings. Analyze the data on the basis of the information in Chap. 3. What conclusions do you draw?

REFERENCES

1. Graneek, M., and J. C. Evans: A Pneumatic Calibrator of High Sensitivity, *Engineer*, p. 62, July 13, 1951.
2. Hildebrand, F. B.: "Introduction to Numerical Analysis," McGraw-Hill Book Company, New York, 1956.
3. Peters, C. G., and W. B. Emerson: Interference Methods for Producing and Calibrating End Standards, *J. Res. Nat. Bur. Std.*, vol. 44, p. 427, 1950.
4. Sokolnikoff, I. S., and R. M. Redheffer: "Mathematics of Physics and Modern Engineering," McGraw-Hill Book Company, New York, 1958.
5. ———, Metrology of Gauge Blocks, *Natl. Bur. Std. (U.S.)*, Circ. 581, April 1957.
6. Fish, V. T., and G. M. Lance: An Accelerometer for Fluidic Control Systems, ASME Paper No. 67-WA/FE-29.
7. Goldstein, S.: A New Type of Pneumatic Triangle-Wave Oscillator, ASME Paper No. 66-WA/AUT-4.
8. Rosenbaum, H. M., and J. S. Gant: A Pneumatic Tape Reader, 2d Cranfield Fluids Conference, E2-13, January 1967.
9. Gant, G. C.: A Fluidic Digital Displacement Indicator, 2d Cranfield Fluidics Conference, E1-1, January 1967.
10. Eckerlin, H. M., and D. A. Small: A Method of Air Gauge Circuit Analysis, *Advan. Fluidics*, ASME, 1967.
11. Moses, H. L., D. A. Small, and G. A. Cotta: Response of a Fluidic Air Gauge, *J. Basic Eng.*, pp. 475–478, September 1969.

SIX

PRESSURE MEASUREMENT

6-1 INTRODUCTION

Pressure is represented as a force per unit area. As such, it has the same units as stress and may, in a general sense, be considered as a type of stress. For our purposes we shall designate the force per unit area exerted by a fluid on a containing wall as the pressure. The forces that arise as a result of strains in solids are designated as stresses and discussed in Chap. 10. Thus, our discussion of pressure measurement is one which is restricted to fluid systems. *Absolute pressure* refers to the absolute value of the force per unit area exerted on the containing wall by a fluid. *Gage pressure* represents the difference between the absolute pressure and the local atmospheric pressure. *Vacuum* represents the amount by which the atmospheric pressure exceeds the absolute pressure. From these definitions we see that the absolute pressure may not be negative and the vacuum may not be greater than the local atmospheric pressure. The three terms are illustrated graphically in Fig. 6-1. It is worthwhile to mention that local fluid pressure may be dependent upon many variables; elevation, flow velocity, fluid density, and temperature are parameters that are of frequent importance.

In the English system of units pressure is usually expressed in pounds per square inch absolute (psia). Gage pressure carrying the same unit is designated with the symbol psig. The standard SI unit for pressure is the newton per square meter (N/m^2) or pascal (Pa). Pressure is frequently expressed in terms of the height of a column of fluid (viz., mercury) which it will support at a temperature

p(absolute)

Positive gage
pressure

Atmospheric pressure

Negative gage
pressure or vacuum

0 — Zero absolute pressure

Figure 6-1 Relationship between pressure terms.

of 20°C. At standard atmospheric pressure this height is 760 mm of mercury having a density of 13.5951 g/cm^3. Some common units of pressure are:

$$1 \text{ atmosphere (atm)} = 14.696 \text{ pounds per square inch absolute}$$

$$= 1.01325 \times 10^5 \text{ newtons per square meter (Pa)}$$

$$= 2116 \text{ pounds-force per square foot (lbf/ft}^2)$$

$$1 \text{ N/m}^2 \equiv 1 \text{ pascal (Pa)}$$

$$1 \text{ atmosphere (atm)} = 760 \text{ millimeters of mercury (mmHg)}$$

$$1 \text{ bar} = 10^5 \text{ newtons per square meter (100 kPa)}$$

$$1 \text{ microbar} = 1 \text{ dyne per square centimeter}$$

$$= 2.089 \text{ pounds-force per square foot}$$

$$= 0.1 \text{ newtons per square meter (0.1 Pa)}$$

$$1 \text{ millimeter of mercury (mmHg)} = 1333.22 \text{ microbar}$$

$$= 133.322 \text{ newtons per square meter (133.3 Pa)}$$

$$1 \text{ micrometer} = 10^{-6} \text{ meters of mercury (}\mu\text{m)}$$

$$= 10^{-3} \text{ millimeters of mercury (mmHg)}$$

$$= 0.133322 \text{ newtons per square meter (0.133 Pa)}$$

$$1 \text{ torr} \equiv 1 \text{ millimeter of mercury (mmHg)}$$

$$1 \text{ inch of mercury} = 70.73 \text{ pounds-force per square foot}$$

$$1 \text{ pound per square inch absolute} = 6894.76 \text{ newtons per square meter (6.894 kPa)}$$

Fluid pressure results from a momentum exchange between the molecules of the fluid and a containing wall. The total momentum exchange is dependent upon the total number of molecules striking the wall per unit time and the average velocity of the molecules. For an ideal gas, it may be shown that the pressure is given by

$$p = \tfrac{1}{3}nmv_{rms}^2 \tag{6-1}$$

where n = molecular density, molecules/unit volume
   m = molecular mass
   v_{rms} = root mean square molecular velocity

It may also be shown that

$$v_{rms} = \sqrt{\frac{3kT}{m}} \tag{6-2}$$

where T = absolute temperature of the gas, K
   $k = 1.3803 \times 10^{-23}$ J/molecule $\cdot$ K (Boltzmann's constant)

Equation (6-1) is a kinetic-theory interpretation of the ideal-gas law. An expression for the pressure in a liquid would not be so simple.

The *mean free path* is defined as the average distance a molecule travels between collisions. For an ideal gas whose molecules act approximately like billiard balls,

$$\lambda = \frac{\sqrt{2}}{8\pi r^2 n} \tag{6-3}$$

where r is the effective radius of the molecule and λ is the mean free path. It is clear that the mean free path increases with a decrease in the gas density. At standard atmospheric pressure and temperature the mean free path is quite small, of the order of 10^{-5} cm. At a pressure of 1 μm, however, the mean free path would be of the order of 1 cm. At very low pressures the mean free path may be significantly greater than a characteristic dimension of the containing vessel. For air the relation for mean free path reduces to

$$\lambda = 8.64 \times 10^{-7} \frac{T}{p} \text{ ft} \qquad T \text{ in } °R \text{ and } p \text{ in lbf/ft}^2 \tag{6-3a}$$

$$\lambda = 2.27 \times 10^{-5} \frac{T}{p} \text{ m} \qquad T \text{ in K and } p \text{ in Pa} \tag{6-3b}$$

A variety of devices are available for pressure measurement, as we shall see in the following sections. Static, i.e., steady-state, pressure is not difficult to measure with good accuracy. Dynamic measurements, however, are much more perplexing because they are influenced strongly by the characteristics of the fluid being studied as well as the construction of the measurement device. In many instances a pressure instrument that gives very accurate results for a static measurement may be entirely unsatisfactory for dynamic measurements. We shall discuss some of the factors that are important for good dynamic response in conjunction with the exposition associated with the different types of pressure-measurement devices.

6-2 DYNAMIC RESPONSE CONSIDERATIONS

The transient response of pressure-measuring instruments is dependent on two factors: (1) the response of the transducer element that senses the pressure, and (2) the response of the pressure-transmitting fluid and the connecting tubing, etc. This latter factor is frequently the one which determines the overall frequency response of a pressure-measurement system, and, eventually, direct calibration must be relied upon for determining this response. An estimate of the behavior may be obtained with the following analysis. Consider the system shown in Fig. 6-2. The fluctuating pressure has a frequency of ω and an amplitude of p_0 and is impressed on the tube of length L and radius r. At the end of this tube is a chamber of volume V where the connection to the pressure-sensitive transducer is made. The mass of fluid vibrates under the influence of fluid friction in the tube, which tends to dampen the motion. If the conventional formula for laminar friction resistance in tube flow is used to represent this friction, the resulting expression for the pressure-amplitude ratio is

$$\left|\frac{p}{p_0}\right| = \frac{1}{\{[1 - (\omega/\omega_n)^2]^2 + 4h^2(\omega/\omega_n)^2\}^{1/2}} \tag{6-4}$$

In this equation, p is the amplitude of the pressure signal impressed on the transducer. The natural frequency ω_n is given by

$$\omega_n = \sqrt{\frac{3\pi r^2 c^2}{4LV}} \tag{6-5}$$

and the damping ratio h is

$$h = \frac{2\mu}{\rho c r^3}\sqrt{\frac{3LV}{\pi}} \tag{6-6}$$

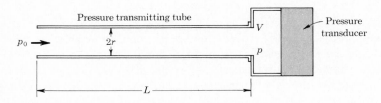

Figure 6-2 Schematic of pressure-transmitting system.

In the above formulas, c represents the velocity of sound in the fluid, μ is the dynamic viscosity of the fluid, and ρ is the fluid density. The phase angle for the pressure signal is

$$\phi = \tan^{-1} \frac{-2h(\omega/\omega_n)}{1 - (\omega/\omega_n)^2} \tag{6-7}$$

The velocity of sound for air may be calculated from

$$c = 49.1 T^{1/2} \text{ ft/s} \qquad \text{with } T \text{ in } °R$$

$$c = 20.04 T^{1/2} \text{ m/s} \qquad \text{with } T \text{ in } K$$

When the tube diameter is very small, as in a capillary, it is possible to produce a very large damping ratio, so that Eq. (6-4) will reduce to the following for frequencies below the natural frequencies:

$$\left| \frac{p}{p_0} \right| = \frac{1}{[1 + 4h^2(\omega/\omega_n)^2]^{1/2}} \tag{6-8}$$

If the transmitting fluid is a gas, the entire system can act as a Helmholtz resonator with a resonant frequency of

$$\omega_n = \left[\frac{\pi r^2 c^2}{V(L + \frac{1}{2}\sqrt{\pi^2 r^2})} \right]^{1/2} \tag{6-9}$$

More complete information on the dynamic response of pressure-measurement systems is given in Refs. [1], [7], and [11].

From both Eqs. (6-4) and (6-8) it is evident that a capillary tube may be used for effective damping of pressure signals. The tube is then said to act as an acoustical filter. The similarity of this system to that described in Sec. 2-7 and the discussion of the seismic instrument in Sec. 11-3 are to be noted. It should be noted also that the actual dynamic response for tube systems is strongly frequency-dependent, and the preceding formulas must be accepted with possible modification in high-frequency ranges. Tijdeman [18] and others [14 to 17] discuss experimental and analytical solutions which are available for such problems. In a dynamic pressure-measurement application one must also consider the frequency response of the pressure transducer and its movement in the overall measurement system. In general, one should try to design the system so that the natural frequency of the transducer is substantially greater than the signal frequency to be measured. Dynamic pressure measurements are particularly applicable to sound-level determinations, as discussed in Chap. 11.

Example 6-1 A small tube, 0.5 mm in diameter, is connected to a pressure transducer through a volume of 3.5 cm^3. The tube has a length of 7.5 cm. Air at 1 atm and 20°C is the pressure-transmitting fluid. Calculate the natural frequency for this system and the damping ratio.

SOLUTION We shall use Eq. (6-9) for this calculation. For air,

$$\rho = \frac{p}{RT} = \frac{1.0132 \times 10^5}{(287)(293)} = 1.205 \text{ kg/m}^3$$

$$c = (20.04)(293)^{1/2} = 343 \text{ m/s}$$

$$\mu = 1.91 \times 10^{-5} \text{ kg/m} \cdot \text{s (Table A-6, Appendex)}$$

Thus
$$\omega_n = \left\{ \frac{\pi(0.25 \times 10^{-3})^2(343)^2}{(3.5 \times 10^{-6})[0.075 + (0.5)(\pi)(0.25 \times 10^{-3})]} \right\}^{1/2}$$

$$= 296 \text{ Hz}$$

The damping ratio is calculated with Eq. (6-6):

$$h = \frac{(2)(1.91 \times 10^{-5})}{(1.205)(343)(0.25 \times 10^{-3})^3} \left[\frac{(3)(0.075)(3.5 \times 10^{-6})}{\pi} \right]^{1/2}$$

$$= 2.96$$

Example 6-2 Calculate the attenuation of a 100-Hz pressure signal in the system of Example 6-1.

SOLUTION For this calculation we employ Eq. (6-4). We have

$$\frac{\omega}{\omega_n} = \frac{100}{296} = 0.338$$

so that

$$\left| \frac{p}{p_0} \right| = \frac{1}{\{[1 - (0.338)^2]^2 + 4(2.96)^2(0.338)^2\}^{1/2}} = 0.457$$

The calibration of pressure transducers for dynamic measurement applications is rather involved; it is discussed in detail in Ref. [19].

6-3 MECHANICAL PRESSURE-MEASUREMENT DEVICES

Mechanical devices offer the simplest means for pressure measurement. In this section we shall examine the principles of some of the more important arrangements.

The fluid manometer is a widely used device for measurement of fluid pressures under steady-state conditions. Consider first the *U-tube manometer* shown in Fig. 6-3. The difference in pressure between the unknown pressure *p* and the atmosphere is determined as a function of the differential height *h*. The density of

Pressure
signal

p_a

A_2

p

p_a

h

A_1

Figure 6-3 U-tube manometer.

the fluid transmitting the pressure p is ρ_f, and the density of the manometer fluid is designated as ρ_m. A pressure balance of the two columns dictates that

$$p_a + \frac{g}{g_c} h\rho_m = p + \frac{g}{g_c} h\rho_f \qquad (6\text{-}10)$$

or

$$p - p_a = \frac{g}{g_c} h(\rho_m - \rho_f) \qquad (6\text{-}11)$$

Equation (6-11) gives the basic principle of the U-tube manometer. It is to be noted that the distance h is measured parallel to the gravitational force and that the differential pressure $p - p_a$ is measured at the location designated by the dashed line. If the location of the pressure source is at a different elevation from this point, there could be an appreciable error in the pressure determination, depending upon the density of the transmitting fluid.

A *well-type manometer* operates in the same manner as the U-tube manometer except that the construction is as shown in Fig. 6-4. In this case the pressure balance of Eq. (6-10) still yields

$$p - p_a = \frac{g}{g_c} h(\rho_m - \rho_f)$$

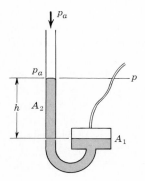

p_a

p_a

p

h

A_2

A_1

Figure 6-4 Well-type manometer.

Figure 6-5 Manometer used as a barometer.

This equation is seldom used, however, because the height h is not the fluid displacement which is normally measured. Typically, the well-type manometer is filled to a certain level at zero-pressure differential conditions. A measurement is then made of the displacement of the small column from this zero level. Designating this displacement by h', we have

$$h'A_2 = (h - h')A_1 \qquad (6\text{-}12)$$

since the volume displacements are the same on both sides of the manometer. Inserting Eq. (6-12) in (6-10) gives

$$p - p_a = \frac{g}{g_c} h'\left(\frac{A_2}{A_1} + 1\right)(\rho_m - \rho_f) \qquad (6\text{-}13)$$

Commercial well-type manometers have the scale for the manometer column graduated so that the user need not apply the area correction factor to the indicated displacement h'. Thus, for an area ratio of $A_2/A_1 = 0.03$, a true reading of 10.0 in for h' would be indicated as 10.3 as a result of the special scale graduation. The indicated value is then substituted for h in Eq. (6-11).

Manometers may be oriented in an inclined position to lengthen the scale and improve readability, or special optical sightglasses and vernier scales may be employed to provide more accurate location and indication of the manometer-fluid height than could be obtained with the naked eye. When mercury is the manometer fluid, variable-reluctance pickups may be used to accurately sense the fluid height. Special metal floats may also afford such a convenience with less dense fluids which are nonconductive.

When a well-type manometer is arranged as in Fig. 6-5, it is commonly called a barometer. The top of the column contains saturated mercury vapor at the local temperature. This saturation pressure is negligible in comparison to atmospheric pressure. The well is exposed to atmospheric pressure. The height h is thus a measure of the absolute atmospheric pressure. When $p_a = 14.696$ psia (1 atm), the height of a column of mercury at 68°F (20°C) would be 760 mm.

The column has a graduated scale fixed in position which requires that the instrument be zeroed for each reading because the level of the well will vary with height of the mercury in the column. The zeroing is accomplished with a screw adjustment which sets the level of the well at a reference position.

6-4 DEAD-WEIGHT TESTER

The dead-weight tester is a device used for balancing a fluid pressure with a known weight. Typically, it is a device which is used for static calibration of pressure gages and is seldom employed for an actual pressure measurement. Our discussion will be concerned only with the use of the dead-weight tester as a calibration device.

Consider the schematic in Fig. 6-6. The apparatus is set up for calibration of the pressure gage G. The chamber and cylinder of the tester are filled with a clean oil by first moving the plunger to its most forward position and then slowly withdrawing it while the oil is poured in through the opening for the piston. The gage to be tested is installed and the piston inserted in the cylinder. The pressure exerted on the fluid by the piston is now transmitted to the gage when the valve is opened. This pressure may be varied by adding weights to the piston or by using different piston-cylinder combinations of varying areas. The viscous friction between the piston and the cylinder in the axial direction may be substantially reduced by rotating the piston-weight assembly while the measurement is taken. As the pressure is increased, it may be necessary to advance the plunger to account for compression of the oil and any entrapped gases in the apparatus. High-pressure–dead-weight testers have a special lever system which is used to apply large forces to the piston.

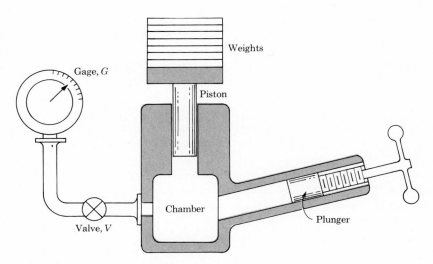

Figure 6-6 Schematic of a dead-weight tester.

The accuracies of dead-weight testers are limited by two factors: (1) the friction between the cylinder and the piston, and (2) the uncertainty in the area of the piston. The friction is reduced by rotation of the piston and use of long enough surfaces to ensure negligible flow of oil through the annular space between the piston and the cylinder. The area upon which the weight force acts is not the area of the piston nor the area of the cylinder; it is some effective area between these two which depends on the clearance spacing and the viscosity of the oil. The smaller the clearance, the more closely the effective area will approximate the cross-sectional area of the piston. It can be shown† that the percentage error due to the clearance varies according to

$$\text{Percent error} \sim \frac{(\rho \ \Delta p)^{1/2} b^3}{\mu D L} \tag{6-14}$$

where ρ = density of the oil
  Δp = pressure differential on the cylinder
  b = clearance spacing
  μ = viscosity
  D = piston diameter
  L = piston length

At high pressures there can be an elastic deformation of the cylinder which increases the clearance spacing and thereby increases the error of the tester.

6-5 BOURDON-TUBE PRESSURE GAGE

Bourdon-tube pressure gages enjoy a wide range of application where consistent, inexpensive measurements of static pressure are desired. They are commercially available in many sizes (1- to 16-in diameter) and accuracies. The Heise gage‡ is an extremely accurate bourdon-tube gage with an accuracy of 0.1 percent of full-scale reading; it is frequently employed as a secondary pressure standard in laboratory work.

The construction of a bourdon-tube gage is shown in Fig. 6-7. The bourdon tube itself is usually an elliptical cross-sectional tube having a "C"-shape configuration. When the pressure is applied to the inside of the tube, an elastic deformation results, which, ideally, is proportional to the pressure. The degree of linearity depends on the quality of the gage. The end of the gage is connected to a spring-loaded linkage, which amplifies the displacement and transforms it to an angular rotation of the pointer. The linkage is constructed so that the mechanism may be adjusted for optimum linearity and minimum hysteresis, as well as to compensate for wear which may develop over a period of time. Electrical-resistance strain gages (Sec. 10-7) may also be installed on the bourdon tube to sense the elastic deformation.

---

† See, for example, Ref. [10], p. 105.
‡ Manufactured by Heise Gage Company, Newton, Conn.

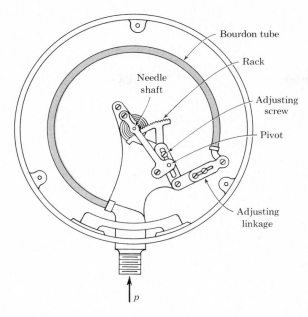

Figure 6-7 Schematic of a bourdon-tube pressure gage.

6-6 DIAPHRAGM AND BELLOWS GAGES

Diaphragm and bellows gages represent similar types of elastic deformation devices useful for many pressure-measurement applications. Consider first the flat diaphragm subjected to the differential pressure $p_1 - p_2$, as shown in Fig. 6-8. The diaphragm will be deflected in accordance with this pressure differential and the deflection sensed by an appropriate displacement transducer. Electrical-resistance strain gages may also be installed on the diaphragm, as shown in Fig. 6-9. The output of these gages is a function of the local strain, which, in turn, may be related to the diaphragm deflection and pressure differential. The deflection generally follows a linear variation with Δp when the deflection is less than one-third the diaphragm thickness. Figure 6-10 compares the deflection characteristics of three diaphragm arrangements as given by Roark [9]. Note that the

Figure 6-8 Schematic of a diaphragm gage.

$\downarrow p_1$

$\uparrow p_2$

Figure 6-9 Diaphragm gage using electrical-resistance strain gages.

$$y_{max} = \frac{3p}{16Et^3} a^4(1 - \mu^2)$$

$$y(r) = \frac{3p(1 - \mu^2)}{16Et^3} (a^2 - r^2)^2$$

(a)

$$y_{max} = \frac{3p(1-\mu^2)}{16Et^3}\left[a^4 + 3b^4 - 4a^2b^2 - 4a^2b^2 \ln(a/b) \right.$$

$$\left. + \frac{16a^2b^2}{a^2-b^2}\left(\ln\frac{a}{b}\right)^2 \right]$$

(b)

$$y_{max} = \frac{3W(1-\mu^2)}{4\pi Et^3}\left[a^2 - b^2 - \frac{4a^2b^2}{a^2-b^2}\left(\ln\frac{a}{b}\right)^2 \right]$$

(c)

Figure 6-10 Deflection characteristics of three diaphragm arrangements, according to Ref. [9]. (a) Edges fixed, uniform load over entire surface; (b) outer edge fixed and supported, inner edge fixed, uniform load over entire actual surface; (c) outer edge fixed and supported, inner edge fixed, uniform load along inner edge.

$\downarrow p_1$

$\uparrow p_2$

Figure 6-11 Corrugated-disk diaphragm.

Displacement
of bellows

Figure 6-12 Schematic of a bellows pressure gage.

first two diaphragms have uniform pressure loading over the entire surface of the disk, while the third type has a load which is applied at the center boss. In all three cases it is assumed that the outer edge of the disk is rigidly fixed and supported. To facilitate linear response over a larger range of deflections than that imposed by the one-third-thickness restriction, the diaphragm may be constructed from a corrugated disk, as shown in Fig. 6-11. This type of diaphragm is most suitable for those applications where a mechanical device is used for sensing the deflection of the diaphragm. Larger deflections are usually necessary with a mechanical amplification device than for electric transducers. A good summary of the properties of corrugated diaphragms is given in Ref. [12].

The bellows gage is depicted schematically in Fig. 6-12. A differential pressure force causes a displacement of the bellows, which may be converted to an electric signal or undergo a mechanical amplification to permit display of the output on an indicator dial. The bellows gage is generally unsuitable for transient measurements because of the larger relative motion and mass involved. The diaphragm gage, on the other hand, may be quite stiff, involves rather small displacements, and is suitable for high-frequency pressure measurements.

The deflection of a diaphragm under pressure may be sensed by a capacitance variation, as shown in Fig. 6-13. Such pressure pickups are well suited for dynamic measurements, since the natural frequency of diaphragms can be rather high. The capacitance pickup, however, involves low sensitivity, and special care must be exerted in the construction of readout circuitry. A schematic diagram of a LVDT-diaphragm differential pressure gage is shown in Fig. 6-14. Commercial models of this type of gage permit measurement of pressures as low as 0.000035 psi (0.25 Pa).

Varying
gap spacing

p

Diaphragm

6-13 Capacitance pressure gage.

Figure 6-14 Schematic of LVDT-diaphragm differential pressure gage. (*Courtesy of Sanborn Company.*)

The natural frequency of a circular diaphragm fixed at its perimeter is given by Hetenyi [5] as

$$f = \frac{10.21}{a^2} \sqrt{\frac{g_c E t^2}{12(1 - \mu^2)\rho}} \; \text{Hz} \tag{6-15}$$

where E = modulus of elasticity, psi or Pa
  t = thickness, in or m
  a = radius of diaphragm, in or m
  ρ = density of material, lbm/in^3 or kg/m^3
  g_c = dimensional conversion constant
    = 385.9 lbm · in/lbf · s^2 or 1.0 kg · m/N · s^2
  μ = Poisson's ratio

Equation (6-15) may be simplified to the following relation for steel diaphragms:

$$f = 1.934 \times 10^6 \, \frac{t}{\pi a^2} \qquad (t \text{ and } a \text{ in in}) \tag{6-16a}$$

$$= 4.912 \times 10^4 \, \frac{t}{\pi a^2} \qquad (t \text{ and } a \text{ in m}) \tag{6-16b}$$

Example 6-3 A diaphragm pressure gage is to be constructed of spring steel ($E = 200$ GN/m^2, $\mu = 0.3$) 5.0 cm in diameter and is to be designed to measure a maximum pressure of 1.4 MPa. Calculate the thickness of the gage required so that the maximum deflection is one-third this thickness. Calculate the natural frequency of this diaphragm.

SOLUTION Using the relation from Fig. 6-10, we have

$$\frac{1}{3} t = \frac{3 \, \Delta p}{16Et^3} \, a^4 (1 - \mu^2)$$

$$t^4 = \frac{(9)(1.4 \times 10^6)(0.025)^4[1 - (0.3)^2]}{(16)(2 \times 10^{11})}$$

$$t = 1.09 \text{ mm}$$

We may calculate the natural frequency from Eq. (6-15):

$$f = \frac{10.21}{(0.025)^2} \left[\frac{(1.0)(2 \times 10^{11})(0.00109)^2}{(12)[1 - (0.3)^2](7800)} \right]^{1/2}$$

$$= 27,280 \text{ Hz}$$

6-7 THE BRIDGMAN GAGE†

It is known that the resistance of fine wires changes with the pressure according to a linear relationship.

$$R = R_1(1 + b \, \Delta p) \qquad\qquad (6\text{-}17)$$

R_1 is the resistance at 1 atm, b is the pressure coefficient of resistance, and Δp is the gage pressure. The effect may be used for measurement of pressures as high as 100,000 atm [4]. A pressure transducer based on this principle is called a Bridgman gage. A typical gage employs a fine wire of Manganin (84% Cu, 12% Mn, 4% Ni) wound in a coil and enclosed in a suitable pressure container. The pressure coefficient of resistance for this material is about 1.7×10^{-7} psi^{-1} $(2.5 \times 10^{-11}$ Pa$^{-1})$. The total resistance of the wire is about 100 Ω, and conventional bridge circuits are employed for measuring the change in resistance. Such gages are subject to aging over a period of time, so that frequent calibration is required; however, when properly calibrated the gage can be used for high-pressure measurement with an accuracy of 0.1 percent. The transient response of the gage is exceedingly good. The resistance wire itself can respond to variations in the megahertz range. Of course, the overall frequency response of the pressure-meaurement system would be limited to much lower values because of the acoustic response of the transmitting fluid. Many of the problems associated with high-pressure measurement are discussed more fully in Refs. [2], [6], and [13].

† P. W. Bridgman, *Proc. Natl. Acad. Sci., U.S.*, vol. 3, p. 10, 1917.

6-8 LOW-PRESSURE MEASUREMENT

The science of low-pressure measurement is a rather specialized field which requires considerable care on the part of the experimentalist. The purpose of our discussion is to call attention to the more prominent types of vacuum instruments and describe the physical principles upon which they operate. For those readers requiring more specialized information, we refer them to the excellent monograph by Dushman and Lafferty [3]. The reader may also consult this reference for information on the various techniques for producing and maintaining a vacuum.

For moderate vacuum measurements the bourdon gage, manometers, and various diaphragm gages may be employed. Our discussion in this section, however, is concerned with the measurement of low pressures which are not usually accessible to the conventional gages. In this sense, we are primarily interested in absolute pressures below 1 torr (1 mmHg, 133 Pa).

6-9 THE McLEOD GAGE†

The McLeod gage is a modified mercury manometer which is constructed as shown in Fig. 6-15. The movable reservoir is lowered until the mercury column drops below the opening O. The bulb B and capillary C are then at the same pressure as the vacuum source p. The reservoir is subsequently raised until the mercury fills the bulb and rises in the capillary to a point where the level in the reference capillary is located at the zero point. The volume of the capillary per unit length is denoted by a, so that the volume of the gas in the capillary is

$$V_c = ay \qquad (6-18)$$

where y is the length of the capillary occupied by the gas.

We designate the volume of the capillary, bulb, and tube down to the opening as V_B. If we assume isothermal compression of the gas in the capillary we have

$$p_c = p \frac{V_B}{V_c} \qquad (6-19)$$

Now, the pressure indicated by the capillary is

$$p_c - p = y \qquad (6-20)$$

where we are expressing the pressure in terms of the height of the mercury column. Combining Eqs. (6-18) to (6-20) gives

$$p = \frac{ay^2}{V_B - ay} = \frac{yV_c}{V_B - ay} \qquad (6-21)$$

† H. McLeod, *Phil. Mag.*, vol. 48, p. 110, 1874.

To vacuum space, p

Reference

0
1
2
3
4
5
6
7
8

Capillary, C

Bulb, B

Opening, O

Movable
reservoir

Figure 6-15 The McLeod gage.

For most cases $ay \ll V_B$ and

$$p = \frac{ay^2}{V_B} \qquad (6\text{-}22)$$

Commercial McLeod gages have the capillary calibrated directly in micrometers. The McLeod gage is sensitive to condensed vapors that may be present in the sample because they can condense upon compression and invalidate Eq. (6-19). For dry gases the gage is applicable from 10^{-2} to 10^2 μm (0.0013 + 13.3 Pa).

Example 6-4 A McLeod gage has $V_B = 100$ cm^3 and a capillary diameter of 1 mm. Calculate the pressure indicated by a reading of 3.00 cm. What error would result if Eq. (6-22) were used instead of Eq. (6-21)?

SOLUTION We have

$$V_c = \frac{\pi(1)^2}{4}(30.0) = 23.6 \text{ mm}^3$$

$$V_B = 10^5 \text{ mm}^3$$

From Eq. (6-21),

$$p = \frac{(23.6)(30.0)}{10^5 - 23.6} = 0.0071 \text{ torr} = 7.1 \ \mu\text{m} \ (0.94 \text{ Pa})$$

The fractional error in using Eq. (6-22) would be

$$\text{Error} = \frac{ay}{V_B} = 2.36 \times 10^{-4}$$

or a negligibly small value.

6-10 PIRANI THERMAL-CONDUCTIVITY GAGE†

At low pressures the effective thermal conductivity of gases decreases with pressure. The Pirani gage is a device that measures the pressure through the change in thermal conductance of the gas. The gage is constructed as shown in Fig. 6-16. An electrically heated filament is placed inside the vacuum space. The heat loss from the filament is dependent upon the thermal conductivity of the gas and the filament temperature. The lower the pressure, the lower the thermal conductivity and consequently the higher the filament temperature for a given electric-energy input. The temperature of the filament could be measured by a thermocouple, but in the Pirani-type gage the measurement is made by observing the variation in resistance of the filament material (tungsten, platinum, etc.). The resistance measurement may be performed with an appropriate bridge circuit. The heat loss from the filament is also a function of the ambient temperature, and, in practice, *two* gages are connected in series, as shown in Fig. 6-17, to compensate for possible variations in the ambient conditions. The measurement gage is evacuated, and both it and the sealed gage are exposed to the same environment conditions. The bridge circuit is then adjusted (through resistance R_2) to produce a null condition. When the test gage is now exposed to the particular pressure conditions, the deflection of the bridge from the null position will be compensated for changes in environment temperature.

Pirani gages require an empirical calibration and are not generally suitable for use at pressures much below 1 μm. The upper limit is about 1 torr (133 Pa), giving an overall range of about 0.1 to 100 Pa. For higher pressures the thermal conductance changes very little with pressure. It must be noted that the heat loss

† M. Pirani, *Verhandl. deut. physik. Ges.*, vol. 8, p. 686, 1906.

To bridge circuit   To vacuum space

Figure 6-16 Schematic of Pirani gage.

Figure 6-17 Pirani-gage arrangement to compensate for change in ambient temperature.

from the filament is also a function of the conduction losses to the filament supports and radiation losses to the surroundings. The lower limit of applicability of the gage is the point where these effects overshadow the conduction into the gas. The transient response of the Pirani gage is poor. The time necessary for the establishment of thermal equilibrium may be of the order of several minutes at low pressures.

6-11 THE KNUDSEN GAGE†

Consider the arrangement shown in Fig. 6-18. Two vanes V along with the mirror M are mounted on the thin-filament suspension. Near these vanes are two heated plates P, each of which is maintained at a temperature T. The separation distance between the vanes and plates is *less than* the mean free path of the surrounding gas. Heaters are installed so that the temperature of the plates is higher than that of the surrounding gas. The vanes are at the temperature of the gas T_g. The molecules striking the vanes from the hot plates have a higher velocity than those leaving the vanes because of the difference in temperature. Thus, there is a net momentum imparted to the vanes which may be measured by observing the angular displacement of the mirror, similar to the technique used in a lightbeam galvanometer. The total momentum exchange with the vanes is a function of molecular density, which, in turn, is related to the pressure and temperature of the gas. An expression for the gas pressure may thus be derived in terms of the temperatures and the measured force. For small temperature differences $T - T_g$ it may be shown that this relation is [3]

$$p = 4F \frac{T_g}{T - T_g} \tag{6-23}$$

where the pressure is in dynes per square centimeter when the force is in dynes. The temperatures are in degrees kelvin.

† M. Knudsen, *Ann. Physik*, vol. 32, p. 809, 1910.

Figure 6-18 Schematic of Knudsen gage.

The Knudsen gage furnishes an absolute measurement of the pressure which is independent of the molecular weight of the gas. It is suitable for use between 10^{-5} and 10 μm (10^{-6} to 1 Pa) and may be used as a calibration device for other gages in this region.

6-12 THE IONIZATION GAGE

Consider the arrangement shown in Fig. 6-19, which is similar to the ordinary triode vacuum tube. The heated cathode emits electrons, which are accelerated by the positively charged grid. As the electrons move toward the grid, they produce

Figure 6-19 Schematic of ionization gage.

ionization of the gas molecules through collisions. The plate is maintained at a negative potential so that the positive ions are collected there, producing the plate current i_p. The electrons and negative ions are collected by the grid, producing the grid current i_g. It is found that the pressure of the gas is proportional to the ratio of plate current to grid current.

$$p = \frac{1}{S} \frac{i_p}{i_g} \qquad (6\text{-}24)$$

where the proportionality constant S is called the sensitivity of the gage. A typical value for nitrogen is $S = 20$ torr^{-1} (2.67 kPa^{-1}), but the exact value must be determined by calibration of the particular gage. The value of S is a function of the tube geometry and the type of gas.

Conventional ionization gages are suitable for measurements between 1.0 and 10^{-5} μm (0.13 to 1.3×10^{-6} Pa), and the current output is usually linear in this range. At higher pressures there is the danger of burning out the cathode. Special types of ionization gages are suitable for measurements of pressures as low as 10^{-12} torr (0.13 nPa). Very precise experimental techniques are required, however, in order to perform measurements at these high vacuums. The interested reader should consult Ref. [3] for additional information.

6-13 THE ALPHATRON†

The Alphatron is a radioactive ionization gage, shown schematically in Fig. 6-20. A small radium source serves as an α-particle emitter. These particles ionize the gas inside the gage enclosure, and the degree of ionization is determined by measuring the voltage output E_o. The degree of ionization is a direct linear

† National Research Corp., Cambridge, Mass.

Figure 6-20 Schematic of Alphatron gage.

function of pressure for a rather wide range of pressures, from 10^{-3} to 10^3 torr (0.1 to 10^5 Pa). The output characteristics, however, are different for each type of gas used. The lower pressure limit of the gage is determined by the length of the mean free path of the alpha particles as compared with the enclosure dimensions. At very low pressures the mean free path becomes so large that very few collisions are probable in the gage, and hence the ionization level is very small. The Alphatron has the advantages that it may be used at atmospheric pressure as well as high vacuum and that there is no heated filament to contend with as in the conventional ionization gage. Consequently, there is no problem of accidentally burning out a filament because of an inadvertent exposure of the gage to high (above 10^{-1} torr or 13 Pa) pressures.

6-14 SUMMARY

Figure 6-21 gives a convenient summary of the pressure ranges for which the gages discussed are normally employed in practice.

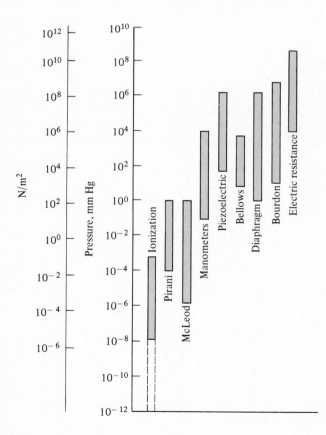

Figure 6-21 Summary of applicable range of pressure gages.

REVIEW QUESTIONS

6-1 Distinguish between gage pressure, absolute pressure, and vacuum.
6-2 To transmit a high-frequency pressure signal, one should select
    (*a*) A short, small-diameter tube
    (*b*) A short, large-diameter tube
    (*c*) A long, small-diameter tube
    (*d*) A long, large-diameter tube
6-3 What are the advantages of the manometer pressure-measurement device?
6-4 What is the advantage of a well-type manometer?
6-5 What are some advantages of the bourdon-tube, diaphragm, and bellows gages?
6-6 Describe the principle of operation of a McLeod gage.
6-7 Describe the Pirani gage.
6-8 When is the Knudsen gage used?
6-9 Describe the ionization gage. How does it differ from the Pirani gage? What disadvantages does it have?

PROBLEMS

6-1 A mercury barometer is constructed like that shown in Fig. 6-5. The column is a glass tube 0.250 in ID and 0.375 in OD, and the well is a glass dish 1.50 in ID. Calculate the percent error which would result if an area correction factor were not used.

6-2 Derive an expression for the radius of a simple diaphragm as shown in Fig. 6-10 using the following restrictions.
    (*a*) The maximum deflection is one-third the thickness.
    (*b*) The maximum deflection must be 100 times as great as the uncertainty in the deflection measurement w_y.
    Assume that the maximum pressure differential Δp is given as well as w_y.

6-3 Determine the factor to convert pressure in inches of water to pounds per square foot.

6-4 The effective radius of an air molecule is about 1.85×10^{-8} cm. Calculate the mean free path at 70°F and the following pressures: 1 atm, 1 torr, 1 μm, 1 inH_2O, and 10^{-3} μm.

6-5 Reduce Eq. (6-3) to an expression for mean free path in terms of pressure in micrometers, temperature in degrees kelvin, molecular weight of the gas, and effective molecular diameter.

6-6 A dynamic pressure measurement is to be made with an apparatus similar to that shown in Fig. 6-2. The appropriate dimensions are:

$$L = 2.50 \text{ in}$$

$$r = 0.005 \text{ in}$$

$$V = 0.10 \text{ in}^3$$

The fluid is air at 70°F and 14.7 psia. Plot the pressure-amplitude ratio versus ω/ω_n according to both Eq. (6-4) and Eq. (6-8).

6-7 Plot the error in Eq. (6-8) versus $r^2 L/V$.

6-8 Calculate the resonant frequency of the system in Prob. 6-6, assuming that it acts as a Helmholtz resonator.

6-9 A well-type manometer uses a special bromide fluid having a specific gravity of 2.95. The well has a diameter of 3.00 in and the tube has a diameter of 0.200 in. The manometer is to be used to measure a differential pressure in a water-flow system. The scale placed alongside the tube has no correction factor for the area ratio of the manometer. Calculate the value of a factor that may be multiplied by the manometer reading in inches to find the pressure differential in pounds per square inch.

6-10 A vacuum gage is to use an LVDT-diaphragm combination like that shown in Fig. 6-14. The LVDT has a sensitivity of 2.5 nm, and the diaphragm is to be constructed of steel ($E = 2 \times 10^{11}$ Pa, $\mu = 0.3$) with a diameter of 15 cm. Calculate the diaphragm thickness in accordance with the restriction that the maximum deflection not exceed one-third this thickness. What is the lowest pressure which may be sensed by this instrument?

6-11 Calculate the natural frequency of the diaphragm in Prob. 6-10.

6-12 A Bridgman gage uses a coil of Manganin wire having a nominal resistance of 100 Ω at atmospheric pressure. The gage is to be used to measure a pressure of 1000 psig with an uncertainty of 0.1 percent. What is the allowable uncertainty in the resistance measurement?

6-13 Suppose the Bridgman gage of Prob. 6-12 is connected to the bridge circuit of Fig. 4-25 so that the gage is R_1 and all resistances are equal to 100 Ω at a pressure of 1 atm. The battery voltage is 4.0 V, and the detector is a high-impedance voltage-measuring device. The bridge is assumed to be in balance at $p = 1$ atm. Calculate the voltage output of the bridge at $p = 1000$ psig.

6-14 Rework Example 6-2 assuming the diameter of the capillary to be 0.2 mm.

6-15 A Knudsen gage is to be designed to operate at a maximum pressure of 1.0 μm. For this application the spacing of the vane and plate is to be less than 0.3 mean free path at this pressure. Calculate the force on the vanes at pressures of 1.0 and 0.01 μm when the gas temperature is 20°C and the temperature difference is 50.0 K.

6-16 A capacitance-diaphragm pressure gage as shown in Fig. 6-13 is to be used to measure pressure differentials as high as 1000 psi at frequencies as high as 15,000 Hz. The diameter of the diaphragm is not to exceed 0.500 in. Calculate the thickness and diameter of a diaphragm to accomplish this (the natural frequency should be at least 30,000 Hz). Choose a suitable gap spacing, and estimate the capacitance-pressure sensitivity of the device. Assume the dielectric constant is that of air.

6-17 A bourdon-tube pressure gage having an internal volume of 1.0 in^3 is used for measuring pressure in a fluctuating air system having frequencies as high as 100 Hz. Design an acoustical filter which will attenuate all frequencies above 20 Hz by 99 percent. Plot the frequency response of this filter.

6-18 An acoustic filter is to be designed to attenuate sharp pressure transients in air above 50 Hz. The volume of air contained in the pressure-transducer cavity is 0.6 in^3, and a capillary tube connects the cavity to the pressure source. If the 50-Hz frequency is to be attenuated by 50 percent, determine the capillary length and diameter for system natural frequencies of (a) 50 Hz, (b) 100 Hz, and (c) 500 Hz.

6-19 A U-tube manometer uses tubes of 0.250 and 0.500 in diameters for the two legs. When subjected to a certain pressure, the difference in height of the two fluid columns is 10.0 inHg. What would have been the reading if both tubes were the same diameter? The measurement is performed on air.

6-20 A pressure signal is fed through a line having an inside diameter of 1.5 mm and a length of 1.5 m. The line is connected to a pressure transducer having a volume of approximately 5 cm^3. Air at 690 kPa and 90°C is the transmitting fluid. Calculate the natural frequency and damping ratio for this system.

6-21 The manometer of Prob. 6-19 uses a fluid having a specific gravity of 1.85. The sensing fluid is water. What is the pressure difference when the difference in heights of the columns is 5.0 in? Assume that both legs of the manometer are filled with water.

6-22 A diaphragm-pressure gage is constructed of spring steel to measure a pressure differential of 1000 psi. The diameter of the diaphragm is 0.5 in. Calculate the diaphragm thickness so that the maximum deflection is one-third the thickness. What is the natural frequency of this diaphragm?

6-23 A Bridgman gage is to be used to measure a pressure of 10,000 psi using a Manganin element having a resistance of 100 Ω at atmospheric pressure. Calculate the resistance of the gage under high-pressure conditions. If the gage is one leg of a bridge whose other legs all have values of exactly 100 Ω, calculate the voltage output of the bridge for a constant-voltage source of 24 V.

6-24 A McLeod gage is available which has a volume V_B of 150 cm^3 and a capillary diameter of 0.3 mm. Calculate the gage reading for a pressure of 30 μm.

6-25 What is the approximate range of mean free paths for air over the range of pressures for which the Knudsen gage is applicable?

6-26 A special high-pressure U-tube manometer is constructed to measure pressure differential in air at 13.8 MPa and 20°C. When an oil having a specific gravity of 0.83 is used as the fluid, calculate the differential pressure in psia that would be indicated by a 135-mm reading.

6-27 A diaphragm like that shown in Fig. 6-10c has $a = 1.0$ in, $b = 0.125$ in, and $t = 0.048$ in and is constructed of spring steel. It is subjected to a total loading of 600 lbf. Calculate the deflection.

REFERENCES

1. Arons, A. B., and R. H. Cole: Design and Use of Piezo-Electric Gages for Measurement of Large Transient Pressures, *Rev. Sci. Instr.*, vol. 21, pp. 31–38, 1950.
2. Bridgman, P. W.: "The Physics of High Pressure," The Macmillan Company, New York, 1931.
3. Dushman, S., and J. M. Lafferty: "Scientific Foundations of Vacuum Technique," 2d ed., John Wiley & Sons, Inc., New York, 1962.
4. Hall, H. T.: Some High Pressure—High Temperature Apparatus Design Considerations, *Rev. Sci. Instr.*, vol. 29, p. 267, 1958.
5. Hetenyi, M. (ed.): "Handbook of Experimental Stress Analysis," John Wiley & Sons, Inc., New York, 1950.
6. Howe, W. H.: The Present Status of High Pressure Measurements, *ISA Journal*, vol. 2, pp. 77 and 109, 1955.
7. Iberall, A. S.: Attenuation of Oscillatory Pressures in Instrument Lines, *Trans. ASME*, vol. 72, p. 689, 1950.
8. Neubert, H. K. P.: "Instrument Transducers," Oxford University Press, Fair Lawn, N. J., 1963.
9. Roark, R. J.: "Formulas for Stress and Strain," 3d ed., McGraw-Hill Book Company, New York, 1954.
10. Sweeney, R. J.: "Measurement Techniques in Mechanical Engineering," John Wiley & Sons, Inc., New York, 1953.
11. Taback, I.: The Response of Pressure Measuring Systems to Oscillating Pressure, *NACA Tech. Note* 1819, February 1949.
12. Wildhack, W. A., R. F. Dresslea, and E. C. Lloyd: Investigation of the Properties of Corrugated Diaphragms, *Trans. ASME*, vol. 79, pp. 65–82, 1957.
13. Giardini, A. A. (ed.): "High Pressure Measurements," Butterworth & Co. (Publishers), Ltd., London, 1963.
14. Karam, J. T., and M. E. Franke: The Frequency Response of Pneumatic Lines, *J. Basic Eng., Trans. ASME*, ser. D. vol. 90, pp. 371–378, 1967.
15. Nichols, N. B.: The Linear Properties of Pneumatic Transmission Lines, *Trans. Instr. Soc. Am.*, vol. 1, pp. 5–14, 1962.
16. Watts, G. P.: The Response of Pressure Transmission Lines, Preprint No. 13.3-1-65, 20th Annual ISA Conf. and Exhibit, Los Angeles, Calif., October 4–7, 1965.
17. Bergh, H., and H. Tijdeman: Theoretical and Experimental Results for the Dynamic Response of Pressure Measuring Systems, *NLR Tech. Rept.* F. 238, 1965.
18. Tijdeman, H.: Remarks on the Frequency Response of Pneumatic Lines, *J. Basic Eng.*, pp. 325–328, June 1969.
19. Schweppe, J. L., et al.: Methods for the Dynamic Calibration of Pressure Transducers, *Natl. Bur. Std. (U.S.), Monograph* 67, 1963.
20. Funk, J. E., D. J. Wood, and S. P. Chao: The Transient Response of Orifices and Very Short Lines, ASME paper 71-WA/FE-14, December 1971.

FLOW MEASUREMENT

7-1 INTRODUCTION

The measurement of fluid flow is important in applications ranging from measurements of blood-flow rates in a human artery to the measurement of the flow of liquid oxygen in a rocket. Many research projects and industrial processes depend upon a measurement of fluid flow to furnish important data for analysis. In some cases extreme precision is called for in the flow measurement, while in other instances only crude measurements are necessary. The selection of the proper instrument for a particular application is governed by many variables, including cost. For many industrial operations the accuracy of a fluid-flow measurement is directly related to profit. A simple example is the gasoline pump at the neighborhood service station; another example is the water meter at home. It is easy to see how a small error in flow measurement on a large natural gas or oil pipeline could make a difference of thousands of dollars over a period of time. Thus, the laboratory scientist is not the only person who is concerned with accurate flow measurement; the engineer in industry is also vitally interested because of the impact which flow measurements may have on the profit and loss statement of the company.

Flow-rate-measurement devices frequently require accurate pressure and temperature measurements in order to calculate the output of the instrument. Chapters 6 and 8 consider these associated measurement topics in detail, and the reader should consult the appropriate sections from time to time to relate specific pressure and temperature measurement devices to the material in the present chapter. We may remark at this time, however, that the overall accuracy of many of the most widely used flow-measurement devices is governed primarily by the accuracy of some pressure or temperature measurement.

Flow rate is expressed in both volume and mass units of varying sizes. Some commonly used terms are

1 gallon per minute (gpm)

$\quad$ = 231 cubic inches per minute (in^3/min)

$\quad$ = 63.09 cubic centimeters per second (cm^3/s)

1 liter

$\quad$ = 0.26417 gallons = 1000 cubic centimeters

1 cubic foot per minute (cfm, or ft^3/min)

$\quad$ = 0.028317 cubic meters per minute = 471.95 cubic centimeters per second

1 standard cubic foot per minute of air at 20°C, 1 atm

$\quad$ = 0.07513 pound-mass per minute

$\quad$ = 0.54579 gram per second

We should alert the reader to the fact that commercial gas flow meters typically specify flow ratings in volume flow rate at standard conditions of 1 atm and 20°C. The units employed are standard cubic feet per minute (scfm) and standard cubic centimeters per minute (sccm).

Our objective in this chapter is to present a broad discussion of flow measurements and indicate the principles of operation of a number of devices that are commonly used. We shall also give the calculation methods that are connected with some of these devices and discuss some methods of flow visualization. In concluding the chapter a tabular comparison of the various methods will be presented, pointing out their range of applicability and expected accuracies.

7-2 POSITIVE-DISPLACEMENT METHODS

The flow rate of a nonvolatile liquid like water may be measured through a direct-weighing technique. The time necessary to collect a quantity of the liquid in a tank is measured, and an accurate measurement is then made of the weight of liquid collected. The average flow rate is thus calculated very easily. Improved accuracy may be obtained by using longer or more precise time intervals or more precise weight measurements. The direct-weighing technique is frequently employed for calibration of water and other liquid flowmeters and thus may be taken as a standard calibration technique. Obviously, it is not suited for transient flow measurements.

Positive-displacement flowmeters are generally used for those applications where consistently high accuracy is desired under steady flow conditions. A typical positive-displacement device is the home water meter shown schematically

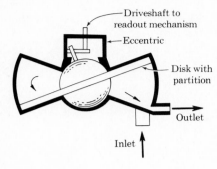

Figure 7-1 Schematic of a nutating-disk meter.

in Fig. 7-1. This meter operates on the nutating-disk principle. Water enters the left side of the meter and strikes the disk, which is eccentrically mounted. In order for the fluid to move through the meter, the disk must "wobble" or nutate about the vertical axis since both the top and the bottom of the disk remain in contact with the mounting chamber. A partition separates the inlet and outlet chambers of the disk. As the disk nutates, it gives direct indication of the volume of liquid which has passed through the meter. The indication of the volumetric flow is given through a gearing and register arrangement which is connected to the nutating disk. The nutating-disk meter may give reliable flow measurements within 1 percent.

Another type of positive-displacement device is the rotary-vane meter shown in Fig. 7-2. The vanes are spring-loaded so that they continuously maintain contact with the casing of the meter. A fixed quantity of fluid is trapped in each section as the eccentric drum rotates, and this fluid eventually finds its way out the exit. An appropriate register is connected to the shaft of the eccentric drum to record the volume of the displaced fluid. The uncertainties of rotary-vane meters are of the order of 0.5 percent, and the meters are relatively insensitive to viscosity since the vanes always maintain good contact with the inside of the casing.

The lobed-impeller meter shown in Fig. 7-3 may be used for either gas- or liquid-flow measurements. The impellers and case are carefully machined so that accurate fit is maintained. In this way, the incoming fluid is always trapped between the two rotors and is conveyed to the outlet as a result of their rotation. The number of revolutions of the rotors is an indication of the volumetric flow rate.

Figure 7-2 Schematic of rotary-vane flowmeter.

Inlet

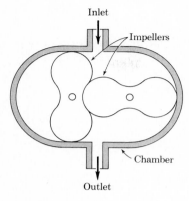

Impellers

Chamber

Outlet

Figure 7-3 Schematic of lobed-impeller flowmeter.

Example 7-1 A lobed-impeller flowmeter is used for measurement of the flow of nitrogen at 20 psia and 100°F. The meter has been calibrated so that it indicates the volumetric flow with an accuracy of ± one-half of 1 percent from 1000 to 3000 cfm. The uncertainties in the gas pressure and temperature measurements are ±0.025 psi and ±1.0°F, respectively. Calculate the uncertainty in a mass flow measurement at the given pressure and temperature conditions.

SOLUTION The mass flow is given by

$$\dot{m} = \rho Q$$

where the density of nitrogen is given by

$$\rho = \frac{p}{R_{N_2} T}$$

Using Eq. (3-2), we obtain the following equation for the uncertainty in the mass flow:

$$\frac{w_{\dot{m}}}{\dot{m}} = \left[\left(\frac{w_Q}{Q} \right)^2 + \left(\frac{w_p}{p} \right)^2 + \left(\frac{w_T}{T} \right)^2 \right]^{1/2}$$

Using the given data,

$$\frac{w_{\dot{m}}}{\dot{m}} = \left[(0.005)^2 + \left(\frac{0.025}{20} \right)^2 + \left(\frac{1}{560} \right)^2 \right]^{1/2} = 5.05 \times 10^{-3}$$

or 0.505 percent. Thus, the uncertainties in the pressure and temperature measurements do not appreciably influence the overall uncertainty in the mass flow measurements.

7-3 FLOW-OBSTRUCTION METHODS

Several types of flowmeters fall under the category of obstruction devices. Such devices are sometimes called *head meters* because a head-loss or pressure-drop measurement is taken as an indication of the flow rate. Let us first consider some of the general relations for obstruction meters. We shall then examine the applicability of these relations to specific devices.

Consider the one-dimensional flow system shown in Fig. 7-4. The continuity relation for this situation is

$$\dot{m} = \rho_1 A_1 u_1 = \rho_2 A_2 u_2 \qquad (7\text{-}1)$$

where u is the velocity. If the flow is adiabatic and frictionless and the fluid is incompressible, the familiar Bernoulli equation may be written

$$\frac{p_1}{\rho_1} + \frac{u_1^2}{2g_c} = \frac{p_2}{\rho_2} + \frac{u_2^2}{2g_c} \qquad (7\text{-}2)$$

where now $\rho_1 = \rho_2$. Solving Eqs. (7-1) and (7-2) simultaneously gives for the pressure drop

$$p_1 - p_2 = \frac{u_2^2 \rho}{2g_c}\left[1 - \left(\frac{A_2}{A_1}\right)^2\right] \qquad (7\text{-}3)$$

and the volumetric flow rate may be written

$$Q = A_2 u_2 = \frac{A_2}{\sqrt{1 - (A_2/A_1)^2}} \sqrt{\frac{2g_c}{\rho}(p_1 - p_2)} \qquad (7\text{-}4)$$

where $Q = \text{ft}^3/\text{s or m}^3/\text{s}$
$A = \text{ft}^2 \text{ or m}^2$
$\rho = \text{lbm/ft}^3 \text{ or kg/m}^3$
$p = \text{lbf/ft}^2 \text{ or N/m}^2$
$g_c = 32.17 \text{ lbm}\cdot\text{ft/lbf}\cdot\text{s}^2 \text{ or } 1.0 \text{ kg}\cdot\text{m/N}\cdot\text{s}^2$

Thus, we see that a channel like the one shown in Fig. 7-4 could be used for a flow measurement by simply measuring the pressure drop $(p_1 - p_2)$ and calculating the flow from Eq. (7-4). No such channel, however, is frictionless, and some

Figure 7-4 General one-dimensional flow system.

losses are always present in the flow. The volumetric flow rate calculated from Eq. (7-4) is the ideal value, and it is usually related to the actual flow rate and an empirical *discharge coefficient C* by the following relation:

$$\frac{Q_{actual}}{Q_{ideal}} = C \tag{7-5}$$

The discharge coefficient is not a constant and may depend strongly on the flow Reynolds number and the channel geometry.

When the flow of an ideal gas is considered, the following equation of state applies:

$$p = \rho R T \tag{7-6}$$

where T is the absolute temperature and R is the gas constant for the particular gas, which can be expressed in terms of the universal gas constant $\mathscr{R}$ and the molecular weight by

$$R = \frac{\mathscr{R}}{M}$$

The value of $\mathscr{R}$ is 8314 J/kg·mol·K or 1545 ft·lbf/lbm·mol·°R. For reversible adiabatic flow the steady-flow energy equation for an ideal gas is

$$c_p T_1 + \frac{u_1^2}{2g_c} = c_p T_2 + \frac{u_2^2}{2g_c} \tag{7-7}$$

where c_p is the specific heat at constant pressure and is assumed constant for an ideal gas. When Eqs. (7-1), (7-6), and (7-7) are combined, there results

$$\dot{m}^2 = 2g_c A_2^2 \frac{\gamma}{\gamma-1} \frac{p_1^2}{R T_1} \left[\left(\frac{p_2}{p_1}\right)^{2/\gamma} - \left(\frac{p_2}{p_1}\right)^{(\gamma+1)/\gamma} \right] \tag{7-8}$$

where the velocity of approach, i.e., the velocity at section 1, is assumed to be very small. This relationship may be simplified to

$$\dot{m} = \sqrt{\frac{2g_c}{R T_1}} A_2 \left[p_2 \, \Delta p - \left(\frac{1.5}{\gamma} - 1\right)(\Delta p)^2 + \cdots \right]^{1/2} \tag{7-9}$$

with $\Delta p = p_1 - p_2$ and $\gamma = c_p/c_v$ is the ratio of specific heats for the gas. Equation (7-9) is valid for $\Delta p < p_1/4$. When $\Delta p < p_1/10$, a further simplification may be made to give

$$\dot{m} = A_2 \sqrt{\frac{2g_c p_2}{R T_1} (p_1 - p_2)} \tag{7-10}$$

where
m = mass flow rate, lbm/s or kg/s
A = area, ft² or m²
g_c = 32.17 lbm·ft/lbf·s² or 1.0 kg·m/N·s²
p = pressure, lbf/ft² or N/m²
R = gas constant, lbf·ft/lbm·°R or N·m/kg·K
T = absolute temperature, °R or K

Note that Eq. (7-10) reduces to Eq. (7-4) when the relation for density from Eq. (7-6) is substituted. Thus, for small values of Δp compared with p_1, the flow of a compressible fluid may be approximated by the flow of an incompressible fluid.

Three typical obstruction meters are shown in Fig. 7-5. The venturi offers the advantages of high accuracy and small pressure drop, while the orifice is considerably lower in cost. Both the flow nozzle and the orifice have relatively high permanent pressure drop. Flow-rate calculations for all three devices are made on the basis of Eq. (7-4) with appropriate empirical constants defined as follows:

$$M = \text{velocity of approach factor} = \frac{1}{\sqrt{1 - (A_2/A_1)^2}} \tag{7-11}$$

$$K = \text{flow coefficient} = CM \tag{7-12}$$

$$\beta = \text{diameter ratio} = \frac{d}{D} = \sqrt{\frac{A_2}{A_1}} \tag{7-13}$$

When flow measurements of a compressible fluid are made, an additional parameter, the *expansion factor* Y, is used. For venturis and nozzles this factor is given by

$$Y_a = \left[\left(\frac{p_2}{p_1} \right)^{2/\gamma} \frac{\gamma}{\gamma - 1} \frac{1 - (p_2/p_1)^{(\gamma - 1)/\gamma}}{1 - (p_2/p_1)} \frac{1 - (A_2/A_1)^2}{1 - (A_2/A_1)^2 (p_2/p_1)^{2/\gamma}} \right]^{1/2} \tag{7-14}$$

while for orifices an empirical expression for Y is given as

$$Y_1 = 1 - \left[0.41 + 0.35 \left(\frac{A_2}{A_1} \right)^2 \right] \frac{p_1 - p_2}{\gamma p_1} \tag{7-15}$$

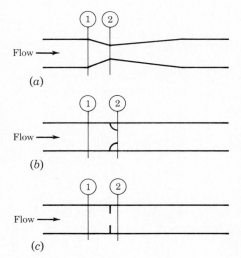

Flow →

(a)

Flow →

(b)

Flow →

(c)

Figure 7-5 Schematic of three typical obstruction meters. (a) Venturi; (b) flow nozzle; (c) orifice.

when either flange taps or vena contracta taps are used. For orifices with pipe taps the following relation applies:

$$Y_2 = 1 - [0.333 + 1.145(\beta^2 + 0.7\beta^5 + 12\beta^{13})]\frac{p_1 - p_2}{\gamma p_1} \tag{7-16}$$

The empirical expansion factors given by Eqs. (7-15) and (7-16) are accurate within ± 0.5 percent for $0.8 < p_2/p_1 < 1.0$. Plots of the expansion factors Y_a and Y_1 are given in Figs. 7-14 and 7-15, respectively.

We thus have the following semiempirical equations, which are conventionally applied to venturis, nozzles, or orifices:

VENTURIS, INCOMPRESSIBLE FLOW:

$$Q_{\text{actual}} = CMA_2 \sqrt{\frac{2g_c}{\rho}} \sqrt{p_1 - p_2} \tag{7-17}$$

NOZZLES AND ORIFICES, INCOMPRESSIBLE FLOW:

$$Q_{\text{actual}} = KA_2 \sqrt{\frac{2g_c}{\rho}} \sqrt{p_1 - p_2} \tag{7-18}$$

The use of the flow coefficient instead of the product CM is merely a matter of convention. When compressible fluids are used, the above equations are modified by the factor Y and the fluid density is evaluated at inlet conditions. We then have

VENTURIS, COMPRESSIBLE FLOW:

$$\dot{m}_{\text{actual}} = YCMA_2 \sqrt{2g_c\,\rho_1(p_1 - p_2)} \tag{7-19}$$

NOZZLES AND ORIFICES, COMPRESSIBLE FLOW:

$$\dot{m}_{\text{actual}} = YKA_2 \sqrt{2g_c\,\rho_1(p_1 - p_2)} \tag{7-20}$$

In Eqs. (7-17) to (7-20) the appropriate units are

$$Q = \text{volume flow rate, ft}^3/\text{s or m}^3/\text{s}$$

$$A = \text{area, ft}^2 \text{ or m}^2$$

$$g_c = 32.17 \text{ lbm} \cdot \text{ft/lbf} \cdot \text{s}^2 \text{ or } 1.0 \text{ kg} \cdot \text{m/N} \cdot \text{s}^2$$

$$\rho = \text{density, lbm/ft}^3 \text{ or kg/m}^3$$

$$p = \text{pressure, lbf/ft}^2 \text{ or N/m}^2$$

Detailed tabulations of the various coefficients have been made in Ref. [1], some of which are presented in Figs. 7-9 through 7-15. Examples 7-2 and 7-3 illustrate the use of these charts for practical calculations.

7-4 PRACTICAL CONSIDERATIONS FOR OBSTRUCTION METERS

The construction of obstruction meters has been standardized by the American Society of Mechanical Engineers [1], [2]. The recommended proportions of venturi tubes are shown in Fig. 7-6. Note that the pressure taps are connected to manifolds which surround the upstream and throat portions of the tube. These manifolds receive a sampling of the pressure all around the sections so that a good average value is obtained. The discharge coefficients for such venturi tubes are shown in Fig. 7-9, with the tolerance limits indicated by the dashed lines. In general, the discharge coefficient is smaller for pipes less than 2 in in diameter, and the approximate behavior is indicated in Fig. 7-10. More precise values of the discharge coefficient for a venturi may be obtained by direct calibration, in which case accuracies of ± 0.5 percent may be obtained fairly easily. The recommended dimensions for ASME flow nozzles are shown in Fig. 7-7, and the discharge coefficients are shown in Fig. 7-11.

The recommended installations for concentric, thin-plate orifices are shown in Fig. 7-8. Note that three standard sets of pressure-tap locations are used:

1. Both pressure taps are installed in the flanges as shown.

D = Pipe diameter inlet and outlet
d = Throat diameter as required
$a = 0.25D$ to $0.75D$ for $4'' \leqq D \leqq 6''$, $0.25D$ to $0.50D$ for $6'' < D \leqq 32''$
$b = d$
$c = d/2$
$\delta = 3/16$ in to $1/2$ in according to D. Annular pressure chamber
  with at least four piezometer vents
$r_2 = 3.5d$ to $3.75d$
$r_1 = 0$ to $1.375D$
$\alpha_1 = 21° \pm 2°$
$\alpha_2 = 5°$ to $15°$

Figure 7-6 Recommended proportions of venturi tubes, according to Ref. [1].

Low β series: $\beta < 0.5$

$r_1 = d$
$r_2 = \frac{2}{3}d$
$L_t = 0.6d$
$1/8'' \gtrless t \gtrless 1/2''$
$1/8'' \gtrless t_2 \gtrless 0.15D$

High β series: $\beta > 0.25$

$r_1 = \frac{1}{2}D$
$r_2 = \frac{1}{2}(D-d)$
$L_t \gtrless 0.6d$ or $L_t \gtrless \frac{1}{3}D$
$2t \gtrless D-(d+1/8'')$
$1/8'' \gtrless t_2 \gtrless 0.15D$

Optional designs
of nozzle outlet

Figure 7-7 Recommended proportions of the ASME long-radius flow nozzle, according to Ref. [1].

2. The inlet pressure tap is located one pipe diameter upstream, and the outlet pressure tap is located one-half diameter downstream of the orifice as measured from the upstream face of the orifice.
3. The inlet pressure tap is located one pipe diameter upstream, and the outlet pressure tap is located at the vena contracta of the orifice as given by Fig. 7-12.

Figure 7-8 Recommended location of pressure taps for use with concentric, thin-plate, square-edged orifices, according to Ref. [1].

Figure 7-9 Discharge coefficients for the venturi tube shown in Fig. 7-6, according to Ref. [1]. Values are applicable for $0.25 < \beta < 0.75$ and $D > 2$ in.

Figure 7-10 Approximate venturi coefficients for various throat diameters, according to Ref. [15].

Figure 7-11 Discharge coefficients for ASME long-radius nozzles shown in Fig. 7-7, according to Ref. [1].

Figure 7-13 gives the values of the orifice flow coefficient for pipe sizes $1\frac{1}{4}$ to 3 in with pressure taps located according to case 2 above. Flow coefficients for other cases are given in Ref. [1].

The various flow coefficients are plotted as a function of Reynolds number, defined by

$$\mathrm{Re} = \frac{\rho u_m d}{\mu} \tag{7-21}$$

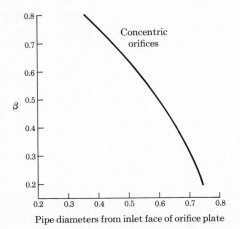

Pipe diameters from inlet face of orifice plate

Figure 7-12 Location of outlet pressure connections for orifices with vena contracta taps, according to Ref. [1].

Ratio of orifice diameter to pipe diameter, β

Figure 7-13 Flow coefficients for concentric orifices in pipes. Pressure taps one diameter upstream and one-half diameter downstream. Applicable for $1.25 < D < 3.00$ in. (*From Ref. [15]*.)

Figure 7-14 Adiabatic expansion factors for use with venturis and flow nozzles as calculated from Eq. (7-14). (*From Ref.* [2].)

where ρ = fluid density
μ = dynamic viscosity
u_m = mean flow velocity
d = diameter *at the particular section for which the Reynolds number is specified*

Note that some charts, viz., Fig. 7-13, base the Reynolds number on upstream conditions, while others, viz., Fig. 7-10, base it on throat conditions. The product ρu_m may be calculated from the mass flow according to

$$\dot{m} = \rho u_m A_c \tag{7-22}$$

where A_c is the cross-sectional area for the flow where u_m is measured. For a circular cross section, $A_c = \pi d^2/4$. Further information on orifice and venturi meters is contained in Refs. [33] to [37].

Figure 7-15 Expansion factors for square-edged orifices with pipe taps as calculated from Eq. (7-16). (*From Ref.* [2].)

Example 7-2 A venturi tube is to be used to measure a maximum flow rate of water of 50 gpm at 70°F. The throat Reynolds number is to be at least 10^5 at these flow conditions. A differential pressure gage is selected, which has an accuracy of 0.25 percent of full scale, and the upper scale limit is to be selected to correspond to the maximum flow rate. Determine the size of the venturi and the maximum range of the differential pressure gage, and estimate the uncertainty in the mass flow measurement at nominal flow rates of 50 and 25 gpm. Use either Fig. 7-9 or Fig. 7-10 to determine the discharge coefficient.

SOLUTION The properties of water are

$$\rho = 62.4 \text{ lbm/ft}^3 = 8.33 \text{ lbm/gal} \qquad \mu = 2.36 \text{ lbm/h} \cdot \text{ft}$$

From the given maximum flow rate and throat Reynolds number we may calculate the maximum allowable throat diameter:

$$\text{Re}_d = \frac{\rho u_m d}{\mu} = \frac{\dot{m}d}{(\pi d^2/4)\mu} = \frac{4\dot{m}}{\pi d\mu} = 10^5$$

The maximum flow rate is

$$\dot{m} = (50)(8.33)(60) = 2.5 \times 10^4 \text{ lbm/h } (3.027 \text{ kg/s})$$

so that

$$d_{\text{max}} = \frac{(4)(2.5 \times 10^4)}{\pi(10^5)(2.36)} = 0.135 \text{ ft} = 1.62 \text{ in } (4.11 \text{ cm})$$

We shall select a venturi with a 1.0-in throat diameter since we have a discharge-coefficient curve for this size in Fig. 7-10. The upstream pipe diameter is taken as 2.0 in. From Fig. 7-10 we estimate the discharge coefficient for this size venturi as 0.976 for $8 \times 10^4 < \text{Re}_d < 3 \times 10^5$. The uncertainty in this coefficient will be taken as ± 0.002 since Fig. 7-10 is a general set of curves. With this selection of venturi size, the maximum throat Reynolds number becomes

$$(\text{Re}_d)_{\text{max}} = (10^5)\left(\frac{1.62}{1.0}\right) = 1.62 \times 10^5$$

The minimum Reynolds number is thus half this value, or 8.1×10^4. The maximum pressure differential may be calculated with Eq. (7-17).

$$Q_{\text{actual}} = CMA_2 \sqrt{\frac{2g_c}{\rho}} \sqrt{\Delta p} \qquad (7\text{-}17)$$

or

$$\frac{(50)(231)}{(60)(1728)} = \frac{(0.976)\pi(1.0)^2}{(4)(144)\sqrt{1 - (\frac{1}{2})^2}} \sqrt{\frac{(2)(32.2)}{62.4}} \sqrt{\Delta p}$$

This yields

$$\Delta p = 948 \text{ psf} = 6.58 \text{ psi } (45.4 \text{ kPa})$$

Let us assume that a differential pressure gage with a maximum range of 1000 psf is at our disposal. In accordance with the problem statement the uncertainty in the pressure reading would be

$$w_{\Delta p} = \pm 2.5 \text{ psf } (119.7 \text{ Pa})$$

When the flow is reduced to 25 gpm, the pressure differential will be *one-fourth* of that at 50 gpm. To estimate the uncertainty in the flow measurement, we shall assume that the dimensions of the venturi are known exactly, as well as the density of the water. For the calculation we utilize Eq. (3-2). The quantities of interest are:

$$\frac{\partial Q}{\partial C} = MA_2 \sqrt{\frac{2g_c}{\rho}} \sqrt{\Delta p}$$

$$\frac{\partial Q}{\partial \Delta p} = \frac{CMA_2}{2\sqrt{\Delta p}} \sqrt{\frac{2g_c}{\rho}}$$

$$w_c = \pm 0.002$$

Thus,

$$\frac{w_Q}{Q} = \left[\left(\frac{w_c}{C} \right)^2 + \frac{1}{4} \left(\frac{w_{\Delta p}}{\Delta p} \right)^2 \right]^{1/2}$$

For $Q = 50$ gpm,

$$\frac{w_Q}{Q} = \left[\left(\frac{0.002}{0.976} \right)^2 + \frac{1}{4} \left(\frac{2.5}{948} \right)^2 \right]^{1/2}$$

$$= 0.002435 \qquad \text{or } 0.2435\%$$

For $Q = 25$ gpm,

$$\frac{w_Q}{Q} = \left[\left(\frac{0.002}{0.976} \right)^2 + \frac{1}{4} \left(\frac{2.5}{984/4} \right)^2 \right]^{1/2}$$

$$= 0.00566 \qquad \text{or } 0.566\%$$

Example 7-3 An orifice with pressure taps one diameter upstream and one-half diameter downstream is installed in a 2.00-in-diameter pipe and used to measure the same flow of water as in Example 7-2. For this orifice, $\beta = 0.50$. The differential pressure gage has an accuracy of 0.25 percent of full scale, and the upper scale limit is selected to correspond to the maximum flow rate. Determine the range of the pressure gage and the uncertainty in the flow-rate measurements at nominal flow rates of 50 and 25 gpm. Assume that the uncertainty in the flow coefficient is ± 0.002.

SOLUTION We first calculate the pipe Reynolds numbers. Using the properties from Example 7-2,

$$\text{Re}_d = \frac{(2.5 \times 10^4)(4)}{\pi(2.0/12)(2.36)} = 8.09 \times 10^4 \qquad \text{at 50 gpm } (5.26 \text{ cm}^3/\text{s})$$

$$\text{Re}_d = 4.05 \times 10^4 \qquad \text{at 25 gpm } (2.63 \text{ cm}^3/\text{s})$$

From Fig. 7-13 the flow coefficient is estimated as

$$K = 0.625 \qquad \text{at 50 gpm}$$

$$K = 0.630 \qquad \text{at 25 gpm}$$

The volumetric flow is

$$Q = \frac{(50)(231)}{(60)(1728)} = 0.1115 \text{ ft}^3/\text{s} \qquad \text{at 50 gpm}$$

$$Q = 0.0558 \text{ ft}^3/\text{s} \qquad \text{at 25 gpm}$$

The nominal values of the differential pressure are then calculated from Eq. (7-18) as

$$0.1115 = (0.625) \frac{\pi(1)^2}{(4)(144)} \sqrt{\frac{(2)(32.2)}{(62.4)}} \sqrt{\Delta p} \qquad \text{at 50 gpm}$$

$$\Delta p = 1037 \text{ psf} = 7.21 \text{ psi (49.7 kPa)} \qquad \text{at 50 gpm}$$

$$\Delta p = 255 \text{ psf} = 1.77 \text{ psi (12.2 kPa)} \qquad \text{at 25 gpm}$$

A suitable differential pressure gage might be one with a maximum range of 1200 psf (57.5 kPa). The same equation for uncertainty applies in this problem as in Example 7-2 except that the flow coefficient K is used instead of the discharge coefficient. Thus,

$$\frac{w_Q}{Q} = \left[\left(\frac{w_K}{K} \right)^2 + \frac{1}{4} \left(\frac{w_{\Delta p}}{\Delta p} \right)^2 \right]^{1/2}$$

with $w_K = 0.002$ and $w_{\Delta p} = (0.0025)(1200) = 3.0 \text{ psf (143.6 Pa)}$.

For $Q = 50$ gpm,

$$\frac{w_Q}{Q} = \left[\left(\frac{0.002}{0.625} \right)^2 + \frac{1}{4} \left(\frac{3.0}{1037} \right)^2 \right]^{1/2}$$

$$= 0.00351 \qquad \text{or } 0.351\%$$

For $Q = 25$ gpm,

$$\frac{w_Q}{Q} = \left[\left(\frac{0.002}{0.630} \right)^2 + \frac{1}{4} \left(\frac{3.0}{255} \right)^2 \right]^{1/2} = 0.00669 \qquad \text{or } 0.669\%$$

7-5 THE SONIC NOZZLE

All the obstruction meters discussed above may be used with gases. When the flow rate is sufficiently high, the pressure differential becomes quite large, and eventually sonic flow conditions may be achieved at the minimum flow area. Under these conditions the flow is said to be "choked," and the flow rate takes on its maximum value for the given inlet conditions. For an ideal gas with constant specific heats it may be shown that the pressure ratio for this choked condition, assuming isentropic flow, is

$$\left(\frac{p_2}{p_1} \right)_{\text{critical}} = \left(\frac{2}{\gamma + 1} \right)^{\gamma/(\gamma - 1)} \qquad (7\text{-}23)$$

This ratio is called the *critical pressure ratio*. Inserting this ratio in Eq. (7-8) gives for the mass flow rate

$$\dot{m} = A_2 p_1 \sqrt{\frac{2g_c}{RT_1}\left[\frac{\gamma}{\gamma+1}\left(\frac{2}{\gamma+1}\right)^{2/(\gamma-1)}\right]^{1/2}} \qquad (7\text{-}24)$$

Equation (7-24) is frequently applied to a nozzle when it is known that the pressure ratio p_2/p_1 is less than the critical value given by Eq. (7-23). Under these conditions the ideal flow is dependent only on the inlet stagnation conditions p_1 and T_1. These conditions are usually easy to measure, so that the sonic nozzle offers a convenient method for measuring gas flow rates. It may be noted, however, that a large pressure drop must be tolerated with the method. *Upstream stagnation conditions* must be used for p_1 and T_1 in the calculation.

The ideal sonic-nozzle flow rate given by Eq. (7-24) must be modified by an appropriate discharge coefficient which is a function of the geometry of the nozzle and other factors. There may be several complicating conditions, but discharge coefficients of about 0.97 are usually observed. A comprehensive survey of critical flow nozzles is presented by Arnberg [3], and the interested reader should consult this discussion for more information. More recent information is given in Refs. [32], [34], and [38].

The flow obstruction devices discussed above require the use of wall pressure taps. (Other devices also require the use of such taps.) Measurements with wall pressure taps can be subject to several influencing factors, which are discussed in detail by Rayle [19]. In general, the diameter of the pressure tap should be small in comparison with the diameter of the pipe.

Example 7-4 A sonic nozzle is to be used to measure a flow of air at 300 psia (2.07 MPa) and 100°F (37.8°C) in a 3-in-diameter pipe. The nominal flow rate is 1 lbm/s (0.46 kg/s). Calculate the throat diameter (nozzle size) such that just critical flow conditions are obtained.

SOLUTION We use Eq. (7-24) for this calculation with $\gamma = 1.4$ for air. The only unknown in this equation is A_2. Thus, we have

$$\dot{m} = CA_2 p_1 \sqrt{\frac{2g_c}{RT_1}\left[\frac{\gamma}{\gamma+1}\left(\frac{2}{\gamma+1}\right)^{2/(\gamma-1)}\right]^{1/2}}$$

$$1 = A_2(300)(144)\sqrt{\frac{(2)(32.2)}{(53.35)(560)}\left[\left(\frac{4.1}{2.4}\right)\left(\frac{2}{2.4}\right)^{2/0.4}\right]^{1/2}}$$

and $A_2 = 0.001078 \text{ ft}^2 = 0.1551 \text{ in}^2 \ (1.0 \text{ cm}^2)$.

The diameter at the throat is

$$d = \sqrt{\frac{4}{\pi}(0.1551)} = 0.444 \text{ in } (1.13 \text{ cm})$$

For the above calculation we have taken the given pressure and temperature as stagnation properties. The temperature would most likely be measured with a stagnation probe, so that $100°F$ is probably the stagnation temperature. The pressure would probably be measured by a static tap in the side of the pipe upstream from the nozzle, so that a static-pressure measurement is most likely the one which will be available. If the upstream pipe diameter is large enough, the static pressure will be very nearly equal to the stagnation pressure, and the error in the above calculation will be small. Let us examine the above situation, assuming that the 300 psia is a static-pressure measurement. The mass flow upstream is

$$m = \frac{p_{1s}}{RT_{1s}} A_1 u_1 \qquad (a)$$

where the subscript s denotes static properties. The velocity upstream may be written in terms of the stagnation temperature as

$$u_1 = \sqrt{2g_c c_p(T_{1_0} - T_{1s})} \qquad (b)$$

Combining Eqs. (a) and (b), we have

$$\dot{m} = \frac{p_{1s}}{RT_{1s}} A_1 \sqrt{2g_c c_p(T_{1_0} - T_{1s})} \qquad (c)$$

Taking $p_{1s} = 300$ psia and $T_{1_0} = 100°F = 560°R$, we may solve Eq. (c) for T_{1s}. The result is

$$T_{1s} \approx 560°R \qquad (d)$$

or the upstream velocity is so small that the stagnation properties are very nearly equal to the static properties. This result may be checked by calculating the upstream velocity from Eq. (a) using the result from Eq. (d). We obtain

$$u_1 = 14.1 \text{ ft/s } (4.3 \text{ m/s})$$

The pressure difference $(p_{1_0} - p_{1s})$ corresponding to this velocity would be only 0.031 psia, while the temperature difference $(T_{1_0} - T_{1s})$ would be $0.017°F$. Both of these values are negligible. It may be noted, however, that if the upstream pipe diameter were considerably smaller, say, 1.0-in diameter, it might be necessary to correct for a difference between the measured static pressure and the stagnation pressure that must be used in Eq. (7-24).

7-6 FLOW MEASUREMENT BY DRAG EFFECTS

The *rotameter* is a very commonly used flow-measurement device and is shown schematically in Fig. 7-16. The flow enters the bottom of the tapered vertical tube and causes the bob or "float" to move upward. The bob will rise to a point in the tube such that the drag forces are just balanced by the weight and buoyancy

Figure 7-16 Schematic of a rotameter.

forces. The position of the bob in the tube is then taken as an indication of the flow rate. The device is sometimes called an *area meter* because the elevation of the bob is dependent on the annular area between it and the tapered glass tube; however, the meter operates on the physical principle of drag, so that we choose to classify it in this category. A force balance on the bob gives

$$F_d + \rho_f V_b \frac{g}{g_c} = \rho_b V_b \frac{g}{g_c} \tag{7-25}$$

where ρ_f and ρ_b are the densities of the fluid and bob, V_b is the total volume of the bob, g is the acceleration of gravity, and F_d is the drag force, which is given by

$$F_d = C_d A_b \frac{\rho_f u_m^2}{2g_c} \tag{7-26}$$

C_d is a drag coefficient, A_b is the frontal area of the bob, and u_m is the mean flow velocity in the annular space between the bob and the tube.

Combining Eqs. (7-25) and (7-26) gives

$$u_m = \left[\frac{1}{C_d} \frac{2g V_b}{A_b} \left(\frac{\rho_b}{\rho_f} - 1 \right) \right]^{1/2} \tag{7-27}$$

or

$$Q = A u_m = A \left[\frac{1}{C_d} \frac{2g V_b}{A_b} \left(\frac{\rho_b}{\rho_f} - 1 \right) \right]^{1/2} \tag{7-28}$$

where A is the annular area and is given by

$$A = \frac{\pi}{4} \left[(D + ay)^2 - d^2 \right] \tag{7-29}$$

D is the diameter of the tube at inlet, d is the maximum bob diameter, y is the vertical distance from the entrance, and a is a constant indicating the tube taper.

The drag coefficient is dependent on the Reynolds number and hence on the fluid viscosity; however, special bobs may be used that have an essentially constant drag coefficient and thus offer the advantage that the meter reading will be essentially independent of viscosity. It may be noted that for many practical meters the quadratic area relation given by Eq. (7-29) becomes nearly linear for the actual dimensions of tube and bob that are used. Assuming such a linear relation, the equation for mass flow would become

$$\dot{m} = C_1 y \sqrt{(\rho_b - \rho_f)\rho_f} \tag{7-30}$$

where C_1 is now an appropriate meter constant.

It is frequently advantageous to have a rotameter that gives an indication that is independent of fluid density; i.e., we wish to have

$$\frac{\partial \dot{m}}{\partial \rho_f} = 0$$

Performing the indicated differentiation, we obtain

$$\rho_b = 2\rho_f \tag{7-31}$$

Figure 7-17 Photograph of a commercial rotameter. (*Courtesy of Fischer and Porter Co.*)

Figure 7-18 Turbine flowmeter. (1) Meter body; (2) inlet-straightening vanes; (3) turbine blades; (4) downstream afterbody to maintain smooth flow; (5) pipe or tube connector; (6) reluctance pickup. (*Courtesy of Fischer and Porter Co.*)

Figure 7-19 Calibration curve for 1-in turbine flowmeter of the type shown in Fig. 7-18. Calibration was performed with water.

and the mass flow is given by

$$\dot{m} = \frac{C_1 y \rho_b}{2} \tag{7-32}$$

Thus, by special bob construction the meter may be used to compensate for density changes in the fluid. The error in Eq. (7-32) is less than 0.2 percent for a fluid-density deviation of 5 percent from that given in Eq. (7-31). A photograph of a commercial rotameter is given in Fig. 7-17.

A popular type of flow-measurement device is the turbine meter shown in Fig. 7-18. As the fluid moves through the meter, it causes a rotation of the small turbine wheel. In the turbine-wheel body a permanent magnet is enclosed so that it rotates with the wheel. A reluctance pickup attached to the top of the meter detects a pulse for each revolution of the turbine wheel. Since the volumetric flow is proportional to the number of wheel revolutions, the total pulse output may be taken as an indication of total flow. The pulse rate is proportional to flow rate, and the transient response of the meter is very good. A flow coefficient K for the turbine meter is defined so that

$$Q = \frac{f}{K} \tag{7-33}$$

where f is the pulse frequency. The flow coefficient is dependent on flow rate and the kinematic viscosity of the fluid v. A calibration curve for a typical meter is given in Fig. 7-19. It may be seen that this particular meter will indicate the flow accurately within ± 0.5 percent over a rather wide range of flow rates. The use of the turbine flowmeter in pulsating flow is discussed in Ref. [31].

Example 7-5 Calculate the range of mass flow rates of liquid ammonia at 20°C for which the turbine meter of Fig. 7-19 would be within $\pm 1/2$ percent. Also determine a flow coefficient for this fluid in terms of cycles per kilogram.

SOLUTION From Fig. 7-19 the range for the $\pm 1/2$ percent calibration is approximately 55 to 700 cycles/s · cSt. From the appendix the properties of ammonia at 20°C are

$$\rho = 612 \text{ kg/m}^3$$

$$v = 0.036 \times 10^{-5} \text{ m}^2/\text{s}$$

The centistoke is defined from

$$1 \text{ stoke (St)} = 10^{-4} \text{ square meter per second (m}^2/\text{s)}$$

$$1 \text{ centistoke (cSt)} = 10^{-6} \text{ square meter per second (m}^2/\text{s)}$$

Therefore
$$v = \frac{0.036 \times 10^{-5}}{10^{-6}} = 0.26 \text{ cSt}$$

The frequency range is then

$$f_{\text{low}} = (55)(0.36) = 19.8 \text{ cycles/s}$$

$$f_{\text{high}} = (700)(0.36) = 252 \text{ cycles/s}$$

Also, 1 gal $= 231$ in^3, so that for ammonia

$$1 \text{ gal} = \frac{(231 \text{ in}^3)(612 \text{ kg/m}^3)}{(39.36 \text{ in/m})^3} = 2.318 \text{ kg}$$

The flow coefficient would then be

$$K = 1092 \text{ cycles/gal} = 471.1 \text{ cycles/kg}$$

and the range of flow rates for the 1/2 percent calibration would be

$$\dot{m}_{\text{low}} = \frac{19.8 \text{ cycles/s}}{471.1 \text{ cycles/kg}} = 0.042 \text{ kg/s}$$

$$\dot{m}_{\text{high}} = \frac{252}{471.1} = 0.535 \text{ kg/s}$$

7-7 HOT-WIRE AND HOT-FILM ANEMOMETERS

The hot-wire anemometer is a device that is often used in research applications to study rapidly varying flow conditions. A fine wire is heated electrically and placed in the flow stream, and the early work of King [5] has shown that the heat-transfer rate from the wire can be expressed in the form

$$q = (a + bu^{0.5})(T_w - T_\infty) \tag{7-34}$$

where T_w = wire temperature
T_∞ = free-stream temperature of fluid
u = fluid velocity
a, b = constants obtained from a calibration of the device

The heat-transfer rate must also be given by

$$q = i^2 R_w = i^2 R_0 [1 + \alpha(T_w - T_0)] \tag{7-35}$$

where i = electric current
R_0 = resistance of the wire at the reference temperature T_0
α = temperature coefficient of resistance

For measurement purposes the hot wire is connected to a bridge circuit as shown in Fig. 7-20. The current is determined by measuring the voltage drop across the standard resistance R_s, and the wire resistance is determined from the bridge circuit. For steady-state measurements the null condition may be used, while an oscillograph output may be used for transient measurements. With i and R_w determined, the flow velocity may be calculated with Eqs. (7-34) and (7-35).

Figure 7-20 Schematic of hot-wire flow-measurement circuit.

Hot-wire probes have been used extensively for measurement of transient flows, especially measurements of turbulent fluctuations. Time constants of the order of 1 ms may be obtained with 0.0001-in-diameter platinum or tungsten wires operating in air.

When the hot wire is to be employed for measurement of rapidly changing flow patterns, full account must be taken of the transient response of both the thermal and electrical resistance characteristics of the wire. Two types of electrical compensation are employed in practice: (1) a constant-current arrangement, where a large resistance is connected in series with the hot wire and a thermal compensating circuit is then applied to the output ac voltage, and (2) a constant-temperature arrangement, where a feedback control circuit is added to vary the current so that the wire temperature remains nearly constant. Response of the wire depends on the angle the flow velocity makes with the wire axis, and techniques have been developed to take this effect into account [6], [44]. The length-to-diameter ratio of the wire (L/d) also has a significant effect on the measurement performance. L/d has a value of about 50 for typical hot wires.

A modification of the hot-wire method consists of a small insulating cylinder that is coated with a thin metallic film. This device is appropriately called a *hot-film probe*, and a sketch of a typical probe (not to scale) is shown in Fig. 7-21 from Ref. [45]. The elements of the probe illustrated in this figure are:

a. Metallic leg of sensor holder, which acts as an electrical terminal
b. Epoxy coating
c. Gold bonding for electrical and mechanical connections
d. Electroplated gold layer, about 5 μm thick
g. Glass tube, about 25 to 100 μm diameter

Figure 7-21 Construction of typical hot-film probe, according to Ref. [45].

The typical coating thicknesses are those employed by one manufacturer, Thermal Systems, Inc., Minneapolis, Minn.

Hot-film probes are extremely sensitive to fluctuations in the fluid velocity and have been used for measurements involving frequencies as high as 50,000 Hz. Sophisticated electronic instrumentation is required for such measurements, and the interested reader is referred to Refs. [6] and [44] for additional information, as well as to the literature of such manufacturers as Thermal Systems and DISA, Inc.

Turbulence Measurements

In turbulent fluid flow, whether in a channel or in a boundary layer, the fluid-velocity components exhibit a random oscillatory character, which depends on the average fluid velocity, fluid density, viscosity, and other variables. The instantaneous-velocity components are expressed as

$$u = \bar{u} + u' \qquad (x \text{ component})$$
$$v = \bar{v} + v' \qquad (y \text{ component})$$
$$w = \bar{w} + w' \qquad (z \text{ component})$$

where the bar quantities are the integrated average-velocity components and the primed quantities represent the fluctuations from the average velocity. Figure

Time t

Figure 7-22 Turbulent fluctuations in direction of flow.

7-22 illustrates a typical plot of turbulent fluctuations. One measure of the *intensity of turbulence* is the root mean square of the fluctuations. For the x component

$$u'_{rms} = (\overline{u'^2})^{1/2} = \left[\frac{1}{T} \int_0^T u'^2 \, dt \right]^{1/2} \tag{7-36}$$

where T is some time period that is large compared to the turbulence scale.

Turbulent fluctuations occur over an extremely broad range. Some typical values are:

Background turbulence in well-designed wind tunnels	0.05%
Turbulence created by grids	0.2–2%
Turbulent wakes	2–5%
Turbulent boundary layers and pipe flows	3–20%
Turbulent jets	20–100%

The average kinetic energy (KE) of turbulence per unit mass may be taken as a measure of the total turbulence intensity of the flow stream. Thus,

$$\frac{\text{Turbulence KE}}{\text{Mass}} = \frac{1}{2} (\overline{u'^2} + \overline{v'^2} + \overline{w'^2}) \tag{7-37}$$

The hot-wire or hot-film anemometer is very useful for turbulence measurements because it can respond to very rapid changes in flow velocity. Two or more wires at one point in the flow can make simultaneous measurements of the fluctuating components. Various mean values and energy spectra can then be computed from the voltage outputs of the probes. In general, these operations can be performed electronically.

7-8 MAGNETIC FLOWMETERS

Consider the flow of a conducting fluid through a magnetic field as shown in Fig. 7-23. Since the fluid represents a conductor moving in the field, there will be an induced voltage according to

$$E = BLu \times 10^{-8} \text{ V} \tag{7-38}$$

where B = magnetic flux density, gauss
u = velocity of the conductor, cm/s
L = length of the conductor, cm

The length of the conductor is proportional to the tube diameter, and the velocity is proportional to the mean flow velocity. The two electrodes detect the induced voltage, which may be taken as a direct indication of flow velocity. The construction of a commercial magnetic flowmeter is shown in Fig. 7-24.

Figure 7-23 Flow of conducting fluid through a magnetic field.

Two types of magnetic flowmeters are used commercially. One type has a nonconducting pipe liner and is used for fluids with low conductivities, like water. The electrodes are mounted so that they are flush with the nonconducting liner and make contact with the fluid. Alternating magnetic fields are normally used with these meters, since the output is low and requires amplification. The second type of magnetic flowmeter is one which is used with high-conductivity fluids,

Figure 7-24 Construction of commercial magnetic flowmeter for measurement of low-conductivity fluids. (*Courtesy of Fischer and Porter Co.*)

principally liquid metals. A stainless-steel pipe is employed in this case, with the electrodes attached directly to the outside of the pipe and diametrically opposed to each other. The output of this type of meter is sufficiently high that it may be used for direct readout purposes. The meter shown in Fig. 7-24 is one which is used for low-conductivity fluids.

7-9 FLOW-VISUALIZATION METHODS

Fluid flow is a complicated subject with many areas that have not yet yielded to precise analytical techniques. Accordingly, flow-measurement problems are not always simple and precise because of the lack of analytical relations to use for calculation and reduction of experimental data. The interpretation of data involving turbulence or measurements of complicated boundary layer, viscous, and stock-wave effects is not easy. Frequently, the flow may be altered as a result of probes that are inserted to measure the pressure, velocity, and temperature profiles, so that the experimentalist is uncertain about the effect that has been measured. Flow visualization by optical methods offers the advantage that when properly executed it does not disturb the fluid stream and thus gives the experimentalist an extra tool to use in conjunction with other measurement devices. In some instances flow-visualization techniques may be employed for rather precise measurements of important flow parameters, while in other cases they may serve only to furnish qualitative information regarding the overall flow behavior. In the following paragraphs we shall discuss the principles of some of the basic flow-visualization schemes and indicate their applications. The interested reader should consult Refs. [8] and [17] for a rather complete survey of the subject. Recent advances are presented in Ref. [47].

Consider the gaseous flow field shown in Fig. 7-25. The flow is in a direction perpendicular to the figure, i.e., the z direction. An incoming light ray is deflected

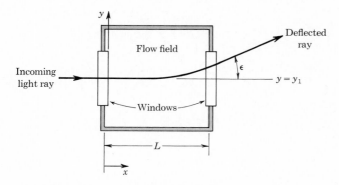

Figure 7-25 Basic optical effects used for flow visualization.

through an angle ϵ as a result of density gradients in the flow. It may be shown†
that the deflection angle for small-density gradients is given as

$$\epsilon = \frac{L}{n_1}\left(\frac{dn}{dy}\right)_{y=y_1} = \frac{L\beta}{\rho_s}\left(\frac{d\rho}{dy}\right)_{y=y_1} \tag{7-39}$$

where L is the width of the flow field, ρ is the local fluid density, ρ_s is a reference
density which is usually taken at standard conditions, and n is the index of
refraction, which for gases may be written as

$$n = \left(1 + \beta\frac{\rho}{\rho_s}\right)n_1 \tag{7-40}$$

β is a dimensionless constant having a value of about 0.000292 for air; n_1 is the
index of refraction outside the flow field and can be taken as very nearly unity in
Eq. (7-39).

According to Eq. (7-39) the angular deflection of the light ray ϵ is proportional to the density gradient in the flow. This is the basic optical effect that is used
for flow-visualization work. It may be noted that the deflection of the light ray is
a measure of the average density gradient integrated over the x coordinate. Thus,
the effect is primarily useful for indicating density variations in two dimensions
(in this case the y and z dimensions) and will average the variations in the third
dimension.

In the following sections we shall discuss several optical methods of flow
visualization for use in gas systems. For liquid flow visualization a typical experimental technique is to add a dye to the liquid in order to study the flow phenomena. Another technique which has received considerable attention for use with
liquids is the so-called hydrogen-bubble method. The technique consists of using
a fine wire, placed in water, as one end of a dc circuit to electrolyze the water.
Very small hydrogen bubbles are generated in the liquid. The motion of the

† See, for example, Ref. [10], p. 156.

Figure 7-26 Voltage pulse circuits for hydrogen-bubble method, according to Ref. [16].

bubbles may be studied by illuminating the flow. The application of this method has been given a very complete description by Schraub, Kline, et al. [16]. Figure 7-26 indicates the voltage-discharge circuits employed in the technique. A variable-input voltage from 10 to 250 V is provided and rectified in the diode circuit. A relay is used to periodically charge and discharge the capacitor, and a variable resistor is provided in the charging circuit to tune the system for optimum performance for each wire diameter and flow condition. In some cases it is necessary to add a small amount of sodium sulfate to the water in order to achieve a satisfactory electrolyte concentration.

7-10 THE SHADOWGRAPH

The shadow technique is a method for direct viewing of flow phenomena. Imagine the flow field as shown in Fig. 7-27 with a density gradient in the y direction. The parallel light rays enter the test section as shown. In the regions where there is no density gradient, the light rays will pass straight through the test section with no deflection. For the regions where a gradient exists, the rays will be deflected. The net effect is that the rays will bunch together after leaving the test section to form bright spots and dark spots. The illumination will depend on the *relative deflection* of the light rays $d\epsilon/dy$, and hence on $d^2\rho/dy^2$. The illumination on a screen placed outside the test section is thus dependent on the second derivative of the density at the particular point.

The shadowgraph is a very simple optical tool, and its effects may be viewed in several everyday phenomena using only the naked eye and local room lighting. The free-convection boundary layer on a horizontal electric hot plate is clearly visible when viewed from the edge. This phenomenon is visible because of the density gradients that result from the heating of the air near the hot surface. It is almost fruitless to try to evaluate local densities using shadow photography; however, the shadowgraph is exceedingly useful for viewing turbulent flow

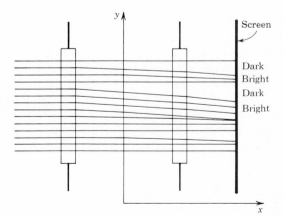

Figure 7-27 Shadowgraph flow-visualization device.

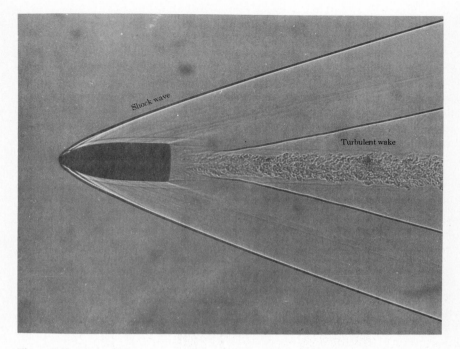

Figure 7-28 Direct shadowgram of flight of a 0.220-caliber Remington Swift shell. Velocity = 1250 m/s; free-stream pressure is atmospheric. Exposure time was 0.13 μs. Note clear definition of shock wave and turbulent wake. (*Courtesy of ARO, Inc., Arnold Air Force Base, Tenn.*)

regions, and the method can be used to establish the location of shock waves with high precision. Figure 7-28 illustrates shock-wave and turbulent-flow phenomena as viewed with a shadowgraph.

7-11 THE SCHLIEREN

While the shadowgraph gives an indication of the second derivative of density in the flow field, the schlieren is a device which indicates the density gradient. Consider the schematic diagram shown in Fig. 7-29. Light from a slit source ab is collimated by the lens L_1 and focused at plane 1 in the test section. After it passes through lens L_2, there is produced an inverted image of the source at the focal plane 2. Lens L_3 then focuses the image of the test section on the screen at plane 3. Now let us consider the imaging process in more detail. The pencils of light originating at point a occupy a different portion of the various lenses from those originating from point b or any other point in the slit source. The regions in which these pencils overlap are shown in Fig. 7-29. Note that *all* pencils of light pass through the image plane cd in the test section and the source image plane $b'a'$. An image of the test section at $d'c'$ is then uniformly illuminated, since the image at $b'a'$ is uniformly illuminated. This means that all points in the plane $b'a'$

are affected in the same manner by whatever fluid effects may take place in the test section.

If the test section is completely uniform in density, the pencils of light appear as shown in Fig. 7-29; a pencil originating at point c is deflected by the same amount as a pencil originating at point d. This is consistent with the observation that all pencils originating in plane cd completely fill the image plane $b'a'$. Now consider the effect of the introduction of an obstruction at plane $b'a'$ under these circumstances. We immediately conclude that such an obstruction would uniformly decrease the illumination on the screen by a factor proportional to the amount of the area $b'a'$ intercepted.

Suppose now that a density gradient exists at the test section focal plane cd. This means that all pencils of light originating in this plane would no longer fill the image plane $b'a'$ completely. If, now, an obstruction is placed at plane $b'a'$, it will intercept more light from some points in the test section plane than from others, resulting in light and dark regions on the screen at plane 3. The obstruction is called a knife edge, and the resultant variation in illumination on the screen is called the *schlieren effect*.

Let us examine the variation in illumination in more detail. In Fig. 7-29b the total height of the source image is y and the portion not intercepted by the knife edge is y_1. Thus, the general illumination on the screen I is proportional to y_1. An angular displacement of a pencil of light in plane 1 is ϵ. This produces a vertical deflection at plane 2 of

$$\Delta y = f_2 \epsilon \qquad (7\text{-}41)$$

where f_2 is the focal length of lens L_2. As a result of this deflection, there is a fractional change in illumination on the screen. The contrast at any point on the

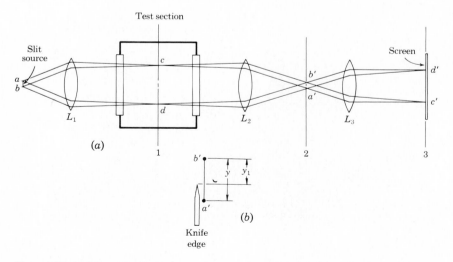

Figure 7-29 (*a*) Schematic of schlieren flow visualization; (*b*) detail of knife edge.

Figure 7-30 Schlieren photograph of 4.45-cm-diameter blunt cone and wake after separation of rear portion. Velocity = 1792 m/s; free-stream pressure = 12.5 kN/m²; cone angle = 18°. (*Courtesy of ARO, Inc., Arnold Air Force Base, Tenn.*)

screen may be defined as the ratio of the fractional change in illumination to the general illumination, or

$$C = \frac{\Delta I}{I} = \frac{\Delta y}{y_1} = \frac{f_2 \epsilon}{y_1} \tag{7-42}$$

The angular deflection is given by Eq. (7-39), so that the expression for contrast may be written

$$C = \frac{f_2 L \beta}{y_1 \rho_s} \left(\frac{d\rho}{dy}\right)_{cd} \tag{7-43}$$

Thus, the contrast on the screen is directly proportional to the density gradient in

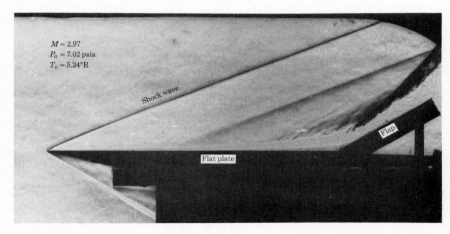

M = 2.97
P_o = 7.02 psia
T_o = 5.24°R

Figure 7-31 Schlieren photograph of flow-separation phenomena. Plate length = 20.3 cm; flap angle = 30°. Note clear definition of shock wave and turbulent flow region at the intersection of the flap and horizontal plate. (*Courtesy of ARO, Inc., Arnold Air Force Base, Tenn.*)

the flow. It may be observed that the contrast may be increased by reducing the distance y_1, that is, by intercepting more light at the source image plane. This also reduces the general illumination, so that the contrast may not be increased indefinitely and a compromise must be accepted. Schlieren photographs are used extensively for location of shock waves and complicated boundary-layer phenomena in supersonic flow systems. Some typical schlieren photographs are shown in Figs. 7-30 and 7-31. In actual practice most schlieren systems use mirrors instead of lenses for reasons of economy.

7-12 THE INTERFEROMETER

The Mach-Zehnder interferometer is the most precise instrument for flow visualization. Consider the schematic representation in Fig. 7-32. The light source is collimated through lens L_1 onto the splitter plate S_1. This plate permits half of the light to be transmitted to mirror M_2 while reflecting the other half toward mirror M_1. Beam 1 passes through the test section, while beam 2 travels an alternative path of approximately equal length. The two beams are brought together again by means of the splitter plate S_2 and eventually focused on the screen. Now, if the two beams travel paths of different *optical lengths*, either because of the geometry of the system or because of the refractive properties of any element of the optical paths, the two beams will be out of phase and will interfere when they are joined together at S_2. There will be alternate bright and dark regions called *fringes*. The number of fringes will be a function of the difference in optical path lengths for the two beams; for a difference in path lengths of one wavelength there will be one fringe, two fringes for a difference of two wavelengths, and so on.

Figure 7-32 Schematic of Mach-Zehnder interferometer.

The interferometer is used to obtain a direct measurement of density variations in the test section. If the density in the test section (i.e., beam 1) is different from that in beam 2, there will be a change in the refractive properties of the fluid medium. If the medium in the test section has the same optical properties as the medium in beam 2, there will be no fringe shifts except those resulting from the geometric arrangement of the apparatus. These fringe shifts may be neutralized by appropriate movements of the mirrors and splitter plates. Then, the appearance of fringes on the screen may be directly related to changes in density in the flow field within the test section by utilizing the following analysis.

The change in optical path in the test section resulting from a change in refractive index is

$$\Delta L = L(n - n_0) \tag{7-44}$$

where L is the thickness of the flow field in the test section. Using Eq. (7-40), the change in optical path may be related to change in density for gases by

$$\Delta L = \beta L \frac{\rho - \rho_0}{\rho_s} \tag{7-45}$$

The number of fringe shifts N is then given by

$$N = \frac{\Delta L}{\lambda} = \frac{\beta L}{\lambda} \frac{\rho - \rho_0}{\rho_s} \tag{7-46}$$

Figure 7-33 Interferometer photograph of a free-convection boundary layer on a vertical flat plate. Fluid is air. Each fringe line represents a line of constant density or a line of constant temperature. (*Photograph courtesy of E. Soehngen.*)

Figure 7-34 Interferometer photograph of the inter-action of free-convection boundary layers on three horizontal heated cylinders. Fluid is air, and each fringe line represents a line of constant temperature. (*Photograph courtesy of E. Soehngen.*)

where λ is the wavelength of the light. In Eq. (7-46) it is to be noted that $\rho - \rho_0$ represents the change in density from the zero-fringe condition. The subscript 0 refers to the zero fringe condition, i.e., conditions in the path followed by beam 2 in Fig. 7-32. ρ_s is the reference density at standard conditions.

The interferometer gives a direct quantitative indication of density changes in the test section, but these changes are represented as integrated values over the entire thickness of the flow field. It is applicable to a wide range of flow conditions ranging from the low-speed (~ 30 cm/s) flow in free-convection boundary layers to shock-wave phenomena in supersonic flow. Figures 7-33 and 7-34 show some typical interferometer photographs. The captions explain the flow phenomena.

We have already given the frequency characteristics of some typical mono-chromatic light sources in Table 5-1. The mercury source is one commonly employed for interferometer work.

7-13 THE LASER DOPPLER ANEMOMETER (LDA)

We have seen how optical flow-visualization methods offer the advantage that they do not disturb the flow during the measurement process. The laser ane-mometer is a device that offers the nondisturbance advantages of optical methods while affording a very precise quantitative measurement of velocities. The device is capable of rapid response and is suitable for measurement of high-frequency turbulence fluctuations.

(a)

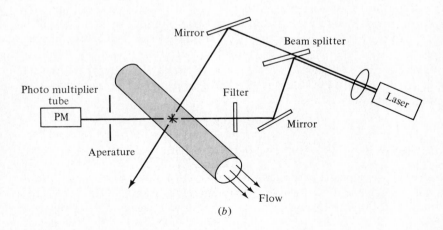

(b)

Figure 7-35 Schematic of laser-anemometer flow-measurement system.

where p_0 is the stagnation pressure, p_∞ is the free-stream static pressure, and M_∞ is the free-stream Mach number given by

$$M_\infty = \frac{u_\infty}{a} \qquad (7\text{-}48)$$

a is the acoustic velocity and may be calculated with

$$a = \sqrt{\gamma g_c RT} \qquad (7\text{-}49)$$

for an ideal gas. It is convenient to express Eq. (7-47) in terms of the dynamic pressure q defined by

$$q = \tfrac{1}{2}\rho u_\infty^2 = \tfrac{1}{2}\gamma p M_\infty^2 \qquad (7\text{-}50)$$

Equation (7-47) thus becomes

$$p_0 - p_\infty = \frac{2q}{\gamma M_\infty^2}\left[\left(1 + \frac{\gamma - 1}{2}M_\infty^2\right)^{\gamma/(\gamma-1)} - 1\right] \qquad (7\text{-}51)$$

This relation may be simplified to

$$p_0 - p_\infty = \frac{2q}{\gamma M_\infty^2}\left(1 + \frac{M_\infty^2}{4} + \frac{2-\gamma}{24}M_\infty^4 + \cdots\right) \qquad (7\text{-}52)$$

when

$$M_\infty^2\left(\frac{\gamma - 1}{2}\right) < 1.$$

For very small Mach numbers Eq. (7-52) reduces to the familiar incompressible flow formula

$$p_0 - p_\infty = \tfrac{1}{2}\rho u_\infty^2 \qquad (7\text{-}53)$$

We thus observe that a measurement of the static and stagnation pressures permits a determination of the flow velocity, by either Eq. (7-53) or Eq. (7-51), depending on the fluid medium.

A basic total-pressure probe may be constructed in several different ways, as shown in Fig. 7-37. In each instance the opening in the probe is oriented in a direction exactly parallel to the flow when a measurement of the total stream pressure is desired. If the probe is inclined at an angle θ to the free-stream velocity, a somewhat lower pressure will be observed. This reduction in pressure is indicated in Fig. 7-37 according to Ref. [7]. Configuration a represents an open-ended tube placed in the flow. Configuration b is called a shielded probe and consists of a venturi-shaped tube placed in the flow with an open-ended tube at the throat of the section to sense the stagnation pressure. It may be noted that this probe is rather insensitive to flow direction. Configuration c represents an open-ended tube with a chamfered opening. The chamfer is about 15°, and the ratio of OD to ID of the tube is about 5. Configuration d represents a tube having a small hole drilled in its side, which is placed normal to the flow direction. This type of probe, as might be expected, is the most sensitive to changes in yaw angle. Also indicated in Fig. 7-37 is a portion of the curve for a Kiel probe,

Figure 7-37 Stagnation pressure response of various probes to changes in yaw angle. (*a*) Open-ended tube; (*b*) channel tube; (*c*) chamfered tube; (*d*) tube with orifice in side. (*From Ref.* [7].)

which is similar in construction to configuration *b* except that a smoother venturi shape is used, as shown in Fig. 7-38. The Kiel probe is the least sensitive to yaw angle.

The measurement of static pressure in a flowstream is considerably more difficult than the measurement of stagnation pressure. A typical probe used for the measurement of both static and stagnation pressures is the *Pitot tube* shown in Fig. 7-39. The opening in the front of the probe senses the stagnation pressure, while the small holes around the outer periphery of the tube sense the static pressure. The static-pressure measurement with such a device is strongly dependent on the distance of the peripheral openings from the front opening as well as on the yaw angle. Figure 7-40 indicates the dependence of the static-pressure indication on the distance from the leading edge of the probe for both blunt subsonic and conical supersonic configurations. To alleviate this condition, the static-pressure holes are normally placed at least eight diameters downstream from the front of the probe. The dependence of the static and stagnation pressures on yaw angle for a conventional Pitot tube is indicated in Fig. 7-41. This device is quite sensitive to flow direction. The probe in Fig. 7-39 is sometimes called a *Pitot static tube* because it measures both static and stagnation pressure.

Figure 7-38 Kiel probe (Model 3696) for measurement of stagnation pressure. (*Courtesy of Airflo Instrument Co., Glastonbury, Conn.*)

Four to eight holes 0.04″
diameter equally spaced

Figure 7-39 Schematic drawing of Pitot tube.

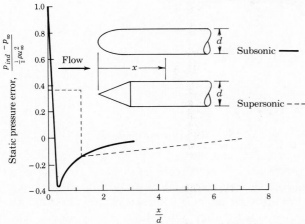

Figure 7-40 Variation of static pressure along standard subsonic and supersonic probe types. (*From Ref.* [*4*].)

Figure 7-41 Variation of static, stagnation, and dynamic pressures with yaw angle for Pitot tube. (*Courtesy of Airflo Instrument Co., Glastonbury, Conn.*)

275

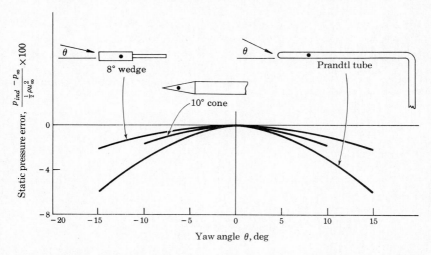

Figure 7-42 Yaw-angle characteristics of various static-pressure probes. (*From Ref.* [*7*].)

The static-pressure characteristics of three types of probes are shown in Figs. 7-42 and 7-43 as functions of Mach number and yaw angle. It may be noted that both the wedge and the Prandtl tube indicate static-pressure values that are too low, while the cone indicates a value that is too high. The wedge is least sensitive to yaw angle. All three probes have two static-pressure holes located 180° apart.

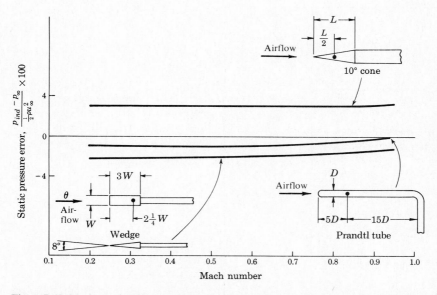

Figure 7-43 Mach-number characteristics of various static-pressure probes. (*From Ref.* [*7*].)

Example 7-7 A Pitot tube is inserted in a flowstream of air at 30°C and 1.0 atm. The dynamic pressure is measured as 1.12 in water when the tube is oriented parallel to the flow. Calculate the flow velocity at that point.

SOLUTION We use Eq. (7-53) for this calculation. The air density is calculated as

$$\rho = \frac{p_\infty}{RT_\infty} = \frac{1.0132 \times 10^5}{(287)(303)} = 1.165 \text{ kg/m}^3$$

We also have

$$p_0 - p_\infty = 1.12 \text{ in water} = 5.82 \text{ psf} = 278.7 \text{ Pa}$$

so that the velocity is

$$u_\infty = \sqrt{\frac{2(p_0 - p_\infty)}{\rho}} = \left[\frac{(2)(278.7)}{1.165}\right]^{1/2}$$

$$= 21.9 \text{ m/s } (71.8 \text{ ft/s})$$

7-16 IMPACT PRESSURE IN SUPERSONIC FLOW

Consider the impact probe shown in Fig. 7-44 which is exposed to a free stream with supersonic flow; that is, $M_1 > 1$. A shock wave will be formed in front of the probe as shown, and the total pressure measured by the probe will not be the free-stream total pressure before the shock wave. It is possible, however, to express the impact pressure at the probe in terms of the free-stream static pressure and the free-stream Mach number. The resulting expression as given in Ref. [10] is

$$\frac{p_\infty}{p_{0_2}} = \frac{\{[2\gamma/(\gamma + 1)]M_1^2 - (\gamma - 1)/(\gamma + 1)\}^{1/(\gamma - 1)}}{\{[(\gamma + 1)/2]M_1^2\}^{\gamma/(\gamma - 1)}} \tag{7-54}$$

where p_∞ is the free-stream static pressure and p_{0_2} is the measured impact pressure behind the normal shock wave. This equation is valid for Reynolds numbers based on probe diameter greater than 400. Equation (7-54) is called the *Rayleigh supersonic Pitot formula*. We see that in order to determine the value of the Mach number it is necessary to have a measurement of the free-stream static pressure. It is possible to make this measurement with special calibrated probes.

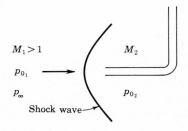

$M_1 > 1$

p_{0_1}

p_∞

Shock wave

M_2

p_{0_2}

Figure 7-44 Impact tube in supersonic flow.

Table 7-1 Operating characteristics of several types of flowmeters, adapted from Ref. [11]

Meter type		Maximum flow range cm^3/s $\times$ 10^{-3}	Useful range ability	Maximum pressure, MPa	Temperature range, °C	Maximum viscosity, cSt (10^{-6} m^2/s)
Obstruction meters	Orifice	Liquid, 0.012–220 Gas 23–43,000	5 : 1	42	−270 to 1100	4000
	Flow nozzle	Liquid 0.031–950 Gas 45–240,000	5 : 1	10	−50 to 800	4000
	Venturi	Liquid 0.03–950 Gas 45–240,000	5 : 1	10	−50 to 800	4000
Drag effects	Glass-tube rotameter	Liquid 0.1 $\times$ 10^{-3}–16 Gas 0.006–330	10 : 1	2	−45 to 200	100
	Metal-tube rotameter	Liquid 0.03–250 Gas 0.2–470	10 : 1	35	−180 to 870	100
	Turbine or propeller	Liquid 0.2 $\times$ 10^{-3}–3100 Gas 0.2–140,000	15 : 1 10 : 1	105	−270 to 540	30
Positive displacement		Clean liquid 0.02–63	10 : 1	1.0	−15 to 150	2000
Magnetic flowmeter		Conductive liquid only 0.1 $\times$ 10^{-3}–3100	20 : 1	4	−130 to 180	1000

(Fluid factors heading spans: Useful range ability, Maximum pressure, Temperature range, Maximum viscosity columns)

7-17 SUMMARY

Comparisons of the operating range, characteristics, and advantages of several flowmeters have been presented by Miesse and Curth [11]. These comparisons are presented in Table 7-1 and may be taken as an appropriate summary of our discussions in this chapter. The reader should bear in mind that the stated accuracies for the various flowmeters may be improved with suitable calibration. Of course, the overall accuracy of a flow-rate determination is also dependent on the accuracy of the readout equipment. As an example, a venturi might be carefully calibrated within 0.5 percent, but if it were used with a crude instrument for differential pressure measurements, a much poorer precision would result than that for which it was calibrated.

				Installation factors				
	Application factors							
Scale	Indication	Standard accuracy, %	Construction materials	Pressure loss, kPa	Line size, cm	Special piping considerations	Typical power requirement	Relative cost
Square root rate	Remote differential pressure	1–2 full-scale differential pressure	Most metals	0.7–200	4–30	Straight pipe: 10 diameters upstream, 3 downstream	None	1.0
Square root rate	Remote differential pressure	1–2 full-scale differential pressure	Bronze, iron steel, stainless steel	0.7–140	2–60	Straight pipe: 10 diameters upstream, 3 downstream	None	1.4
Square root rate	Remote differential pressure	1–2 full-scale differential pressure	Bronze, iron, steel, stainless steel, plastic	0.7–100	2–60	Straight pipe: 10 diameters upstream, 3 downstream	None	1.5
Linear rate	Local or remote electric or pneumatic	1–2 full-scale	Most metals, plastics, and ceramics	0.05–7	0.6–10	Vertical only	None	1.0
Linear rate	Local or remote electric or pneumatic	1–2 full-scale	Most metals, plastics, and ceramics	0.3–70	1.2–30	Vertical only	None	1.2
Linear total	Remote electric	0.5 reading	Aluminum, stainless steel	14–70	0.3–90	Any position	115 V, 60 Hz, 50 W	5.0
Linear total	Local or remote electric	1 reading	Bronze, iron, steel, stainless steel	7–140	1.2–15	Horizontal recommended	None	1.0
Linear rate	Remote electric	0.5–1 full-scale	Plastic, stainless steel with non-conductive liner	Very small	0.2–200	Any position	115 V, 60 Hz, 200 W	6.0

REVIEW QUESTIONS

7-1 What basic methods are used for calibration of flow-measurement devices?

7-2 What is meant by positive-displacement flowmeter?

7-3 What are the relative advantages of the venturi, orifice, and flow-nozzle meters?

7-4 What type of flow-measurement accuracy would you expect with an orifice; a venturi?

7-5 What is a sonic nozzle? How is it used? What are its advantages and disadvantages?

7-6 Why is a rotameter called a drag meter? Could it also be called an area meter?

7-7 What is the utility of the hot-wire anemometer?

7-8 Distinguish between the shadowgraph, schlieren, and interferometer flow-visualization techniques. What basic flow variable is measured in each technique?

7-9 Upon what does the sensitivity of the schlieren depend?

7-10 What is the primary advantage of the laser anemometer?

7-11 What is a Pitot tube?

7-12 Why must the tube be tapered in order for a rotameter to indicate flow rate?

7-13 What particular flow-measurement situations are adapted to the hot-wire anemometer?

7-14 How would one go about calibrating a venturi for liquid measurement? Could the calibration data be adapted to gas flows? If so, how?

PROBLEMS

7-1 Using Eq. (7-8) as a starting point, obtain Eqs. (7-9) and (7-10). Subsequently, calculate the error in these equations when $\Delta p = p_1/4$ for Eq. (7-9) and $\Delta p = p_1/10$ for Eq. (7-10).

7-2 In the compressible flow equation for a venturi the velocity of approach factor M cancels with a like term in the expansion factor Y_a. Construct a graphical plot which will indicate the error that would result if the approach factor given by Eq. (7-11) were used in conjunction with Eq. (7-8). Use the parameter $\Delta p/\gamma p_1$ as the abscissa for this plot.

7-3 A venturi is to be used for measuring the flow of air at 300 psia and 80°F. The maximum flow rate is 1.0 lbm/s. The minimum flow rate is 30 percent of this value. Determine the size of the venturi such that the throat Reynolds number is not less than 10^5. Calculate the differential pressure across the venturi for mass flows of 0.3, 0.5, 0.7, and 1.0 lbm/s. Assume $\beta = 0.5$ for the venturi.

7-4 Work Prob. 7-3 for an orifice with pressure taps one diameter upstream and one-half diameter downstream.

7-5 Work Prob. 7-3 for an ASME long-radius flow nozzle.

7-6 A sonic nozzle is to be used to measure a gas flow with an uncertainty of 1 percent. Assuming that the nozzle throat area and discharge coefficients are known exactly, derive an expression for the required relationship between the uncertainties in the stagnation pressure and temperature measurements. Which of the two measurements, pressure or temperature, is more likely to be the controlling factor?

7-7 Calculate the throat area for the sonic nozzle in Example 7-4 if the stated pressure is static pressure upstream, the temperature is total temperature upstream, and the pipe diameter is 1.0 in.

7-8 Show that the parameter that governs the use of a linear relation for a rotameter as in Eq. (7-30) is

$$\frac{ay}{D}$$

provided that $d \approx D$ to the extent that

$$1 - \left(\frac{d}{D}\right)^2 \ll \frac{ay}{D}$$

Under these restrictions plot the error resulting from the linear approximation as a function of the parameter ay/D. Discuss the physical significance of this analysis, and interpret its meaning in terms of specific design recommendations for rotameters.

7-9 Verify that the error in flow rate for a rotameter calculated from Eq. (7-32) is less than 0.2 percent for density variations of ± 5 percent when the float density is designated according to Eq. (7-31).

7-10 The turbine flowmeter whose calibration is shown in Fig. 7-19 is to measure a nominal flow rate of water of 2.5 gpm at 60°F. A single value of the meter constant K is to be used in the data reduction. What deviations from the nominal flow rate are allowable in order that the nominal value of K is accurate within ± 0.25 percent?

7-11 A rotameter is to be designed to measure a maximum flow of 10 gpm of water at 70°F. The bob has a 1-in diameter and has a total volume of 1 in³. The bob is constructed so that the density is given in accordance with Eq. (7-31). The total length of the rotameter tube is 13 in, and the diameter of the tube at inlet is 1.0 in. Determine the tube taper for drag coefficients of 0.4, 0.8, and 1.20. Plot the flow rate versus distance from the entrance of the tube for each of these drag coefficients.

Determine the meter constant for use in Eq. (7-32), and estimate the error resulting from the use of this relation instead of the exact expression in Eq. (7-28).

7-12 A rotameter is to be used for measurement of the flow of air at 100 psia and 70°F. The maximum flow rate is 0.03 lbm/s, the inlet diameter of the meter is 1 in, and the length of the meter is 12 in. The bob is constructed so that its density is five times that of the air and its volume is 1 in³. Calculate the tube taper for drag coefficients of 0.4, 0.8, and 1.2, and determine the meter constant for use in Eq. (7-30). Plot the error resulting from the use of Eq. (7-30) as a function of flow rate.

7-13 Derive an expression for the product of density and velocity across a hot-wire anemometer in terms of the wire resistance, the current through the wire, and the empirical constants a and b. Subsequently obtain an expression for the uncertainty in this product as a function of the measured quantities.

7-14 The sensitivity of a schlieren system is defined as the fractional deflection obtained at the knife edge per unit angular deflection of a light ray at the test section. Show that this sensitivity may be calculated with

$$S = \frac{f_2}{y_1}$$

Derive an expression for the contrast in terms of the sensitivity, the density gradient, and the test section width.

7-15 Show that the sensitivity of an interferometer, defined as the number of fringe shifts per unit change of density, may be written as

$$S = \frac{\beta L}{\lambda \rho_s}$$

Show that the maximum sensitivity of an interferometer, defined as the number of fringe shifts per unit change in Mach number, will be about 30 when $L = 15$ cm, $\lambda = 5400$ Å, $\beta = 0.000293$, and the stagnation density is that of air at standard conditions.

7-16 Calculate the temperatures corresponding to the four fringes nearest to the plate surface in Example 7-6.

7-17 The velocity in turbulent tube flow varies approximately as

$$\frac{u}{u_c} = \left(1 - \frac{r}{r_0}\right)^{1/7}$$

where u_c is the velocity at the center of the tube and r_0 is the tube radius. An experimental setup using air in a 30-cm-diameter tube is used to check this relation. The air temperature is 20°C, and pressure is 1.0 atm. The maximum flow velocity is 15 m/s, and a Pitot tube is used to traverse the flow and obtain a measurement of the velocity distribution. Measurements are taken at radii of 0, 5, 10, and 12 cm, and the uncertainty in the dynamic-pressure measurement is ± 5 Pa. Using the above relation and Eq. (7-53), calculate the nominal velocity and dynamic pressure at each radial location. Then calculate the uncertainty in the velocity measurement at each location based on the uncertainty in the pressure measurement. Assuming that the above relation does represent the true velocity profile, calculate the uncertainty which could result in a mass flow determined from the experimental data. The mass flow would be obtained by performing the integration

$$\dot{m} = \int_0^{r_0} 2\pi r \rho u \, dr$$

7-18 Calculate the dynamic pressure measured by a Pitot tube in a flowstream of water at 20°C moving at a velocity of 3 m/s.

7-19 Using the supersonic Pitot-tube formula given by Eq. (7-54), obtain an expression for the uncertainty in the Mach number as a function of the percent uncertainty in the pressure ratio (p_∞/p_{0_2}) and the nominal value of the Mach number. Plot the percent uncertainty in the Mach number as a function of Mach number for a constant uncertainty in (p_∞/p_{0_2}) of 1 percent. Assume $\gamma = 1.4$.

7-20 A venturi with throat and upstream diameters of 8 and 16 in is used to measure the flow of water at 70°F. The flow rate is controlled by a motorized valve downstream from the venturi. The valve is operated so that a constant differential pressure of 12 inHg is maintained across the venturi. Suppose someone informs you that this type of control scheme is not very effective because it does not account for possible changes in temperature of the water. Reply to this criticism by plotting the error in the flow rate as a function of water temperature, taking the flow at 70°F as the reference value.

7-21 An obstruction meter is used for the measurement of the flow of moist air at low velocities. Suppose that the flow rate is calculated taking the density as that of dry air at 90°F. Plot the error in this flow rate as a function of relative humidity.

7-22 A small venturi is used for measuring the flow of water in a $\frac{1}{2}$-in-diameter line. The venturi has a throat diameter of $\frac{1}{4}$ in and is constructed according to ASME specifications. What minimum flow rate at 70°F should be used in order that the discharge coefficient remain in the flat portion of the curve? What pressure reading, in inches of mercury, would be experienced for this flow rate?

7-23 The venturi of Prob. 7-22 is used to measure the flow of air at 100 psia and 120°F (upstream flow conditions). The throat pressure is measured as 80 psia. Calculate the flow rate.

7-24 An orifice is to be used to indicate the flow rate of water in a 1.25-in-diameter line. The orifice diameter is 0.50 in. What pressure reading will be experienced on the orifice for a line-flow velocity of 10 ft/s? What would be the flow rate for a pressure reading of twice this value?

7-25 A small sonic nozzle having a throat diameter of 0.8 mm is used to measure and regulate the flow in a 7.5-cm-diameter line. The upstream pressure on the nozzle is varied in accordance with the demand requirements. The downstream pressure is always low enough to ensure sonic flow at the throat. What is the flow rate for upstream conditions at 20°C and 1.0 MPa.

7-26 An impact tube is used to measure the Mach number of a certain air flow in a wind tunnel. The static pressure is 3 psia, and the impact pressure at the probe is 116 kPa. What is the Mach number of the flow ($\gamma = 1.4$ for air)? If the free-stream air temperature is $-40°C$, what is the flow velocity?

7-27 A Pitot tube is used to measure the velocity of an air stream at 20°C and 1.0 atm. If the velocity is 2.5 m/s, what is the dynamic pressure in newtons per square meter? What is the uncertainty of the velocity measurement if the dynamic pressure is measured with a manometer having an uncertainty of 5 Pa?

7-28 A venturi is used to measure the flow of liquid Freon 12 at 20°C. The throat diameter is 1.2 cm and the inlet diameter is 2.4 cm. Calculate the pressure drop reading if the throat Reynolds number is 10^5.

7-29 Repeat Prob. 7-28 for a concentric orifice with the same diameter ratio.

7-30 A flow rate of 1 kg/s of air at 30 atm and 20°C is to be measured with a venturi and an orifice. Select appropriate size devices and specify suitable pressure instrumentation for each.

7-31 The flow rate in Prob. 7-30 is to be measured with a sonic nozzle. Calculate the exit diameter of the nozzle.

7-32 A rotameter is to be used to measure the flow rate of liquid Freon 12 at 20°C. You have at your disposal several specially constructed bobs of varying densities. What density would you select?

7-33 A test is to be conducted of the fuel economy of a certain automobile. The test is to be conducted under road conditions at varying speeds. The manufacturer states that the car will get 8.0 km/liter of fuel at a steady speed of 80 km/h. Select appropriate flow measurement device(s) and specify the necessary pressure and temperature measurement devices for a speed range of 30 to 90 km/h. Analyze your choice and estimate the uncertainties which may result in the final results of fuel economy. Give alternate suggestions for the measurements and show what result they would have on the uncertainty.

REFERENCES

1. "Fluid Meters, Their Theory and Application," 6th ed., ASME, New York, 1971.
2. "Flowmeter Computation Handbook," ASME, New York, 1961.

3. Arnberg, B. T.: Review of Critical Flowmeters for Gas Flow Measurements, *Trans. ASME*, vol. 84D, p. 447, 1962.
4. Gracey, W.: Measurement of Static Pressure on Aircraft, *NACA Tech. Note* 4184, November 1957.
5. King, L. V.: On the Convection of Heat from Small Cylinders in a Stream of Fluid, with Applications to Hot-Wire Anemometry, *Phil. Trans. Roy. Soc. London*, vol. 214, no. 14, p. 373, 1914.
6. Kovasznay, L. S. G.: Hot-Wire Method, in "Physical Measurements in Gas Dynamics and Combustion," p. 219, Princeton University Press, Princeton, N.J., 1954.
7. Krause, L. N., and C. C. Gettelman: Considerations Entering into the Selection of Probes for Pressure Measurement in Jet Engines, *ISA Proc.*, vol. 7, p. 134, 1952.
8. Ladenburg, R. W. (ed.): "Physical Measurements in Gas Dynamics and Combustion," Princeton University Press, Princeton, N.J., 1954.
9. Laurence, J. C., and L. G. Landes: Application of the Constant Temperature Hot-Wire Anemometer to the Study of Transient Air Flow Phenomena, *J. Instr. Soc. Am.*, vol. 1, no. 12, p. 128, 1959.
10. Liepmann, H. W., and A. Roshko: "Elements of Gas Dynamics," John Wiley & Sons, Inc., New York, 1957.
11. Miesse, C. C., and O. E. Curth: How to Select a Flowmeter, *Prod. Eng.*, p. 35, May 8, 1961.
12. Purdy, K. R., T. W. Jackson, and C. W. Gorton: Viscous Fluid Flow under the Influence of a Resonant Acoustic Field, *J. Heat Transfer*, vol. 86C, p. 97, February 1964.
13. Schulze, W. M., G. C. Ashby, Jr., and J. R. Erwin: Several Combination Probes for Surveying Static and Total Pressure and Flow Direction, *NACA Tech. Note* 2830, November 1952.
14. Smith, E., R. H. Reed, and H. D. Hodges: The Measurement of Low Air Speeds by the Use of Titanium Tetrachloride, *Texas Eng. Exp. Sta. Research Rpt.* 25, May 1951.
15. Tuve, G. L.: "Mechanical Engineering Experimentation," McGraw-Hill Book Company, New York, 1961.
16. Schraub, F. A., S. J. Kline, et al.: Use of Hydrogen Bubbles for Quantitative Determination of Time-Dependent Velocity Fields in Low Speed Water Flows, ASME Paper No. 64-WA/FE-20, December 1964.
17. "Flow Visualization Symposium," ASME, New York, 1960.
18. Dean, R. C. (ed.): "Aerodynamic Measurements," Eagle Press, 1953.
19. Rayle, R. E.: "An Investigation of the Influence of Orifice Geometry on Static Pressure Measurements," *S.M. thesis*, Dept. Mech. Eng., Massachusetts Institute of Technology, 1949. (See also Ref. 18.)
20. Yeh, Y., and H. Z. Cummins: Localized Fluid Flow Measurements with an He-Ne Laser Spectrometer, *Appl. Phys. Lett.*, vol. 4, no. 10, pp. 176–178, May 1964.
21. Forman, J. W., E. W. George, and R. D. Lewis: Measurement of Localized Flow Velocities in Gases with a Laser Doppler Flowmeter, *Appl. Phys. Lett.*, vol. 7, no. 4, pp. 77–78, August 1965.
22. Forman, J. W., R. D. Lewis, J. R. Thornton, and H. J. Watson: Laser Doppler Velocimeter for Measurement of Localized Flow Velocities in Liquids, *Proc. IEEE*, vol. 54, pp. 424–425, March 1966.
23. Berman, N. S., and V. A. Santos: Laminar Velocity Profiles in Developing Flows Using a Laser Doppler Technique, *AIChEJ.*, vol. 15, no. 3, pp. 323–327, May 1969.
24. Goldstein, R. J., and D. K. Kreid: Measurements of Laminar Flow Development in a Square Duct Using a Laser Doppler Flowmeter, *J. Appl. Mech.*, vol. 34, no. 4, pp. 813–818, December 1967.
25. Goldstein, R. J., and W. F. Hagen: Turbulent Flow Measurements Utilizing the Doppler Shift of Scattered Laser Radiation, *Phys. Fluids*, vol. 10, pp. 1349–1352, June 1967.
26. Welch, N. E., and W. J. Tomme: The Analysis of Turbulence from Data Obtained with a Laser Velocimeter, *AIAA Paper* 67–179, January 1967.
27. Welch, N. E., and R. H. Hines: The Practical Application of the Laser Anemometer for Fluid Flow Measurements, presented at the ISA Electro-Optical Systems Conf., New York, N.Y., September 1969.

28. Hodges, A. E., and T. N. Pound: Further Development of a Smoke Producer Using Vaporized Oil, *Aerodynamics Note* 233, Australian Defense Scientific Service Aeronautical Res. Labs., November 1964.
29. Benedict, R. P., J. S. Wyler, and G. B. Brandt: The Effect of Edge Sharpness on the Discharge Coefficient of an Orifice, *J. Engr. Power*, p. 576, 1975.
30. Wyler, J. S., and R. P. Benedict: Comparisons between Throat and Pipe Wall Tap Nozzles, *J. Engr. Power*, p. 569, 1975.
31. Lee, W. F., M. J. Kirik, and J. A. Bonner: Gas Turbine Flowmeter Measurement of Pulsating Flow, *J. Engr. Power*, p. 531, 1975.
32. Szaniszlo, A. J.: Experimental and Analytical Sonic Nozzle Discharge Coefficients for Reynolds Numbers up to 8×10^6, *J. Engr. Power*, p. 521, 1975.
33. Smith, R. V., and J. T. Leang: Evaluations of Correlations for Two-Phase Flowmeters, Three Current—One New, *J. Engr. Power*, p. 589, 1975.
34. Benedict, R. P., and R. D. Schulte: A Note on the Critical Pressure Ratio across a Fluid Meter, ASME Paper No. 72-WA/FM-1, December 1972.
35. Head, V. P.: Improved Expansion Factors for Nozzles, Orifices, and Variable Area Meters, ASME Paper No. 73-WA/FM-1, December 1973.
36. Halmi, D.: Metering Performance Investigation and Substantiation of the Universal Venturi Tube, Parts I and II, ASME papers No. 73-WA/FM-3, 4, December 1973.
37. Miller, R. W., and O. Kneisel: A Comparison between Orifice and Flow Nozzle Laboratory Data and Published Coefficients, ASME Paper No. 73-WA/FM-5, December 1973.
38. Arnberg, B. T., C. L. Britton, and W. F. Seidl: Discharge Coefficient Correlations for Circular Arc Venturi Flowmeters at Critical (Sonic) Flow, ASME Paper No. 73-WA/FM-8, December 1973.
39. Goldstein, R. J.: Measurement of Fluid Velocity by Laser Doppler Techniques, *Appl. Mech. Rev.*, vol. 27, p. 753, 1974.
40. Orloff, K. L., G. R. Grant, and W. D. Gunter: Laser Velocimeter for Simultaneous Two-Dimensional Velocity Measurements, NASA Tech. Brief, 73-10267, 1973.
41. Farmer, W. M.: Determination of a Third Orthogonal Velocity Component Using Two Rotationally Displaced Laser Doppler Velocimeter Systems, *Appl. Opt.*, vol. 11, p. 770, 1972.
42. Durst, F., A. Melling, and J. H. Whitelaw: "Principles and Practice of Laser Doppler Anemometry," Academic Press, Inc., New York, 1976.
43. Benedict, R. P.: "Fundamentals of Temperature, Pressure, and Flow Measurement," John Wiley & Sons, Inc., New York, 1969.
44. Richards, B. E., ed.: "Measurement of Unsteady Fluid Dynamic Phenomena," Chap. 6, Hot-Wire and Hot-Film Anemometers, Hemisphere Publishing, 1976.
45. James, D. F., and A. J. Acosta: The Laminar Flow of Dilute Polymer Solutions around Circular Cylinders, *J. Fluid Mech.*, vol. 42, p. 269, 1970.
46. Goldstein, R. J., and D. K. Kreid: The Laser Doppler Anemometer in "Measurements in Heat Transfer," pp. 541–574, McGraw-Hill Book Company, New York, 1976.
47. Asanuma, T. (ed.): "Flow Visualization," Hemisphere Publishing Co., 1979.
48. Goldstein, R. J.: "Fluid Mechanics Measurements," Hemisphere Publishing Co., 1983.

EIGHT

THE MEASUREMENT OF TEMPERATURE

8-1 INTRODUCTION

To most people, temperature is an intuitive concept which tells whether a body is "hot" or "cold." In the exposition of the second principle of thermodynamics, temperature is related to heat, for it is known that heat flows only from a high temperature to a low temperature, in the absence of other effects. In the kinetic theory of gases and statistical thermodynamics it is shown that temperature is related to the average kinetic energy of the molecules of an ideal gas. Further extensions of statistical thermodynamics show the relationship between temperature and the energy levels in liquids and solids. We shall not be able to discuss the many theoretical aspects of the concept of temperature but may only note that it is important in every branch of physical science; hence the experimental engineer should be familiar with the methods employed in temperature measurement. Detailed discussions of the thermodynamic meaning of temperature are given in Refs. [8], [9], [12], and [18].

Since pressure, volume, electrical resistance, expansion coefficients, etc., are all related to temperature through the fundamental molecular structure, they change with temperature, and these changes can be used to measure temperature. Calibration may be achieved through comparison with established standards, as discussed in Chap. 2. The International Temperature Scale serves to *define* temperature in terms of observable characteristics of materials.

8-2 TEMPERATURE SCALES

The two temperature scales in use are the Fahrenheit and Celsius scales. These scales are based on a specification of the number of increments between the freezing point and boiling point of water at standard atmospheric pressure. The

K	°C	°F	°R
2273.16	2000	3632	4091.69
1773.16	1500	2732	3191.69
1273.16	1000	1832	2291.69
773.16	500	932	1391.69
673.16	400	752	1211.69
573.16	300	572	1031.69
473.16	200	392	851.69
373.16	100	212.0	671.69
273.16	0	32.0	491.69
233.16	−40	−40	419.69
173.16	−100	−148	311.69

Figure 8-1 Relationship between Fahrenheit and Celsius temperature scales.

Celsius scale has 100 units between these points, while the Fahrenheit scale has 180 units. The absolute Celsius scale is called the Kelvin scale, while the absolute Fahrenheit scale is termed the Rankine scale. Both absolute scales are so defined that they will correspond as closely as possible with the absolute thermodynamic temperature scale. The zero points on both absolute scales represent the same physical state, and the ratio of two values is the same, regardless of the absolute scale used; i.e.,

$$\left(\frac{T_2}{T_1}\right)_{\text{Rankine}} = \left(\frac{T_2}{T_1}\right)_{\text{Kelvin}} \tag{8-1}$$

The boiling point of water at 1 atm is arbitrarily taken as 100° on the Celsius scale and 212° on the Fahrenheit scale. The relationship between the scales is indicated in Fig. 8-1. It is evident that the following relations apply:

$$°F = 32.0 + \tfrac{9}{5}°C \tag{8-2a}$$

$$°R = \tfrac{9}{5}\,K \tag{8-2b}$$

8-3 THE IDEAL-GAS THERMOMETER

The behavior of an ideal gas at low pressures furnishes the basis for a temperature-measurement device that may serve as a secondary experimental standard. The ideal-gas equation of state is

$$pV = mRT \tag{8-3}$$

Unknown
temperature
T

Volume
V

Pressure
gage

Figure 8-2 Schematic of ideal-gas thermometer.

where V is the volume occupied by the gas, m is the mass, and R is the gas constant for the particular gas, given by

$$R = \mathscr{R}/M$$

where $\mathscr{R}$ is the universal gas constant, having a value of 8314.5 J/kg · mol · K, and M is the molecular weight of the gas. For the gas thermometer a fixed volume is filled with gas and exposed to the temperature to be measured, as shown in Fig. 8-2. At the temperature T the gas-system pressure is measured. Next, the volume is exposed to a standard reference temperature (as discussed in Sec. 2-4), and the pressure is measured under these conditions. According to Eq. (8-3), at constant volume,

$$T = T_{\text{ref}}\left(\frac{p}{p_{\text{ref}}}\right)_{\text{const vol}} \tag{8-4}$$

Now suppose that some of the gas is removed from the volume and the pressure measurements are repeated. In general, there will be a slight difference in the pressure ratio in Eq. (8-4) as the quantity of gas is varied. However, regardless of the gas used, the series of measurements may be repeated and the results plotted as in Fig. 8-3. When the curve is extrapolated to zero pressure, the true temperature as defined by the ideal-gas equation of state will be obtained. A gas thermometer may be used to measure temperatures as low as 1 K by extrapolation.

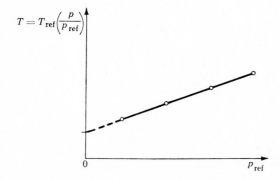

$T = T_{\text{ref}}\left(\frac{p}{p_{\text{ref}}}\right)$

0

p_{ref}

Figure 8-3 Results of measurements with ideal-gas thermometer.

8-4 TEMPERATURE MEASUREMENT BY MECHANICAL EFFECTS

Several temperature-measurement devices may be classified as mechanically operative. In this sense we shall be concerned with those devices operating on the basis of a change in mechanical dimension with a change in temperature.

The liquid-in-glass thermometer is one of the most common types of temperature-measurement devices. The construction details of such an instrument are shown in Fig. 8-4. A relatively large bulb at the lower portion of the thermometer holds the major portion of the liquid, which expands when heated and rises in the capillary tube, upon which are etched appropriate scale markings. At the top of the capillary tube another bulb is placed to provide a safety feature in case the temperature range of the thermometer is inadvertently exceeded. Alcohol and mercury are the most commonly used liquids. Alcohol has the advantage that it has a higher coefficient of expansion than mercury, but it is limited to low-temperature measurements because it tends to boil away at high temperatures. Mercury cannot be used below its freezing point of $-38.78°F$ ($-37.8°C$). The size of the capillary depends on the size of the sensing bulb, the liquid, and the desired temperature range for the thermometer.

In operation, the bulb of the liquid-in-glass thermometer is exposed to the environment whose temperature is to be measured. A rise in temperature causes the liquid to expand in the bulb and rise in the capillary, thereby indicating the

Figure 8-4 Schematic of a mercury-in-glass thermometer.

Figure 8-5 The bimetallic strip.

temperature. It is important to note that the expansion registered by the thermometer is the *difference* between the expansion of the liquid and the expansion of the glass. The difference is a function not only of the heat transfer to the bulb from the environment, but also of the heat conducted into the bulb from the stem; the more the stem conduction relative to the heat transfer from the environment, the larger the error. To account for such conduction effects the thermometer is usually calibrated for a certain specified depth of immersion. High-grade mercury-in-glass thermometers have the temperature scale markings engraved on the glass along with a mark which designates the proper depth of immersion. Very precise mercury-in-glass thermometers may be obtained from the National Bureau of Standards with calibration information for each thermometer.

Mercury-in-glass thermometers are generally applicable up to about 600°F (315°C), but their range may be extended to 1000°F (538°C) by filling the space above the mercury with a gas like nitrogen. This increases the pressure on the mercury, raises its boiling point, and thereby permits the use of the thermometer at higher temperatures.

A very widely used method of temperature measurement is the bimetallic strip. Two pieces of metal with different coefficients of thermal expansion are bonded together to form the device shown in Fig. 8-5. When the strip is subjected to a temperature higher than the bonding temperature, it will bend in one direction; when it is subjected to a temperature lower than the bonding temperature, it will bend in the other direction. Eskin and Fritze [3] have given calculation methods for bimetallic strips. The radius of curvature r may be calculated as

$$r = \frac{t\{3(1 + m)^2 + (1 + mn)[m^2 + (1/mn)]\}}{6(\alpha_2 - \alpha_1)(T - T_0)(1 + m)^2} \qquad (8\text{-}5)$$

where t = combined thickness of the bonded strip
      m = ratio of thicknesses of low- to high-expansion materials
      n = ratio of moduli of elasticity of low- to high-expansion materials
      α_1 = lower coefficient of expansion
      a_2 = higher coefficient of expansion
      T = temperature
      T_0 = initial bonding temperature

Table 8-1 Mechanical properties of some commonly used thermal materials

Material	Thermal coefficient of expansion per °C	Modulus of elasticity psi	GN/m²
Invar	1.7×10^{-6}	21.4×10^6	147
Yellow brass	2.02×10^{-5}	14.0×10^6	96.5
Monel 400	1.35×10^{-5}	26.0×10^6	179
Inconel 702	1.25×10^{-5}	31.5×10^6	217
Stainless-steel type 316	1.6×10^{-5}	28×10^6	193

The thermal-expansion coefficients for some commonly used materials are given in Table 8-1.

Example 8-1 A bimetallic strip is constructed of strips of yellow brass and Invar bonded together at 30°C. Each has a thickness of 0.3 mm. Calculate the radius of curvature when the strip is subjected to a temperature of 100°C.

SOLUTION We use Eq. (8-5) with properties from Table 8-1.

$$T - T_0 = 100 - 30 = 70°C$$

$$m = 1.0$$

$$n = \frac{147}{96.5} = 1.52$$

$$\alpha_1 = 1.7 \times 10^{-6}\,°C^{-1} \qquad \alpha_2 = 2.02 \times 10^{-5}\,°C^{-1}$$

$$t = (2)(0.3 \times 10^{-3}) = 0.6 \times 10^{-3} \text{ m}$$

Thus,

$$r = \frac{(0.6 \times 10^{-3})[(3)(2)^2 + (1 + 1.52)(1 + 1/1.52)]}{6(2.02 - 0.17)(10^{-5})(70)(2)^2}$$

$$= 0.312 \text{ m}$$

Fluid-expansion thermometers represent one of the most economical, versatile, and widely used devices for industrial temperature-measurement applications. The principle of operation is indicated in Fig. 8-6. A bulb containing a liquid, gas, or vapor is immersed in the environment. The bulb is connected by means of a capillary tube to some type of pressure-measuring device, such as the bourdon gage shown. An increase in temperature causes the liquid or gas to expand, thereby increasing the pressure on the gage; the pressure is thus taken as an indication of the temperature. The entire system consisting of the bulb, capillary, and gage may be calibrated directly. It is clear that the temperature of the capillary tube may influence the reading of the device because some of the volume of fluid is contained therein. If an equilibrium mixture of liquid and vapor is used in the bulb, however, this problem may be alleviated, provided that

Figure 8-6 Fluid expansion thermometer.

the bulb temperature is always higher than the capillary-tube temperature. In this circumstance the fluid in the capillary will always be in a subcooled liquid state, while the pressure will be uniquely specified for each temperature in the equilibrium mixture contained in the bulb.

Capillary tubes as long as 200 ft (60 m) may be used with fluid-expansion thermometers. The transient response is primarily dependent on the bulb size and the thermal properties of the enclosed fluid. Highest response may be achieved by using a small bulb connected to some type of electric-pressure transducer through a short capillary.

8-5 TEMPERATURE MEASUREMENT BY ELECTRICAL EFFECTS

Electrical methods of temperature measurement are very convenient because they furnish a signal that is easily detected, amplified, or used for control purposes. In addition, they are usually quite accurate when properly calibrated and compensated.

Electrical-Resistance Thermometer

One quite accurate method of temperature measurement is the electrical-resistance thermometer. It consists of some type of resistive element, which is exposed to the temperature to be measured. The temperature is indicated through a measurement of the change in resistance of the element. Several types of materials may be used as resistive elements, and their characteristics are given in Table 8-2. The linear temperature coefficient of resistance α is defined by

$$\alpha = \frac{R_2 - R_1}{R_1(T_2 - T_1)} \tag{8-6}$$

Table 8-2 Resistance-temperature coefficients and resistivity at 20°C†

Substance	$\alpha(°C^{-1})$	$\rho(\mu\Omega \cdot cm)$
Nickel	0.0067	6.85
Iron (alloy)	0.002 to 0.006	10
Tungsten	0.0048	5.65
Aluminum	0.0045	2.65
Copper	0.0043	1.67
Lead	0.0042	20.6
Silver	0.0041	1.59
Gold	0.004	2.35
Platinum	0.00392	10.5
Mercury	0.00099	98.4
Manganin	± 0.00002	44
Carbon	-0.0007	1400
Electrolytes	-0.02 to -0.09	Variable
Semiconductor (thermistors)	-0.068 to $+0.14$	10^9

† According to Lion [6].

where R_2 and R_1 are the resistances of the material at temperatures T_2 and T_1, respectively. The relationship in Eq. (8-6) is usually applied over a narrow temperature range such that the variation of resistance with temperature approximates a linear relation. For wider temperature ranges the resistance of the material is usually expressed by a quadratic relation

$$R = R_0(1 + aT + bT^2) \tag{8-7}$$

where R = resistance at temperature T

R_0 = resistance at reference temperature T_0

a, b = experimentally determined constants

It may be noted that the platinum resistance thermometer is used for the International Temperature Scale between the oxygen point and the antimony point, as described in Chap. 2.

Various methods are employed for construction of resistance thermometers, depending on the application. In all cases care must be taken to ensure that the resistance wire is free of mechanical stresses and so mounted that moisture cannot come in contact with the wire and influence the measurement.

The resistance measurement may be performed with some type of bridge circuit, as described in Chap. 4. For steady-state measurements a null condition will suffice, while transient measurements will usually require the use of a deflection bridge. One of the primary sources of error in the electrical-resistance thermometer is the effect of the resistance of the leads which connect the element to the bridge circuit. Several arrangements may be used to correct for this effect, as shown in Fig. 8-7. The Siemen's three-lead arrangement is the simplest type of corrective circuit. At balance conditions the center lead carries no current, and the effect of the resistance of the other two leads is canceled out. The Callender

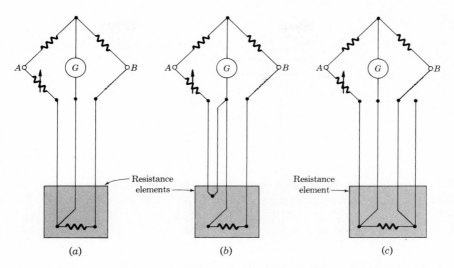

Figure 8-7 Methods of correcting for lead resistance with electrical-resistance thermometer. (*a*) Siemen's three-lead arrangement; (*b*) Callender four-lead arrangement; (*c*) floating-potential arrangement. Power connections made at *A* and *B*.

four-lead arrangement solves the problem by inserting two additional lead wires in the adjustable leg of the bridge so that the effect of the lead wires on the resistance thermometer is canceled out. The floating-potential arrangement in Fig. 8-7c is the same as the Siemen's connection, but an extra lead is inserted. This extra lead may be used to check the equality of lead resistance. The thermometer reading may be taken in the position shown, followed by additional readings with the two right and left leads interchanged, respectively. Through this interchange procedure the best average reading may be obtained and the lead error minimized.

Example 8-2 A platinum resistance thermometer is used at room temperature. Assuming a linear temperature variation with resistance, calculate the sensitivity of the thermometer in ohms per degrees Fahrenheit.

SOLUTION The meaning of a linear variation of resistance with temperature is

$$R = R_0[1 + \alpha(T - T_0)]$$

where R_0 is the resistance at the reference temperature T_0. The sensitivity is thus

$$S = \frac{dR}{dT} = \alpha R_0$$

R_0 depends on the length and size of the resistance wire. At room temperature, $\alpha = 0.00392°\text{C}^{-1} = 0.00218°\text{F}^{-1}$ for platinum.

Thermistors

The *thermistor* is a semiconductor device that has a negative temperature coefficient of resistance, in contrast to the positive coefficient displayed by most metals. Furthermore, the resistance follows an exponential variation with temperature instead of a polynomial relation like Eq. (8-7). Thus, for a thermistor,

$$R = R_0 \exp\left[\beta\left(\frac{1}{T} - \frac{1}{T_0}\right)\right] \tag{8-8}$$

where R_0 is the resistance at the reference temperature T_0 and β is an experimentally determined constant. The numerical value of β varies between 3500 and 4600 K, depending on the thermistor material and temperature. The resistivities of three thermistor materials as compared with platinum are given in Fig. 8-8 according to Ref. [1]. A typical static voltage-current curve is shown in Fig. 8-9, while a typical set of transient voltage-current characteristics is illustrated in Fig. 8-10. The numbers on the curve in Fig. 8-9 designate the degrees Celsius rise in temperature above ambient temperature for the particular thermistor.

The thermistor is an extremely sensitive device, and consistent performance within 0.01°C may be anticipated with proper calibration. A rather nice feature of the thermistor is that it may be used for temperature compensation of electric circuits. This is possible because of the negative-temperature characteristic that it exhibits, so that it can be used to counteract the increase in resistance of a circuit with a temperature increase.

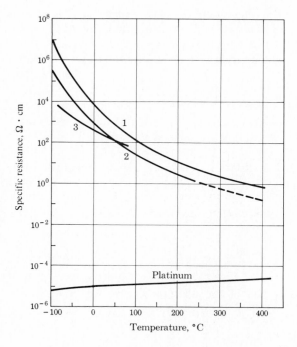

Figure 8-8 Resistivity of three thermistor materials compared with platinum, according to Ref. [1].

Figure 8-9 Static voltage-current curve for a typical thermistor, according to Ref. [1].

A bridge amplifier circuit which may be employed with resistance-thermometer or thermistor devices is shown in Fig. 8-11. In this arrangement R_T is the resistance of the thermometer and is expressed as

$$R_T = R_x = \Delta R$$

For linear operation we must have $R_T \gg R_1$ and R_B is approximately $R_1/10$. The amplifier voltage gain is $A = 1 + R_2/R_1$. If these conditions are satisfied the output voltage can be expressed as

$$v_o = \frac{A v_b \alpha \, \Delta T}{4} \tag{8-9}$$

where α is the temperature coefficient of resistance and ΔT is the temperature difference from balanced conditions, i.e., from $\Delta R = 0$.

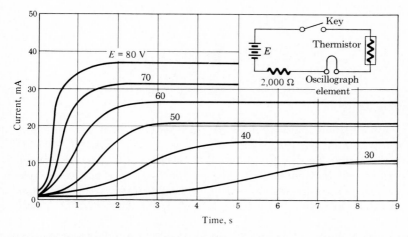

Figure 8-10 Typical set of transient voltage-current curves for a thermistor, according to Ref. [1]. Circuit for measurement is shown in insert.

Figure 8-11 Bridge amplifier circuit, according to Ref. [24].

Example 8-3 Calculate the temperature sensitivity for thermistor No. 1 in Fig. 8-8 at 100°C. Express the result in ohm-centimeters per degree Celsius. Take $\beta = 4120$ K at 100°C.

SOLUTION The sensitivity is obtained by differentiating Eq. (8-8).

$$S = \frac{dR}{dT} = R_0 \exp\left[\beta\left(\frac{1}{T} - \frac{1}{T_0}\right)\right]\left(\frac{-\beta}{T^2}\right)$$

We wish to express the result in resistivity units; thus the resistivity at 100°C is inserted for R_0. Also,

$$T = T_0 = 100°C = 373 \text{ K}$$

so that

$$S = -\rho_{100°C}\frac{4120}{(373)^2}$$

$$= -\frac{(110)(4120)}{(373)^2} = -3.26 \ \Omega \cdot \text{cm/°C}$$

Thermoelectric Effects

The most common electrical method of temperature measurement uses the thermocouple. When two dissimilar metals are joined together as in Fig. 8-12, an emf will exist between the two points A and B, which is primarily a function of the junction temperature. This phenomenon is called the *Seebeck effect*. If the two materials are connected to an external circuit in such a way that a current is drawn, the emf may be altered slightly owing to a phenomenon called the *Peltier effect*. Further, if a temperature gradient exists along either or both of the materials, the junction emf may undergo an additional slight alteration. This is

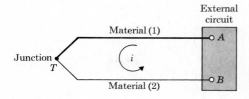

Figure 8-12 Junction of two dissimilar metals indicating thermoelectric effect.

called the *Thomson effect*. There are, then, three emfs present in a thermoelectric circuit: the Seebeck emf, caused by the junction of dissimilar metals; the Peltier emf, caused by a current flow in the circuit; and the Thomson emf, which results from a temperature gradient in the materials. The Seebeck emf is of prime concern since it is dependent on junction temperature. If the emf generated at the junction of two dissimilar metals is carefully measured as a function of temperature, then such a junction may be utilized for the measurement of temperature. The main problem arises when one attempts to measure the potential. When the two dissimilar materials are connected to a measuring device, there will be another thermal emf generated at the junction of the materials and the connecting wires to the voltage-measuring instrument. This emf will be dependent on the temperature of the connection, and provision must be made to take account of this additional potential.

Two rules are available for analysis of thermoelectric circuits:

1. If a third metal is connected in the circuit as shown in Fig. 8-13, the net emf of the circuit is not affected as long as the new connections are at the same temperature. This statement may be proved with the aid of the second law of thermodynamics and is known as the *law of intermediate metals*.
2. Consider the arrangements shown in Fig. 8-14. The simple thermocouple circuits are constructed of the same materials but operate between different temperature limits. The circuit in Fig. 8-14a develops an emf of E_1 between temperatures T_1 and T_2; the circuit in Fig. 8-14b develops an emf of E_2 between temperatures T_2 and T_3. The *law of intermediate temperatures* states that this same circuit will develop an emf of $E_3 = E_1 + E_2$ when operating between temperatures T_1 and T_3, as shown in Fig. 8-14c.

It may be observed that all thermocouple circuits must involve at least two junctions. If the temperature of one junction is known, then the temperature of the other junction may be easily calculated using the thermoelectric properties of

Figure 8-13 Influence of a third metal in a thermoelectric circuit; law of intermediate metals.

$$T_1 \underset{E_1}{\diagdown} T_2 \qquad T_2 \underset{E_2}{\diagdown} T_3 \qquad T_1 \underset{E_3 = E_1 + E_2}{\diagdown} T_3$$

(a)   (b)   (c)

Figure 8-14 Circuits illustrating the law of intermediate temperatures.

the materials. The known temperature is called the *reference temperature.* A common arrangement for establishing the reference temperature is the ice bath shown in Fig. 8-15. An equilibrium mixture of ice and air-saturated distilled water at standard atmospheric pressure produces a known temperature of $32°F$. When the mixture is contained in a Dewar flask, it may be maintained for extended periods of time. Note that the arrangement in Fig. 8-15a maintains both thermocouple wires at a reference temperature of $32°F$, whereas the arrangement in Fig. 8-15b maintains only one at the reference temperature. The system in Fig. 8-15a would be necessary if the binding posts at the voltage-measuring instrument were at different temperatures, while the connection in Fig. 8-15b would be satisfactory if the binding posts were at the same temperature. To be effective the system in Fig. 8-15a must have copper binding posts; i.e., the binding posts and leads must be of the same material.

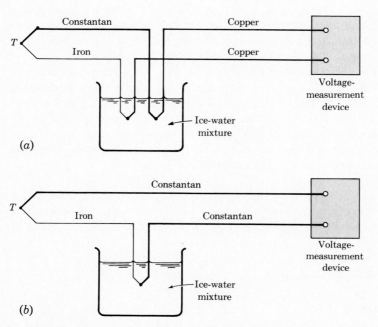

Figure 8-15 Conventional methods for establishing reference temperature in thermocouple circuit. Iron-constantan thermocouple illustrated.

It is common to express the thermoelectric emf in terms of the potential generated with a reference junction at 32°F (0°C). Standard thermocouple tables have been prepared on this basis, and a summary of the output characteristics of the most common thermocouple combinations is given in Table 8-3. These data are shown graphically in Fig. 8-16, along with the behavior of some of the more exotic thermocouple materials. The output voltage E of a simple thermocouple circuit is usually written in the form

$$E = AT + \tfrac{1}{2}BT^2 + \tfrac{1}{3}CT^3 \tag{8-10}$$

where T is the temperature in degrees Celsius and E is based on a reference junction temperature of 0°C. The constants A, B, and C are dependent on the thermocouple material. Powell [19] gives an extensive discussion of the manufacture of materials for thermocouple use, inhomogeneity ranges, and power series

Table 8-3 Thermal emf in absolute millivolts for commonly used thermocouple combinations† (Reference junction at 0°C)

Temperature		Copper vs. constantan (T)	Chromel vs. constantan (E)	Iron vs. constantan (J)	Chromel vs. alumel (K)	Platinum vs. platinum 10% rhodium (S)
°F	°C					
−300	−184.4	−5.341	−8.404	−7.519	−5.632	
−250		−4.745	−7.438	−6.637	−5.005	
−200	−128.9	−4.419	−6.471	−5.760	−4.381	
−150		−3.365	−5.223	−4.623	−3.538	
−100	−73.3	−2.581	−3.976	−3.492	−2.699	
−50		−1.626	−2.501	−2.186	−1.693	
0	−17.8	−0.674	−1.026	−0.885	−0.692	−0.092
50		0.422	0.626	0.526	0.412	0.064
100	37.8	1.518	2.281	1.942	1.520	0.221
150		2.743	4.075	3.423	2.667	0.408
200	93.3	3.967	5.869	4.906	3.819	0.597
250		5.307	7.788	6.425	4.952	0.807
300	148.9	6.647	9.708	7.947	6.092	1.020
350		8.085	11.728	9.483	7.200	1.247
400	204.4	9.523	13.748	11.023	8.314	1.478
450		11.046	15.844	12.564	9.435	1.718
500	260.0	12.572	17.942	14.108	10.560	1.962
600		15.834	22.287	17.178	12.865	2.472
700	371.1	19.095	26.637	20.253	15.178	2.985
800			31.108	23.338	17.532	3.524
1000	537.8		40.056	29.515	22.251	4.609
1200			48.927		26.911	5.769
1500	815.6		62.240		33.913	7.514
1700					38.287	8.776
2000	1093.3				44.856	10.675
2500	1371.1				54.845	14.018
3000	1648.9					17.347

† According to Powell et al. [23].

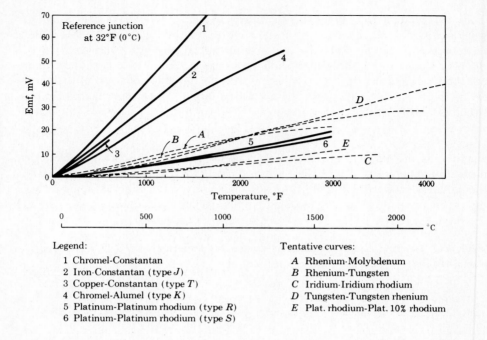

Figure 8-16 Emf temperature relations for thermocouple materials, positive electrode listed first. (*Courtesy of Honeywell, Inc.*)

Legend:

1 Chromel-Constantan
2 Iron-Constantan (type J)
3 Copper-Constantan (type T)
4 Chromel-Alumel (type K)
5 Platinum-Platinum rhodium (type R)
6 Platinum-Platinum rhodium (type S)

Tentative curves:

A Rhenium-Molybdenum
B Rhenium-Tungsten
C Iridium-Iridium rhodium
D Tungsten-Tungsten rhenium
E Plat. rhodium-Plat. 10% rhodium

Table 8-4 Thermoelectric sensitivity $S = dE/dT$ of thermoelement made of materials listed against platinum, $\mu V\,^{\circ}C^{-1}$†

(Reference junction kept at a temperature of $0\,^{\circ}C$)

Bismuth	-72	Silver	6.5
Constantan	-35	Copper	6.5
Nickel	-15	Gold	6.5
Potassium	-9	Tungsten	7.5
Sodium	-2	Cadmium	7.5
Platinum	0	Iron	18.5
Mercury	0.6	Nichrome	25
Carbon	3	Antimony	47
Aluminum	3.5	Germanium	300
Lead	4	Silicon	440
Tantalum	4.5	Tellurium	500
Rhodium	6	Selenium	900

† According to Lion [6].

relationships for thermoelectric voltages of various standard thermocouples. The NBS publication which gives these data is Ref. [23].

The sensitivity, or thermoelectric power, of a thermocouple is given by

$$S = \frac{dE}{dT} = A + BT + CT^2 \qquad (8\text{-}11)$$

Table 8-4 gives the approximate values of the sensitivity of various materials relative to platinum at $0°C$. A summary of the operating range and characteristics of the most common thermocouple materials is given in Figs. 8-16 and 8-17.

If computer processing of thermocouple data is to be performed, the power series relations, like Eq. (8-10) or the extensive collection in Refs. [14] and [23], will certainly be of value. But a firm word of caution must be give here. When one buys a roll of commercial thermocouple wire there are different grades available (at different prices, of course). Precision-grade wire will usually follow the NBS tables by $\pm 1°F$. A "commercial" grade might not be better than $\pm 2°F$. Therefore, the experimentalist should not be lulled into security by a 10-digit computer printout based on the NBS equations. If better precision is required, samples of the thermocouple wire should be calibrated directly against known temperature standards. The individual can then use this calibration data to determine the constants in Eq. (8-10). For precise data analysis it is the specific calibration which must be employed for calculations.

The need for precise calibration is particularly acute when two thermocouples are used for a small-differential temperature measurement. For example, suppose two "precision" thermocouples having an uncertainty of $\pm 1°F$ each are used for a temperature difference measurement of $10°F$. Using the uncertainty analysis of Chap. 3 the uncertainty in the temperature difference would be

$$w(\Delta T) = [(1)^2 + (1)^2]^{1/2} = 1.414°F$$

or *14 percent*! We must note that if both thermocouples come from the same roll they should have essentially the same characteristics and this result would be highly unlikely. Also, the accuracy of the thermocouples for a direct *temperature-difference* measurement will be much better than for an *absolute* temperature measurement.

The output of thermocouples is in the millivolt range and may be measured by the D'Arsonval movement dc millivoltmeter. The voltmeter is basically a current-sensitive device, and hence the meter reading will be dependent on both the emf generated by the thermocouple and the total circuit resistance, including the resistance of the connecting wires. The complete system, consisting of the thermocouple lead wires and millivoltmeter, may be calibrated directly to furnish a reasonably accurate temperature determination. For all precision measurements the thermocouple output is determined with a potentiometer circuit similar to the one described in Sec. 4-4. Potentiometers are available as either manually balanced types like the one shown in Fig. 8-18 or automatically balanced instruments of the indicating or recording type. For very precise laboratory work a microvolt potentiometer is normally used. These instruments can read potentials within 1 μV. It may be noted that the resistance of the lead wires is of no consequence when a potentiometer is used at balance conditions since the current flow is zero in the thermocouple circuit.

Obviously, there are a large number of electronic voltmeters which are suitable for thermocouple measurements. Many provide a digital output which can be used for direct computer processing of the data.

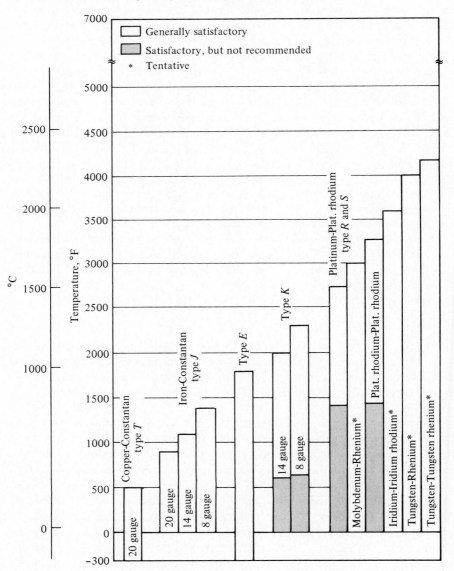

Figure 8-17 Summary of operating range of thermocouples. See also Fig. 8-16. (*Courtesy of Honeywell, Inc.*)

Figure 8-18 Photograph of a typical manually balanced thermocouple potentiometer. (*Courtesy of Honeywell, Inc.*)

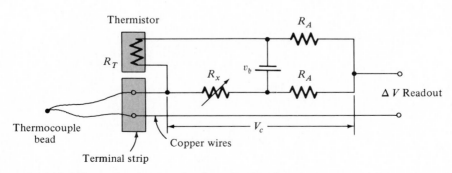

Figure 8-19 Reference junction compensation using thermistor.

The problem of the reference junction can be alleviated with the circuit shown in Fig. 8-19. A thermistor is placed in thermal contact with the terminal strip to which the thermocouple wires are attached. The voltage v_b and temperature coefficient of the thermistor must be adjusted to that V_c will match the thermocouple temperature coefficient in millivolts/degree. The value of R_x is adjusted so that the voltage output ΔV is zero at 0°C. A convenient value for R_A is about 1 kΩ.

Example 8-4 An iron-constantan thermocouple is connected to a potentiometer whose terminals are at 75°F. The potentiometer reading is 3.59 mV. What is the temperature of the thermocouple junction?

SOLUTION The thermoelectric potential corresponding to 75°F is obtained from Table 8-3 as

$$E_{75} = 1.234 \text{ mV}$$

The emf of the thermocouple based on a 32°F (0°C) reference temperature is thus

$$E_T = 1.234 + 3.59 = 4.824 \text{ mV}$$

From Table 8-3 the corresponding temperature is 197°F.

In order to provide a more sensitive circuit, thermocouples are occasionally connected in a series arrangement as shown in Fig. 8-20. Such an arrangement is called a *thermopile*, and for the three-junction situation, the output would be three times that of a single thermocouple arrangement *provided* the temperatures of the hot and cold junctions are uniform. The thermopile arrangement is useful for obtaining a substantial emf for measurement of a small temperature difference

Figure 8-20 Thermopile connection.

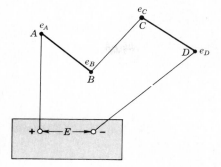

Figure 8-21 Series connection of thermocouples.

between the two junctions. In this way a relatively inexpensive instrument may be used for voltage measurement, whereas a microvolt potentiometer might otherwise be required. When a thermopile is installed, it is important to ensure that the junctions are electrically insulated from one another. We have seen that the typical thermocouple measures the difference in temperature between a certain unknown point and another point designated as the reference temperature. The circuit could just as well be employed for the measurement of a differential temperature. For small differentials, the thermopile circuit is frequently used to advantage. Now consider the series thermocouple arrangement shown in Fig. 8-21. The four junctions are all maintained at different temperatures and connected in series. Since there are an even number of junctions, it is not necessary to install a reference junction because the same type of metal is connected to both terminals of the potentiometer. If we note that the current will flow from plus to minus and assume that junction A produces a potential drop in this direction, then junctions B and D will produce a potential drop in the opposite direction and junction C will generate a potential drop in the same direction as junction A. Thus the total emf measured at the potentiometer terminals is

$$E = e_A - e_B + e_C - e_D \qquad (8\text{-}12)$$

The reading will be zero when all the junctions are at the same temperature and will take on some other value at other conditions. Note, however, that the emf of this series connection is *not* indicative of any particular temperature. It is *not* representative of an average of the junction temperatures.

The parallel connection in Fig. 8-22 may be used for obtaining the average temperature of a number of points. Each of the four junctions may be at a different temperature and hence will generate a different emf. The bucking potential furnished by the potentiometer will be the average of the four junction potentials. There can be a small error in this reading, however, because there is a small current flow in the lead wires as a result of the difference in potential between the junctions. Thus, the resistance of the lead wires will influence the reading to some extent.

A more suitable way of obtaining an average temperature is to use the thermopile circuit in Fig. 8-20. Each of the "hot" junctions may be at a different

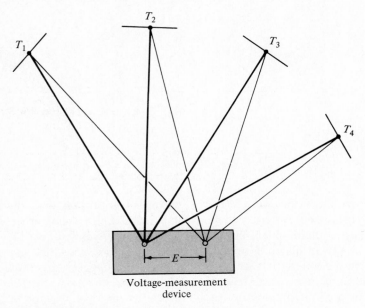

Figure 8-22 Parallel connection of thermocouples.

temperature, while all the "cold" junctions may be maintained at a fixed reference value. The average emf is then given by

$$E_{av} = \frac{E}{n} \qquad (8\text{-}13)$$

where n is the number of junction pairs and E is the total reading of the thermopile. The average temperature corresponding to the average emf given in Eq. 8-13 may then be determined.

From the above discussion it is clear that the thermocouple measures the temperature at the last point of electric contact of the two dissimilar materials. Consider the situations shown in Fig. 8-23. The thermocouple installation in Fig. 8-23a is made so that only the junction bead makes contact with the metal plate whose temperature is to be measured. The installation in Fig. 8-23b allows contact at two points. If there is a temperature gradient in the metal plate, then

(a) (b) (c)

Figure 8-23 Installation of thermocouples on a metal plate. (a) Only junction bead contacts plate; (b) contact at two points; (c) contact at bead and along wires.

the emf of the thermocouple will be indicative of the average of the temperatures of these two points. In Fig. 8-23c only the junction bead contacts the metal plate, but the two thermocouple wires are in electric contact a short distance away from the plate. The temperature indicated by the thermocouple will be that temperature at the shorted electric contact.

Quartz-Crystal Thermometer

A novel and highly accurate method of temperature measurement is based on the sensitivity of the resonant frequency of a quartz crystal to temperature change. When the proper angle of cut is used with the crystal, there is a very linear correspondence between the resonant frequency and temperature. Commercial models for the device utilize electronic counters and digital readout for the frequency measurement. For absolute temperature measurements usable sensitivities of 0.001°C are claimed for the device. Since the measurement process relies on a frequency measurement, the device is particularly insensitive to noise pickup in connecting cables. The device has a normal operating range of -40 to 230°C (-40 to 440°F) and a frequency-temperature relationship of about 100 Hz/°C.

Liquid Crystal Thermography

Cholesteric liquid crystals, formed from esters of cholesterol, exhibit an interesting response to temperature. Over a reproducible temperature range the liquid crystal will exhibit all colors of the visible spectrum. The phenomenon is reversible and repeatable. By varying the particular formulation, one can make cholesteric liquid crystals operate from just below 0°C to several hundred degrees Celsius. The *event temperature range*, or temperature at which the color changes are displayed, can vary from 1 to 50°C. Thus, the liquid crystals afford the opportunity for rather precise temperature indication through observation of the color changes.

To prevent deterioration of the crystals, they can be coated with polyvinyl alcohol, producing encapsulated liquid crystals which are available either as a water-based slurry or precoated on a blackened substrate of paper or Mylar. A description of the application of liquid crystals to heat-transfer measurements is given in Refs. [21] and [22].

Example 8-5 The thermopile arrangement shown in Fig. 8-20 uses copper-constantan thermocouples with $T_1 = 300°F$ and $E = 3.2$ mV for the three-junction pairs. Calculate the value of T_2.

SOLUTION We first calculate the thermoelectric sensitivity at 300°F using the values in Table 8-3.

$$S_{300} = \left(\frac{dE}{dT}\right)_{300} \approx \frac{E_{350} - E_{250}}{350 - 250} = \frac{8.085 - 5.307}{100} = 0.0278 \text{ mV/°F}$$

The thermoelectric emf generated by each junction pair is

$$E = \frac{3.2}{3} = 1.067 \text{ mV}$$

so that the temperature difference is

$$T_2 - T_1 = \frac{E}{S} = \frac{1.067}{0.0278} = 38.4°\text{F}$$

and
$$T_2 = 300 + 38.4 = 338.4°\text{F}$$

Example 8-6 The thermoelectric effect has been used for electric-power generation. Calculate the thermoelectric sensitivity of a device using bismuth and tellurium as the dissimilar materials, and estimate the maximum voltage output for a 100°F temperature difference at approximately room temperature using one junction.

SOLUTION The sensitivity is calculated from the data of Table 8-4 as

$$S = S_{\text{tellurium}} - S_{\text{bismuth}}$$
$$= 500 - (-72) = 572 \ \mu\text{V}/°\text{C}$$

The voltage output for a 100°F temperature difference is calculated as

$$E = S \ \Delta T = (572 \times 10^{-6})(\tfrac{5}{9})(100)$$
$$= 3.18 \times 10^{-2} \text{ V}$$

8-6 TEMPERATURE MEASUREMENT BY RADIATION

In addition to the methods described in the preceding sections, it is possible to determine the temperature of a body through a measurement of the thermal radiation emitted by the body. Two methods are commonly employed for measurement: (1) optical pyrometry and (2) emittance determination. Before we discuss these methods, it is necessary to describe the nature of thermal radiation.

Thermal radiation is electromagnetic radiation emitted by a body as a result of its temperature. This radiation is distinguished from other types of electromagnetic radiation such as radio waves and X rays, which are not propagated as a result of temperature. Thermal radiation lies in the wavelength region from about 0.1 to 100 μm (1 μm = 10^{-6} m). The total thermal radiation emitted by a *blackbody* (ideal radiation) is given as

$$E_b = \sigma T^4 \tag{8-14}$$

where σ = Stefan-Boltzmann constant
$\quad = 0.1714 \times 10^{-8} \text{ Btu/h} \cdot \text{ft}^2 \cdot °\text{R}^4$
$\quad = 5.669 \times 10^{-8} \text{ W/m}^2 \cdot \text{K}^4$
E_b = emissive power, Btu/h·ft² or W/m²
T = absolute temperature, °R or K

The emissive power of the blackbody varies with wavelength according to the Planck distribution equation:

$$E_{b\lambda} = \frac{C_1 \lambda^{-5}}{e^{C_2/\lambda T} - 1} \tag{8-15}$$

where $E_{b\lambda}$ = monochromatic blackbody emissive power
$\quad\quad\quad = \text{Btu/h} \cdot \text{ft}^2 \cdot \mu\text{m}$
$\quad\quad\quad = \text{W/m}^2 \cdot \mu\text{m}$
$\quad\quad \lambda = \text{wavelength, } \mu\text{m}$
$\quad\quad T = \text{temperature, } °\text{R or K}$
$\quad\quad C_1 = 1.187 \times 10^8 \text{ Btu} \cdot \mu\text{m}^4/\text{h} \cdot \text{ft}^2$
$\quad\quad\quad = 3.743 \times 10^8 \text{ W} \cdot \mu\text{m}^4/\text{m}^2$
$\quad\quad C_2 = 2.5896 \times 10^4 \text{ } \mu\text{m} \cdot °\text{R}$
$\quad\quad\quad = 1.4387 \times 10^4 \text{ } \mu\text{m} \cdot \text{K}$

A plot of Eq. 8-15 is given in Fig. 8-24 for two temperatures.

When thermal radiation strikes a material surface, the following relation applies:

$$\alpha + \rho + \tau = 1 \tag{8-16}$$

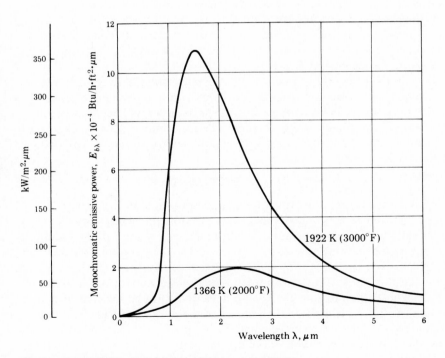

Figure 8-24 Blackbody emissive power for two temperatures.

where α = absorptivity or the fraction of the incident radiation absorbed
ρ = reflectivity or the fraction reflected
τ = transmissivity or the fraction transmitted

For most solid materials, $\tau = 0$, so that

$$\alpha + \rho = 1 \qquad (8\text{-}17)$$

The emissivity ϵ is defined as

$$\epsilon = \frac{E}{E_b} \qquad (8\text{-}18)$$

where E is the emissive power of an actual surface and E_b is the emissive power of a blackbody at the same temperature. Kirchhoff's identity furnishes the additional relationship

$$\epsilon = \alpha \qquad (8\text{-}19)$$

under conditions of thermal equilibrium. A gray body is one for which the emissivity is constant for all wavelengths; i.e.,

$$\epsilon_\lambda = \frac{E_\lambda}{E_{b\lambda}} = \epsilon \qquad (8\text{-}20)$$

Actual surfaces frequently exhibit highly variable emissivities over the wavelength spectrum. Figure 8-25 illustrates the distinctive features of blackbody and gray-body radiation. For purposes of analysis the real surface is frequently approximated by a gray body having an emissivity equal to the average total emissivity of the real surface as defined by Eq. (8-18).

Let us now consider the measurement of temperature through the use of optical pyrometry. This method refers to the identification of the temperature of a surface with the color of the radiation emitted. As a surface is heated, it becomes dark red, orange, and finally white in color. The maximum points in the blackbody radiation curves shift to shorter wavelengths with increase in temperature according to Wien's law.

$$\lambda_{\max} T = 5215.6 \ \mu\text{m} \cdot {}^\circ\text{R}(2897.6 \ \mu\text{m} \cdot \text{K}) \qquad (8\text{-}21)$$

where $\lambda_{\max}$ is the wavelength at which the maximum points in the curves in Fig. 8-24 occur. The shift in these maximum points explains the change in color as a body is heated; i.e., higher temperatures result in a concentration of the radiation in the shorter-wavelength portion of the spectrum. The temperature-measurement problem consists of a determination of the variation of temperature with color of the object. For this purpose an instrument is constructed as shown schematically in Fig. 8-26. The radiation from the source is viewed through the lens and filter arrangement. An absorption filter at the front of the device reduces the intensity of the incoming radiation so that the standard lamp may be operated at a lower level. The standard lamp is placed in the optical path of the incoming radiation. By an adjustment of the lamp current, the color of the filament may be made to

Figure 8-25 Comparison of emissive powers of blackbody, ideal gray body, and actual surfaces.

Figure 8-26 Schematic of optical pyrometer.

Filament
too cold

Null
condition

Filament
too hot

Figure 8-27 Appearance of lamp filament in eyepiece of optical pyrometer.

match the color of the incoming radiation. The red filter is installed in the eyepiece to ensure that comparisons are made for essentially monochromatic radiation, thus eliminating some of the uncertainties resulting from variation of radiation properties with wavelength.

Figure 8-27 illustrates the appearance of the lamp filament as viewed from the eyepiece. When balance conditions are achieved, the filament will seem to disappear in the total incoming radiation field. Temperature calibration is made in terms of the lamp heating current.

The temperature of a body may also be measured by determining the total emitted energy from the body and then calculating the temperature from

$$E = \epsilon \sigma T^4 \tag{8-22}$$

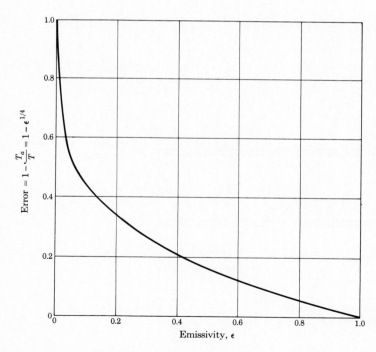

Figure 8-28 Temperature error due to nonblackbody surface conditions.

In order to determine the temperature, the emissivity of the material must be known, so that

$$T = \left(\frac{E}{\epsilon\sigma}\right)^{1/4} \tag{8-23}$$

The *apparent* blackbody temperature is the value as calculated from Eq. (8-23) with $\epsilon = 1$, or

$$T_a = \left(\frac{E}{\sigma}\right)^{1/4} \tag{8-24}$$

If the apparent temperature is taken as the measured value, the *error* in temperature due to nonblackbody conditions is thus

$$\text{Error} = \frac{T - T_a}{T} = 1 - \frac{T_a}{T} = 1 - \epsilon^{1/4} \tag{8-25}$$

Figure 8-28 gives this error as a function of emissivity.

Several methods are available to measure the emitted thermal energy from a body, and these methods will be discussed in Chap. 12. For now, it is important to realize that temperature may be calculated from the above equations once this measurement is made.

In practice the optical pyrometer is the more widely used of the two radiation temperature methods since it is relatively inexpensive and portable and the determination does not depend strongly on the surface properties of the material. The measurement of radiant energy from a surface can be quite accurate, however, when suitable instruments are employed. If the surface emissive properties are accurately known, this measurement can result in a very accurate determination of temperature. It will be noted that a total radiation measurement forms the basis for the upper limit of the International Temperature Scale, as discussed in Chap. 2.

The interested reader should consult Refs. [2], [4], and [7] for more information on thermal radiation and temperature measurements by the above methods.

Example 8-7 The energy emitted from a piece of metal is measured, and the temperature is determined to be 1050°C, assuming a surface emissivity of 0.82. It is later found that the true emissivity is 0.75. Calculate the error in the temperature determination.

SOLUTION The emitted energy is given by

$$q/A = \epsilon\sigma T^4$$

when $T = 1050°C = 1323$ K, $\epsilon = 0.82$. We wish to calculate the value of the true temperature T' such that

$$q/A = \epsilon'\sigma(T')^4$$

where $\epsilon' = 0.75$. Thus,

$$(0.82)(1323)^4 = (0.75)(T')^4$$

and

$$T' = (1323)\left(\frac{0.82}{0.75}\right)^{1/4} = 1352 \text{ K}$$

so that the temperature error is

$$\Delta T = 1352 - 1323 = 29°\text{C}$$

8-7 EFFECT OF HEAT TRANSFER ON TEMPERATURE MEASUREMENT

A heat-transfer process is associated with all temperature measurements. When a thermometer is exposed to an environment, the temperature is determined in accordance with the total heat-energy exchange with the temperature-sensing element. In some instances the temperature of the thermometer can be substantially different from the temperature which is to be measured. In this section we wish to discuss some of the methods that may be used to correct the temperature readings. It may be noted that the errors involved are classified as fixed errors.

Heat transfer may take place as a result of one or more of three modes: conduction, convection, or radiation. In general, all three modes must be taken into account in analyzing a temperature-measurement problem. Conduction is described by Fourier's law:

$$q = -kA \frac{\partial T}{\partial x} \tag{8-26}$$

where k = thermal conductivity
   A = area through which the heat transfer takes place
   q = heat-transfer rate in the direction of the decreasing temperature gradient

If a temperature gradient exists along a thermometer, heat may be conducted into or out of the sensing element in accordance with this relation.

Convection heat transfer is described in accordance with Newton's law of cooling:

$$q = hA(T_s - T_\infty) \tag{8-27}$$

where h = convection heat-transfer coefficient
   A = surface area exchanging heat with the fluid
   T_s = surface temperature
   T_∞ = fluid temperature

The radiation heat transfer between two surfaces is proportional to the difference in absolute temperatures to the fourth power according to the Stefan-Boltzmann law of thermal radiation:

$$q_{1-2} = \sigma F_G F_\epsilon (T_1^4 - T_2^4) \tag{8-28}$$

where F_G is a geometric factor and F_ϵ is a factor that describes the radiation properties of the surfaces.

Consider the temperature-measurement problem illustrated in Fig. 8-29. A thermocouple junction is installed in the flat plate whose temperature is to be measured. The plate is exposed to a convection environment on both sides, and the thermocouple wires are exposed to this same environment. The thermocouple wires are covered with insulating material as shown. If the plate temperature is higher than the convection environment, heat will be conducted out along the thermocouple wire and the temperature of the junction will be lower than the true plate temperature. An analytical solution to this problem has been presented by Schneider [11], and the results are given in Fig. 8-30. This solution neglects radiation heat-transfer effects.

The nomenclature for the parameters in Fig. 8-30 is as follows:

h_1, h_2 = convection heat-transfer coefficients on the sides of the plate as indicated, Btu/h · ft² · °F or W/m² · °C

h_t = convection heat-transfer coefficient from each of the thermocouple wires, Btu/h · ft² · °F or W/m² · °C

k = thermal conductivity of the plate material, Btu/h · ft · °F or W/m · °C

δ = plate thickness, ft or m

T_f = fluid temperature surrounding the thermocouple wire, °F or °C

T_i = temperature indicated by the thermocouple, °F or °C

T_p = true plate temperature (temperature a large distance away from the thermocouple junction), °F or °C

$r_s = \sqrt{2} r_t$, where r_t is the radius of each of the thermocouple wires, ft or m

$$m = \left(\frac{h_1 + h_2}{k\delta} \right)^{1/2} \qquad \text{ft}^{-1} \text{ or m}^{-1} \tag{8-29}$$

$$K = \sqrt{2} \pi (k_A^{1/2} + k_B^{1/2}) r_t^{3/2} \left(\frac{1}{h_t} + \frac{\delta_i}{k_i} \right)^{-1/2} \qquad \text{Btu/h · °F or W/°C} \tag{8-30}$$

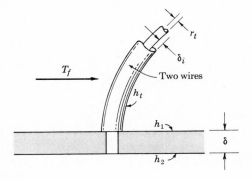

Figure 8-29 Schematic of general thermocouple installation in a flat plate.

where k_A and k_B are the thermal conductivities of the two thermocouple materials, δ_i is the thickness of the thermocouple insulation, and k_i is the thermal conductivity of the insulation.

If the solid is a relatively massive one, the temperature correction can be made in accordance with calculations developed by Hennecke and Sparrow [16] and presented in graphical form in Fig. 8-31. In this figure k is the thermal conductivity of the solid, r is the radius of the wire or $\sqrt{2}$ times the radius for two wires, L is the length of the thermocouple leads, h_s is the convection heat-transfer coefficient between the solid and the fluid, kA is the equivalent conductivity area product for axial conduction in the wire, and R is the radial thermal resistance of the wire, insulation, and convection to the fluid. For a wire of radius r_w, covered by a layer of insulation having an outside radius r_i and thermal conductivity k_i, the value of R is calculated from

$$R = \frac{1}{h2\pi r_i} + \frac{\ln (r_i/r_w)}{2\pi k_i} \tag{8-31}$$

where h is the convection heat-transfer coefficient from the outside of the insulation to the fluid. We should note that the length L is sufficiently long in most applications to make the hyperbolic tangent term unity in the abscissa of Fig. 8-31.

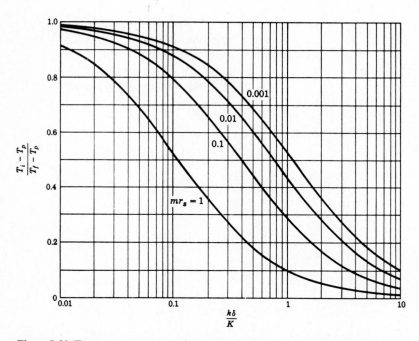

Figure 8-30 Temperature-compensation curves for installation in Fig. 8-29. (*From Ref. [11]*.)

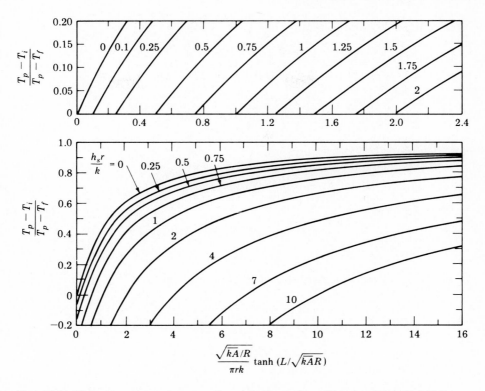

Figure 8-31 Temperature-correction curves for relatively massive solids, according to Ref. [16].

Example 8-8 A copper-constantan thermocouple is attached to a $\frac{1}{16}$-in-thick stainless-steel plate [$k = 35$ Btu/h·ft·°F] as shown in Fig. 8-29. The diameter of the wires is 0.04 in, and the following estimates are made of the significant heat-transfer parameters:

$$h_1 = 4.0 \text{ Btu/h·ft}^2 \cdot °F$$

$$h_2 = 2.5 \text{ Btu/h·ft}^2 \cdot °F$$

$$h_t = 20 \text{ Btu/h·ft}^2 \cdot °F$$

$$k_A = 14 \text{ Btu/h·ft·°F} \quad \text{(constantan)}$$

$$k_B = 220 \text{ Btu/h·ft·°F} \quad \text{(copper)}$$

$$T_f = 500°F$$

The thermocouple wires are coated with an electrically insulating lacquer, which may be considered negligibly thin insofar as heat transfer is concerned. The plate is cooled on the side opposite the thermocouple installation, and the thermocouple indicates a temperature of 300°F. Calculate the true plate temperature.

SOLUTION We observe that

$$\delta = \tfrac{1}{16} \text{ in} = 0.0071 \text{ ft}$$

$$r_t = 0.02 \text{ in} = 0.00167 \text{ ft}$$

$$\delta_i \approx 0$$

$$T_i = 300°F$$

From Eq. (8-29)

$$m = \left[\frac{4.0 + 2.5}{(35)(0.0142)} \right]^{1/2} = 3.62$$

From Eq. (8-30)

$$K = \sqrt{2}\pi(14^{1/2} + 220^{1/2})(0.00167)^{3/2}(\tfrac{1}{20} + 0)^{-1/2}$$

$$= 0.0251$$

The parameters for use with Fig. 8-30 are thus

$$mr_s = \sqrt{2}(0.00167)(3.62) = 0.00856$$

$$\frac{k\delta}{K} = \frac{(35)(0.0071)}{(0.0251)} = 9.9$$

Using these values, we obtain from Fig. 8-30

$$\frac{T_i - T_p}{T_f - T_p} = 0.08$$

or $\qquad 300 - T_p = (0.08)(500 - T_p) \qquad$ and $\qquad T_p = 282.6°F$

The thermocouple conduction error is $T_i - T_p = 17.4°F$. This error could be reduced by using smaller thermocouple wires or by insulating the wires.

Example 8-9 A thermocouple wire having an effective diameter of 1.5 mm is attached to a ceramic solid having the properties $\rho = 2500 \text{ kg/m}^3$, $c = 0.7 \text{ kJ/kg}\cdot°C$, and $k = 0.9 \text{ W/m}\cdot°C$. The thermocouple has an effective conductivity of 80 W/m · °C. The wire is very long and essentially bare, with a convection coefficient of 250 W/m² · °C. The convection coefficient h_s is 20 W/m² · °C. Calculate the true plate temperature when the thermocouple indicates 200°C and the fluid temperature is 90°C.

SOLUTION We make use of Fig. 8-31 and calculate the quantities

$$R = \frac{1}{h2\pi r_i} = \frac{1}{(250)(2\pi)(0.75 \times 10^{-3})} = 0.849$$

$$\overline{kA} = (80)\pi(0.75 \times 10^{-3})^2 = 1.414 \times 10^{-4}$$

Because $L \to \infty$, $\tanh(\infty) \to 1.0$.

$$\frac{\sqrt{kA/R}}{\pi r k} = \frac{[(1.414 \times 10^{-4})/0.849]^{1/2}}{\pi(0.75 \times 10^{-3})(0.9)} = 6.085$$

$$\frac{h_s r}{k} = \frac{(250)(0.75 \times 10^{-3})}{0.9} = 0.208$$

From Fig. 8-31 we read

$$\frac{T_p - T_i}{T_p - T_f} = 0.8$$

and, with $T_i = 200°C$, $T_f = 90°C$, the true plate temperature is

$$T_p = 640°C$$

This installation results in a very large error.

Example 8-10 Repeat Example 8-9 for installation on a block of aluminum with $k = 200$ W/m · °C.

SOLUTION We recalculate the appropriate parameters as

$$\frac{\sqrt{kA/R}}{\pi r k} = \frac{[(1.414 \times 10^{-4})/0.849]^{1/2}}{\pi(0.75 \times 10^{-3})(200)} = 0.0274$$

$$\frac{h_s r}{k} = \frac{(250)(0.75 \times 10^{-3})}{200} = 9.36 \times 10^{-4}$$

Then, from Fig. 8-31,

$$\frac{T_p - T_i}{T_p - T_f} \approx 0$$

and

$$T_p \approx 200°C$$

Of course, this indicates the extreme importance of the conductivity of the solid to the measurement error.

Now let us consider the general problem of the temperature measurement of a gas stream and the influence of radiation on this measurement. The situation is illustrated in Fig. 8-32. The temperature of the thermometer is designated as T_t,

$$q_{rad} = \sigma\epsilon(T_t^4 - T_s^4)$$

Gas
T_g

Surrounding, T_s

T_t

Figure 8-32 Schematic illustrating influence of radiation on temperature of thermometer.

the true temperature of the gas is T_g, and the effective radiation temperature of the surroundings is T_s. If it is assumed that the surroundings are very large, then the following energy balance may be made:

$$hA(T_g - T_t) = \sigma A \epsilon (T_t^4 - T_s^4) \qquad (8\text{-}32)$$

where h = convection heat-transfer coefficient from the gas to the thermometer
      A = surface area of the thermometer
      ϵ = surface emissivity of the thermometer

Equation (8-32) may be used to determine the true gas temperature.

In practice, the radiation error in a temperature measurement is reduced by placing a radiation shield around the thermometer, which reflects most of the radiant energy back to the thermometer. A simple radiation-shield arrangement is shown in Fig. 8-33. The environment is assumed to be very large, and we wish to find the true gas temperature knowing the indicated temperature T_t and the other heat-transfer parameters, such as the heat-transfer coefficients and emissivities. The thermometer size may vary considerably, from the rather substantial dimensions of a mercury-in-glass thermometer to a tiny thermistor bead embedded in the tip of a hypodermic needle. An energy balance that neglects conduction effects may be made for both the thermometer and the shield using a radiation-network analysis to obtain expressions for the radiation heat transfer.†
The results of the analysis are

THERMOMETER

$$h_t(T_g - T_t) = \frac{\epsilon_t}{1 - \epsilon_t} (E_{bt} - J_t) \qquad (8\text{-}33)$$

† An explanation of the radiation-network method is given in Ref. [5].

Figure 8-33 A simple radiation-shield arrangement for a thermometer.

SHIELD

$$2h_s(T_g - T_s) = \frac{E_{bs} - J_t}{(1/F_{ts})(A_s/A_t) + 1/\epsilon_s - 1} + \frac{E_{bs} - E_{be}}{1/\epsilon_s - 1 + 1/F_{se}} \tag{8-34}$$

where

$$J_t = \frac{\{E_{bt}[\epsilon_t/(1 - \epsilon_t)] + F_{te} E_{be}\}A_t/A_s + E_{bs}/[(1/F_{ts})(A_s/A_t) + 1/\epsilon_s - 1]}{(A_t/A_s)[F_{te} + \epsilon_t/(1 - \epsilon_t)] + 1/[(1/F_{ts})(A_s/A_t) + 1/\epsilon_s - 1]} \tag{8-35}$$

ϵ_t and ϵ_s are the emissivities of the thermometer and shield, respectively, h_t and h_s are the convection heat-transfer coefficients from the gas to the thermometer and shield, respectively, A_t is the surface area of the thermometer for convection and radiation, A_s is the surface area of the radiation shield *on each side*, and the blackbody emissive powers E_b are given by

$$E_{bt} = \sigma T_t^4 \tag{8-36}$$

$$E_{bs} = \sigma T_s^4 \tag{8-37}$$

$$E_{be} = \sigma T_e^4 \tag{8-38}$$

The radiation-shape factors F are defined as

> F_{ts} = fraction of radiation that leaves the thermometer and arrives at the shield
> F_{se} = fraction of radiation that leaves the outside surface of shield and gets to the environment (this fraction is 1.0)
> F_{te} = fraction of radiation that leaves the thermometer and gets to the environment

The analysis that arrives at Eqs. (8-33) and (8-34) has been simplified somewhat by assuming that all the radiation exchange between the shield and environment may be taken into account by adjusting the value of F_{se} to include radiation from the inside surface of the shield. Some type of convection heat-transfer analysis must be used to determine the values of h_t and h_s, depending on whether natural or forced convection is involved, etc. An iterative solution must usually be performed on Eqs. (8-33) and (8-34) except in special cases. Some useful charts are given in Ref. [15] for correcting temperature measurements for the effects of conduction and radiation.

In general, the determination of convection heat-transfer coefficients requires very detailed consideration of the fluid dynamics of the problem, and empirical relations must often be used for the calculation. For more information the reader should consult Refs. [2] and [5]. Some frequently used formulas for calculating convection heat transfer are summarized in Table 8-5.

Table 8-5 Frequently used convection heat-transfer formulas, adapted from Holman [5]

Physical situation	Type of fluid	Range of validity	Heat-transfer relation	Fluid properties evaluated at
Forced convection over flat plate, plate heated over entire length	Gas or liquid	Laminar: $Re_x < 5 \times 10^5$ Turbulent: $Re_x > 5 \times 10^5$ Laminar: $Re_L < 5 \times 10^5$ Turbulent: $Re_L > 5 \times 10^5$	$Nu_x = 0.332\ Re_x^{1/2}\ Pr^{1/3}$ $St_x\ Pr^{2/3} = 0.0296\ Re_x^{-1/5}$ $\overline{Nu}_L = 0.664\ Re_L^{1/2}\ Pr^{1/3}$ $\overline{Nu}_L = (0.037\ Re_L^{0.8} - 850)\ Pr^{1/3}$	Film temperature T_f
Forced convection in smooth circular tube	Gas or liquid	$Re_d > 3000$ $Re_d < 2100$ and $Re_d\ Pr\left(\dfrac{d}{L}\right) > 0$	$\overline{Nu}_d = 0.023\ Re_d^{0.8}\ Pr^n$ $n = 0.4$ for heating; $n = 0.3$ for cooling $\overline{Nu}_d = 1.86(Re_d\ Pr)^{1/3}\left(\dfrac{d}{L}\right)^{1/3}\left(\dfrac{\mu}{\mu_w}\right)^{0.14}$	Average bulk temperature of fluid
Forced-convection crossflow over cylinder	Gas or liquids	$40 < Re_d < 4000$ $4000 < Re_d < 40{,}000$ $40{,}000 < Re_d < 400{,}000$	$\overline{Nu}_d = 0.683\ Re_d^{0.466}\ Pr^{1/3}$ $\overline{Nu}_d = 0.193\ Re_d^{0.618}\ Pr^{1/3}$ $\overline{Nu}_d = 0.0266\ Re_d^{0.805}\ Pr^{1/3}$	Film temperature T_f Film temperature T_f
Forced convection over spheres	Gas	$17 < Re_d < 70{,}000$	$\overline{Nu}_d = 0.37\ Re_d^{0.6}$	Film temperature T_f
	Liquids	$1 < Re_d < 200{,}000$	$\overline{Nu}_d\ Pr^{-0.3}\left(\dfrac{\mu}{\mu_w}\right)^{0.25} = 1.2 + 0.5\ Re_d^{0.54}$	Free-stream temperature T_∞

Free convection from vertical flat plate	Gas or liquid	$10^4 < Gr_L\ Pr < 10^9$	$\overline{Nu}_L = 0.59(Gr_L\ Pr)^{1/4}$	Film temperature T_f
		$10^9 < Gr_L\ Pr < 10^{12}$	$\overline{Nu}_L = 0.10(Gr_L\ Pr)^{1/3}$	Film temperature T_f
Free convection from horizontal cylinders	Gas or liquid	$10^4 < Gr_d\ Pr < 10^9$	$\overline{Nu}_d = 0.53(Gr_d\ Pr)^{1/4}$	Film temperature T_f
		$10^9 < Gr_d\ Pr < 10^{12}$	$\overline{Nu}_d = 0.13(Gr_d\ Pr)^{1/3}$	Film temperature T_f

Definition of symbols: All quantities in consistent set of units so that Nu, Re, Pr, Gr, and St are dimensionless.

$$\overline{Nu}_d = \frac{\overline{h}d}{k}; \quad \overline{Nu}_L = \frac{\overline{h}L}{k}; \quad Nu_x = \frac{h_x x}{k}; \quad Re_d = \frac{\rho u_\infty d}{\mu} \text{ for flow over cylinders or spheres}; \quad Re_d = \frac{\rho u_m d}{\mu} \text{ for flow in tubes}; \quad Re_x = \frac{\rho u_\infty x}{\mu}; \quad Re_L = \frac{\rho u_\infty L}{\mu};$$

$$Pr = \frac{c_p \mu}{k}; \quad Gr_L = \frac{\rho^2 g\beta(T_w - T_x)L^3}{\mu^2}; \quad Gr_d = \frac{\rho^2 g\beta(T_w - T_\infty)d^3}{\mu^2}; \quad St_x = \frac{h_x}{\rho u_\infty c_p};$$

where

d = diameter of tube
g = acceleration of gravity
$\overline{h}$ = average heat-transfer coefficient over entire surface
h_x = local heat-transfer coefficient on flat plate
k = thermal conductivity of fluid
L = total length or height of flat plate or length of tube

T_f = film temperature $(T_w + T_\infty)/2$
T_w = wall or surface temperature
T_∞ = free-stream temperature
u_m = mean fluid velocity in tube
u_∞ = free-stream velocity past flat plate, cylinder, or sphere

x = distance from leading edge of flat plate
β = volume coefficient of expansion of fluid
μ = dynamic viscosity of fluid
μ_w = viscosity evaluated at wall temperature
ρ = fluid density

Example 8-11 A mercury-in-glass thermometer is placed inside a cold room in a frozen-food warehouse to measure the change in air temperature when the door is left open for extended periods of time. In one instance the thermometer reads 1°C, while the automatic temperature-control system for the room indicates that the *wall temperature* of the room is -10°C. The convection heat-transfer coefficient for the thermometer is estimated at 10 W/m² · °C, and $\epsilon = 0.9$ for glass. Estimate the true air temperature.

SOLUTION We use Eq. (8-32) to calculate the true air temperature T_g:

$$h(T_g - T_t) = \epsilon\sigma(T_t^4 - T_s^4)$$

$$(10)(T_g - 274) = (5.669 \times 10^{-8})(0.9)(274^4 - 263^4)$$

Thus, $$T_g = 278.3 \text{ K} = 5.3°C$$

We may note, of course, the rather substantial difference between the indicated air temperature of 1°C and the actual temperature of 5.3°C.

8-8 TRANSIENT RESPONSE OF THERMAL SYSTEMS

When a temperature measurement is to be made under non-steady-state conditions, it is important that the transient response characteristics of the thermal system be taken into account. Consider the simple system shown in Fig. 8-34. The thermometer is represented by the mass m and specific heat c and is suddenly exposed to a convection environment of temperature T_∞. The convection heat-transfer coefficient between the thermometer and fluid is h, and radiation heat transfer is assumed to be negligible. It is also assumed that the thermometer is substantially uniform in temperature at any instant of time; i.e., the thermal conductivity of the thermometer material is sufficiently large in comparison with the surface conductance because of the convection heat-transfer coefficient. The energy balance for the transient process may be written as

$$hA(T_\infty - T) = mc \frac{dT}{d\tau} \tag{8-39}$$

The solution of Eq. (8-39) gives the temperature of the thermometer as a function of time:

$$\frac{T - T_\infty}{T_0 - T_\infty} = e^{(-hA/mc)\tau} \tag{8-40}$$

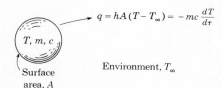
$$q = hA(T - T_\infty) = -mc \frac{dT}{d\tau}$$

T, m, c

Surface area, A

Environment, T_∞

Figure 8-34 Simple thermal system exposed to a sudden change in ambient temperature.

where T_0 is the thermometer temperature at time zero. Equation (8-40) represents the familiar exponential decay behavior described in Sec. 2-8. The time constant for the system in Fig. 8-34 is

$$\tau = \frac{mc}{hA} \tag{8-41}$$

When the heat-transfer coefficient is sufficiently large, there may be substantial temperature gradients within the thermometer itself and a different type of analysis must be used to determine the temperature variation with time. The interested reader should consult Ref. [5] for more information on this subject.

To indicate the approximate response times for semi-infinite solids, Figs. 8-35

Figure 8-35 Surface temperature rise for semi-infinite insulating solids subjected to constant surface heat fluxes between 1 W/cm² and 10 kW/cm². Signal levels to be expected from platinum and gold resistance thermometers and chromel-alumel thermocouples are also indicated. (*From Ref.* [*20*].)

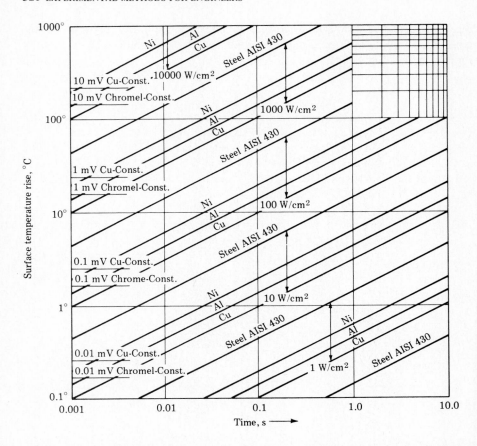

Figure 8-36 Surface temperature rise for semi-infinite metallic solids subjected to constant heat fluxes between 1 W/cm² and 10 kW/cm². Signal levels to be expected from chromel-constantan and copper-constantan thermocouples are also indicated. (*From Ref.* [*20*].)

and 8-36 show the surface temperature rises to be expected from different heat flux inputs. The approximate signal outputs from resistance thermometers and different thermocouple materials are also indicated. These figures were developed by Schultz and Jones [20] and may be used to estimate when transient effects become important.

8-9 THERMOCOUPLE COMPENSATION

Suppose a thermocouple is used to measure a transient temperature variation. The response of the thermocouple will be dependent on several factors, as outlined above, and will follow a variation like that of Eq. (8-40) when subjected to a step change in environment temperature. If a compensating electric network is applied to the system, it is possible to increase the frequency response of the thermocouple. The disadvantage of the compensating network is that it reduces the thermocouple output; however, if the measuring instrument is sufficiently sensitive, this problem is not too critical.

A typical thermocouple-compensation network is shown in Fig. 8-37. The thermocouple *input voltage* is represented by E_i, and the *output voltage* is represented by E_o. This particular network will attenuate low frequencies more than high frequencies and has a frequency response to a step input that is approximately opposite to that of a thermocouple. The network and thermocouple may thus be used in combination to produce a more nearly flat response over a wider frequency range. The output response of the network is given by

$$\frac{E_o}{E_i} = \alpha \frac{1 + j\omega\tau_c}{1 + \alpha j\omega\tau_c} \tag{8-42}$$

where $\alpha = \dfrac{R}{R + R_c}$

$\tau_c = R_c C$

ω = frequency of the input signal

$j = \sqrt{-1}$

The amplitude response is given by the absolute value of the function in Eq. (8-42). Thus,

$$\left(\frac{E_o}{E_i}\right)_{amplitude} = \alpha \sqrt{\frac{1 + \omega^2\tau_c^2}{1 + \alpha^2\omega^2\tau_c^2}} \tag{8-43}$$

The high-frequency compensation of the network is improved as the value of α is decreased, but this brings about a decreased output since the steady-state output is

$$\left(\frac{E_o}{E_i}\right)_{steady\ state} = \frac{R}{R + R_c} = \alpha \tag{8-44}$$

In practice, the compensating circuit is usually designed so that τ_c is equal to the time constant for the thermocouple. Variable resistors may be used to change the compensation with a change in thermal environment conditions.

Figure 8-37 Typical thermocouple-compensation network.

The overall problem we have described above is one of compensating a thermocouple output so that higher-frequency response may be obtained. The step response of the thermocouple given by Eq. (8-40) will only hold, of course, for the conditions described. Therefore, the compensating network of Fig. 8-37 will achieve the desired objectives in varying degrees, depending on the exact thermal environment to which the thermocouple is exposed. Let us examine the thermocouple and compensating network response in more detail. Figure 8-38 shows the thermocouple response to a step change in temperature of the environment T_∞. We assume that the voltage output of the thermocouple is directly proportional to temperature over this range so that the voltage output would also follow a similar curve. In Fig. 8-39 we have plotted the output voltage of the compensating network for a step-voltage input and for $\alpha = 0.1$. The thermocouple response is obtained by solving the electric-network equivalent of the thermocouple. In that network the thermal capacitance is analogous to electric capacitance, and the convection heat-transfer coefficient is analogous to thermal or electric resistance. If Figs. 8-38 and 8-39 are added together, we obtain the plot shown in Fig. 8-40 where the response is shown for different values of the ratio τ_c/τ. For $\tau_c/\tau > 1$, the response curve "overshoots" the steady-state voltage ratio of 0.1. It is clear that the curve $\tau = \tau_c$ offers a broader frequency response than the uncompensated curve. Careful inspection will also show that some overshoot can be desirable. The curve for $\tau_c/\tau = 1.1$ indicates a substantial extension of the frequency response within a plus or minus range of about 5 percent. Excessive overshoot, of course, gives a decided nonlinear response. Optimal dynamic response of the network will be obtained when $\tau_c/\tau \approx 1.1$, provided the thermocouple input behaves nicely. In practice, variable resistors will be provided in the circuits to adjust the compensation in accordance with the actual variations in the thermal environment.

Figure 8-38 Thermocouple response to step change in environment temperature.

Figure 8-39 Response of the network in Fig. 8-37 for a step voltage input with $\alpha = 0.1$.

We note again that the foregoing discussion pertains to a step function input, and the actual input may be markedly different. In practice, one should make the thermocouple as small as possible and calibrate against some known transient temperature. Adjustments to the network can be made to obtain the best calibration. Further information on transient response of thermocouples is given in Ref. [17].

Figure 8-40 Combined response of thermocouple and compensating network to step change in environment temperature.

Example 8-12 A thermocouple bead has the approximate shape of a sphere 1.5 mm in diameter. The properties may be taken as those of iron ($\rho = 7900$ kg/m^3, $c = 450$ J/kg · °C). Suppose such a bead is exposed to a convection environment where $h = 75$ W/m^2 · °C. Estimate the time constant for the thermocouple.

SOLUTION We use Eq. (8-41) to calculate the time constant with

$$m = \rho V = \rho \tfrac{4}{3} \pi r^3$$

$$A = 4 \pi r^2$$

so that

$$\tau = \frac{(7900)(450)(0.75 \times 10^{-3})}{(3)(75)} = \frac{\rho \tfrac{4}{3} \pi r^3 c}{h 4 \pi r^2}$$

$$= 11.85 \text{ s}$$

Example 8-13 Design a thermocouple-compensation network like that shown in Fig. 8-37 having a steady-state attenuation of a factor of 10. Let the network have the same time constant as the thermocouple bead in Example 8-12.

SOLUTION We take

$$\alpha = \frac{R}{R + R_c} = 0.1 \qquad \text{and} \qquad \tau_c = R_c C = 11.85$$

We must arbitrarily choose a value for one of the resistances. Let us choose a value of R_c which will make C a reasonable number. Take

$$R_c = 1 \text{ M}\Omega = 10^6 \ \Omega$$

Then $C = 11.85/10^6 = 1.185 \times 10^{-5} = 11.85 \ \mu$F. Using the above value of R_c we obtain

$$R = 0.111 \text{ M}\Omega$$

A more practical set of values would be

$$C = 10 \ \mu\text{F}$$

$$R_c = 1.185 \text{ M}\Omega$$

$$R = 0.132 \text{ M}\Omega$$

8-10 TEMPERATURE MEASUREMENTS IN HIGH-SPEED FLOW

The measurements of temperatures in high-speed gas-flow streams are sometimes difficult because of the fact that a stationary probe must measure the temperature of the gas that is brought to rest at the probe surface. As the gas velocity is reduced to zero, the kinetic energy is converted to thermal energy and evidenced as an increase in temperature. If the gas is brought to rest adiabatically, the resulting stagnation temperature is given by

$$T_0 = T_\infty + \frac{u_\infty^2}{2c_p g_c} \qquad (8\text{-}45)$$

where T_0 = stagnation temperature

T_∞ = free-stream or *static* temperature

u_∞ = flow velocity

c_p = constant-pressure specific heat of the gas

The stagnation temperature of the gas may also be expressed in terms of the Mach number as

$$\frac{T_0}{T_\infty} = 1 + \frac{\gamma - 1}{2} M^2 \qquad (8\text{-}46)$$

where $\gamma = c_p/c_v$ and has a value of 1.4 for air. An inspection of Eqs. (8-45) and (8-46) shows that the stagnation temperature is substantially the same as the static temperature for low-speed flow. The difference in these temperatures is only 4°F (2.2°C) for a velocity of 226 ft/s (68.9 m/s) in air.

For the actual case of a probe inserted in a high-speed flow stream, the temperature indicated by the probe will *not*, in general, be equal to the stagnation temperature given by Eq. (8-45). The measured temperature is called the *recovery temperature* and is strongly dependent on the probe configuration. The *recovery factor* is defined by

$$r = \frac{T_r - T_\infty}{T_0 - T_\infty} \qquad (8\text{-}47)$$

where T_r is the recovery temperature. In practice, the recovery factor must be determined from a calibration of the probe at different flow conditions. It is usually in the range $0.75 < r < 0.99$. One of the important objectives of a probe design is to achieve a configuration which produces a near constant recovery factor over a wide range of flow velocities.

A typical high-speed temperature probe is shown in Fig. 8-41 according to Winkler [13]. The gas stream enters the front of the probe and is diffused to a

Figure 8-41 High-speed temperature probe, according to Winkler [13].

Table 8-6 Summary of characteristics of temperature-measurement devices

Device	Applicable temperature range		Approximate accuracy		Transient response	Cost	Remarks
	°F	°C	°F	°C			
Liquid-in-glass thermometer:							
a. Alcohol	−90 to 150	−70 to 65	±1	±0.5	Poor	Low	Used as inexpensive low-temperature thermometers
b. Mercury	−35 to 600	−40 to 300	±0.5	±0.25	Poor	Variable	Accuracy of ±0.1°F (0.05°C) may be obtained with special calibrated thermometers
c. Gas-filled mercury	−35 to 1000	−40 to 550	±0.5	±0.25	Poor	Variable	
Fluid-expansion thermometer:							
a. Liquid or gas	−150 to 1000	−100 to 550	±2	±1	Poor	Low	Widely used for industrial temperature measurements
b. Vapor-pressure	20 to 400	−6 to 200	±2	±1	Poor	Low	
Bimetallic strip	−100 to 1000	−70 to 550	±0.5	±0.25	Poor	Low	Widely used in simple temperature-control devices
Electrical-resistance thermometer	−300 to 1800	−180 to 1000	±0.005†	±0.0025	Fair to good, depending on size of element	Readout equipment can be rather expensive for high-precision work	Most accurate of all methods
Thermistor	−100 to 500	−70 to 250	±0.02†	±0.01	Very good	Low, but readout equipment may be expensive for high-precision work	Useful for temperature-compensation circuits; thermistor beads may be obtained in very small sizes

Copper-constantan thermocouple	−300 to 650	−180 to 350	±0.5	±0.25	Good, depends on wire size	Low	Superior in reducing atmospheres
Iron-constantan thermocouple	−300 to 1200	−180 to 650	±0.5	±0.25	Good, depends on wire size	Low	Resistant to oxidation at high temperatures
Chromel-alumel thermocouple	−300 to 2200	−180 to 1200	±0.5	±0.25	Good, depends on wire size	Low	
Platinum-platinun 10% rhodium thermocouple	0 to 3000	−15 to 1650	±0.5	±0.25	Good, depends on wire size	High	Low output; most resistant to oxidation at high temperatures; accuracy of ±0.15°F (0.07°C) may be obtained in carefully controlled conditions
Optical pyrometer	1200 up	650 up	±20	±10	Poor	Medium	Widely used for measurement of industrial furnace temperatures
Radiation pyrometer	0 up	−15 up	±0.5°C at low ranges, 2.5 to 10°C at high temperatures; depends on blackbody conditions and type of pyrometer; see Chap. 12		Good, depending on type of pyrometer	Medium to high	Increased applications resulting from new high-precision devices being developed

† Accuracy is that which *may* be achieved. Inexpensive versions of the devices may not give the optimum accuracy.

lower velocity in the enclosing shield arrangement. The flow subsequently leaves the probe through the side ventholes. The shield also serves to reduce radiation losses from the thermocouple-sensing element. The recovery factor for this probe is essentially 1.0.

8-11 SUMMARY

A summary of the range and application of the various temperature-measurement devices is given in Table 8-6. It will be noted that many of the devices overlap in their range of application. Thus, the selection of a device for use in a particular application is often a matter of preference. In many cases, the selection is based on a consideration of the type of readout equipment that is already available at the installation where the device is to be used.

REVIEW QUESTIONS

8-1 Distinguish between the Fahrenheit and Celsius temperature scales.

8-2 Upon what principle does the mercury-in-glass thermometer operate?

8-3 Describe the resistance characteristics of thermistors.

8-4 State the law of intermediate metals for thermocouples.

8-5 What is the Seebeck effect?

8-6 State the law of intermediate temperatures for thermocouples.

8-7 Why is a reference temperature necessary when using thermocouples?

8-8 Why does the resistance of thermocouple wires not influence a temperature reading when a potentiometer is used?

8-9 In general, why does heat transfer influence the accuracy of a temperature measurement?

8-10 In general, would one classify the response of thermal systems as fast, slow, or intermediate?

PROBLEMS

8-1 Calculate the temperature at which the Fahrenheit and Celsius scales coincide.

8-2 A certain mercury-in-glass thermometer has been calibrated for a prescribed immersion depth. The thermometer is immersed too much, such that the extra depth is equal to a distance of 10° on the scale. The true temperature reading may be calculated with

$$T_{true} = T_{ind} - 0.000088(T_{ind} - T_{amb})D$$

where T_{ind} = indicated temperature
   T_{amb} = ambient temperature of the exposed stem
   D = extra immersion depth of the thermometer past the correct mark

Calculate the thermometer error for an indicated temperature of 210°F and an ambient temperature of 70°F.

8-3 The bimetallic strip material of Example 8-1 is to be used in an on-off–temperature-control device, which will operate at a nominal temperature of 200°F. Calculate the deflection at the end of a 4-in strip for deviations of ±1 and 2°F from the nominal temperature.

8-4 Suppose the bimetallic strip in Prob. 8-3 is used to indicate temperatures of 150, 200, and 300°F. The tip of the strip is connected to an appropriate mechanism, which amplifies the deflection so that it may be read accurately. The uncertainty in the length of the strip is ±0.01 in, and the uncertainty in the thickness of each material is ±0.0002 in. The perpendicular deflection from the 100°F position is taken as an indication of the temperature. Calculate the uncertainty in each of the above temperatures, assuming zero uncertainty in all the material properties; i.e., calculate the uncertainty in the deflection at each of these temperatures.

8-5 A bimetallic strip of yellow brass and Monel 400 is bonded at 120°F. The thickness of the yellow brass is 0.014 ± 0.0002 in, and the thickness of the Monel is 0.010 ± 0.0001 in. The length of the strip is 5 in. Calculate the deflection sensitivity defined as the deflection per degree Fahrenheit temperature difference. Estimate the uncertainty in this deflection sensitivity.

8-6 The specific volume of mercury is given by the relation

$$v = v_0(1 + aT + bT^2)$$

where T is in degrees Celsius and

$$a = 0.1818 \times 10^{-3}$$
$$b = 0.0078 \times 10^{-6}$$

A high-temperature thermometer is constructed of a Monel-400 tube having an inside diameter of 0.8 mm ± 5 nm. After the inside is evacuated, mercury is placed on the inside such that a column height of 10 cm ± 0.25 mm is achieved when the thermometer temperature is 260°C. If the tube is to be used for a temperature measurement, calculate the uncertainty at 260°C if the uncertainty in the height measurement is ±0.25 mm.

8-7 A fluid-expansion thermometer uses liquid Freon ($C = 0.22$ Btu/lbm · °F. $\rho = 85.2$ lbm/ft³) enclosed in a 0.25-in-ID copper ($C = 0.092$, $\rho = 560$) cylinder with a wall thickness of 0.032 in. The thermometer is exposed to a crossflow of air at 400°F, 15 psia, and 20 ft/s. Estimate the time constant for such a thermometer. Neglect all effects of the capillary tube. Repeat for a crossflow of liquid water at 180°F and 20 ft/s. Recalculate the time constant for both of these situations for a stainless-steel bulb ($C = 0.11$, $\rho = 490$).

8-8 It is desired to measure a temperature differential of 5°F using copper-constantan thermocouples at a temperature level of 200°F. A millivolt recorder is available for the emf measurement which has an uncertainty of 0.004 mV. Precision thermocouple wire is available, which may be assumed to match the characteristics in Table 8-3 exactly. How many junction pairs in a thermopile must be used in order that the uncertainty in the temperature differential measurement does not exceed 0.05°F? Neglect all errors due to heat transfer.

8-9 A thermopile of chromel-alumel with four junction pairs is used to measure a differential temperature of 4.0°F at a temperature level of 400°F. The sensitivity of the thermocouple wire is found to match that in Table 8-3 within ±0.5 percent, and the millivolt potentiometer has an uncertainty of 0.002 mV. Calculate the uncertainty in the differential temperature measurement.

8-10 For a certain thermistor, $\beta = 3420$ K, and the resistance at 200°F is known to be 1010 ± 3 Ω. The thermistor is used for a temperature measurement, and the resistance is measured as 2315 ± 4 Ω. Calculate the temperature and the uncertainty.

8-11 Calculate the thermoelectric sensitivity for iron-constantan and copper-constantan at 0°C. Plot the error which would result if these values were assumed constant over the temperature range from −50 to +600°F.

8-12 A chromel-alumel thermocouple is exposed to a temperature of 1560°F. The potentiometer is used as the cold junction, and its temperature is estimated to be 83°F. Calculate the emf indicated by the potentiometer.

8-13 Four iron-constantan thermocouple junctions are connected in series. The temperatures of the four junctions are 200, 300, 100, and 32°F. Calculate the emf indicated by a potentiometer.

8-14 A millivolt recorder is available with a total range of 0 to 10 mV and an accuracy of ± 0.25 percent of full scale. Calculate the corresponding temperature ranges for use with iron-constantan, copper-constantan, and chromel-alumel thermocouples, and indicate the temperature accuracies for each of these thermocouples.

8-15 When a material with spectral emissivity less than unity is viewed with an optical pyrometer, the apparent temperature will be somewhat less than the true temperature and will depend on the wavelength at which the measurements are made. The error is given by

$$\frac{1}{T} - \frac{1}{T_a} = \frac{\lambda}{C_2} \ln \epsilon_\lambda$$

where T = true temperature, K
T_a = apparent temperature, K
ϵ_λ = spectral emissivity at the particular value of λ
C_2 = constant from Eq. (8-15)

For a measurement at 0.665 μm and an apparent temperature of 1300°C, plot the error as a function of ϵ_λ.

8-16 A radiant energy measurement is made to determine the temperature of a hot block of metal. The emitted energy from the surface of the metal is measured as 28 ± 0.4 kW/m^2 and the surface emissivity is estimated as $\epsilon = 0.90 \pm 0.05$. Calculate the surface temperature of the metal, and estimate the uncertainty.

8-17 A radiometer is used to measure the radiant energy flux from a material having a temperature of 280 ± 0.5°C. The emissivity of the material is 0.95 ± 0.03. The radiometer is then used to view a second material having exactly the same geometric shape and orientation as the first material and an emissivity of 0.72 ± 0.05. The uncertainty in the radiant energy flux measurement is ± 225 W/m^2 for both measurements, and the second measurement is 2.23 times as large as the first measurement. Calculate the temperature of the second material and the uncertainty.

8-18 Consider the thermometer-shield arrangement shown in Fig. 8-33. Assuming that the shield area is very large compared with the thermometer area and that all radiation leaving the thermometer is intercepted by the shield, calculate the true gas temperature using the following data:

$$h_t = 2.0 \text{ Btu/h} \cdot \text{ft}^2 \cdot \text{°F}$$

$$h_s = 1.5 \text{ Btu/h} \cdot \text{ft}^2 \cdot \text{°F}$$

$$\epsilon_t = 0.8$$

$$\epsilon_s = 0.2$$

$$T_t = 600\text{°F}$$

$$T_s = 100\text{°F}$$

(It is assumed from the statement of the problem that $F_{ts} = 1.0$, $F_{se} = 1.0$, and $F_{te} = 0$.) What temperature would the thermometer indicate if the shield were removed? Calculate the error reduction owing to the use of the radiation shield.

8-19 Rework Prob. 8-18 assuming that $A_s/A_t = 100$ and that the radiation shape factors are given as $F_{ts} = 0.9$, $F_{te} = 0.1$, and $F_{se} = 1.1$ (recall the assumption pertaining to F_{se}).

8-20 A platinum resistance thermometer is placed in a duct to measure the temperature of an airflow stream. The thermometer is placed inside a cylindrical shell 6 mm in diameter which has a polished outside surface with $\epsilon = 0.08 \pm 0.02$. The airstream velocity is known to be 3 m/s, and the pressure is 1.0 atm. The thermometer indicates a temperature of 115°C. The duct wall temperature is measured at 193°C. Calculate the true air temperature, and estimate the uncertainty in this temperature. Assume that the uncertainty in the convection heat-transfer coefficient calculated with the appropriate formula from Table 8-5 is ± 15 percent. Assume that the uncertainty in the resistance-thermometer measurement is ± 0.03°C.

8-21 Repeat Prob. 8-20, but assume that the cylinder is a chrome-plated copper rod with a thermo-couple embedded in the surface. The uncertainty in the thermocouple measurement is assumed to be $\pm 0.3°C$.

8-22 A small copper sphere is constructed with a thermocouple embedded in the center and is used to measure the air temperature in an oven. The walls of the oven are at $1200 \pm 20°F$. The emissivity of the copper surface is 0.57 ± 0.04, and the temperature indicated by the thermocouple is $1015 \pm 1.0°F$. The convection heat-transfer coefficient is 5.0 Btu/h·ft²·°F $\pm$ 15 percent. Calculate the true air temperature and the uncertainty. Calculate the true air temperature if the surface of the sphere had been chromeplated with $\epsilon = 0.06 \pm 0.02$. Estimate the uncertainty in this circumstance.

8-23 The thermocouple wires in Example 8-8 are coated with an insulation material having a thickness of 0.003 in and a thermal conductivity of 0.1 Btu/h·ft·°F. Calculate the true plate temperature, assuming the other conditions are the same as in the example.

8-24 A $\frac{1}{32}$-in constantan strip is heated by the gas stream on one side and cooled on the other side as shown. Two bare 0.004-in-diameter copper wires are attached to the strip to indicate the temperature difference in the transverse direction. A potentiometer indicates 0.15 ± 0.001 mV, and the sensitivity of copper constantan may be taken as 0.0279 ± 0.0001 mV/°F. Estimates of the significant heat-transfer parameters are as follows:

$$h_1 = 3.5 \text{ Btu/h·ft}^2 \cdot °F$$

$$h_2 = 2.5 \text{ Btu/h·ft}^2 \cdot °F$$

$$h_{t_1} = 15 \text{ Btu/h·ft}^2 \cdot °F$$

$$h_{t_2} = 10 \text{ Btu/h·ft}^2 \cdot °F$$

$$k_{cu} = 220 \text{ Btu/h·ft} \cdot °F$$

$$k_{con} = 14 \text{ Btu/h·ft} \cdot °F$$

Calculate the true temperature difference, and estimate the uncertainty.

$T_f = 200°F$

h_{t_1}

h_1

h_{t_2}

$\frac{1}{32}''$

Hot gas

Figure Prob. 8-24

8-25 Calculate the time constant for the copper-rod thermometer of Prob. 8-21. Consider only the convection heat transfer from the gas in calculating this value.

8-26 A thermocouple is placed inside a $\frac{1}{8}$-in-OD, $\frac{1}{16}$-in-ID copper tube. The tube is then placed inside a furnace whose walls are at $1500°F$. The air temperature in the furnace is $1200°F$. Calculate the temperature indicated by the thermocouple, and estimate the time τ to obtain

$$\frac{T - T_i}{T_t - T_i} = 0.5$$

where T_i = initial temperature of the thermocouple before it is placed in the furnace
T_t = indicated temperature after a long time
T = temperature at time τ

Assume $\epsilon = 0.78$ for copper.

8-27 A certain thermocouple has a time constant of 1.2 s. Design a compensation network like that shown in Fig. 8-37 having a steady-state attenuation factor of 8. Select the value of the resistance so that the capacitor has an even multiple value.

8-28 A certain high-speed temperature probe having a recovery factor of 0.98 ± 0.01 is used to measure the temperature of air at Mach 3.00. The thermocouple installed in the probe is accurate within $\pm 1.0°C$ and indicates a temperature of 380°C. Calculate the free-stream temperature and the uncertainty.

8-29 A 4-in-square copper plate is suspended vertically in a room, and a thermocouple is attached to measure the plate temperature. The walls of the room are at 110°F, and the thermocouple indicates a temperature of 95°F. Assuming the copper has an emissivity of 0.7, calculate the air temperature in the room. Determine the convection heat-transfer coefficient from Table 8-5. If the uncertainty in the surface emissivity is ± 0.05, what is the resulting uncertainty in the calculated air temperature?

8-30 Repeat Prob. 8-29 for polished copper having an emissivity of $\epsilon = 0.05 \pm 0.01$.

8-31 A thermometer having an emissivity of 0.9 is placed in a large room and indicates a temperature of 25°C. The walls of the room are at 35°C and the convection heat-transfer coefficient is 5.0 $W/m^2 \cdot °C$. Calculate the true temperature of the air.

8-32 A heat flux of 100 W/cm^2 is imposed on a slab of AISI 430 steel. What temperature rise would be expected in a time of 0.1 s?

8-33 A heat flux of 1 W/cm^2 is imposed on a slab of silicon rubber. What temperature rise would be expected in 1 s?

8-34 A heat flux of 10 W/cm^2 is imposed on a slab of copper. What temperature rise would be expected in 1 s?

8-35 A stainless-steel sphere having a diameter of 3.0 mm is imbedded with a copper-constantan thermocouple and suddenly exposed to air at 50°C with a convection coefficient of 20 $W/m^2 \cdot °C$. The initial temperature of the sphere is 20°C. Plot the output of the thermocouple as a function of time. What is the time constant?

8-36 A thermopile is to be constructed for measuring small temperature differences. Devise arrangements to measure temperature differences of 1°, 3°, and 5°C using copper-constantan and cromel-constantan thermocouples. Discuss the uncertainties which may be encountered.

8-37 Two platinum resistance thermometers are used to measure a temperature difference of 8°C. Discuss the uncertainties which may be involved. Would a thermopile arrangement be better and if so, why?

REFERENCES

1. Becker, J. A., C. B. Green, and G. L. Pearson: Properties and Uses of Thermistors, *Trans. AIEE*, vol. 65, pp. 711–725, November 1946.
2. Eckert, E. R. G., and R. M. Drake: "Heat and Mass Transfer," 2d ed., McGraw-Hill Book Company, New York, 1959.
3. Eskin, S. G., and J. R. Fritze: Thermostatic Bimetals, *Trans. ASME*, vol. 62, pp. 433–442, July 1940.
4. Hackforth, H. L.: "Infrared Radiation," McGraw-Hill Book Company, New York, 1960.
5. Holman, J. P.: "Heat Transfer," 5th ed., McGraw-Hill Book Company, New York, 1981.
6. Lion, K. S.: "Instrumentation in Scientific Research," McGraw-Hill Book Company, New York, 1959.
7. Siegel, R., and J. R. Howell: "Thermal Radiation Heat Transfer," 2d ed., McGraw-Hill Book Company, 1981.
8. Obert, E. F.: "Concepts of Thermodynamics," McGraw-Hill Book Company, New York, 1960.
9. Rossini, F. D. (ed.): "Thermodynamics and Physics of Matter," Princeton University Press, Princeton, N.J., 1955.

10. Sears, F. W.: "Thermodynamics," Addison-Wesley Publishing Company, Inc., Reading, Mass., 1953.
11. Schneider, P. J.: "Conduction Heat Transfer," Addison-Wesley Publishing Company, Inc., Reading, Mass., 1955.
12. Tribus, M.: "Thermostatics and Thermodynamics," D. Van Nostrand Company, Inc., Princeton, N.J., 1961.
13. Winkler, E. M.: Design and Calibration of Stagnation Temperature Probes for Use at High Supersonic Speeds and Elevated Temperatures, *J. Appl. Phys.*, vol. 25, p. 231, 1954.
14. ———, Reference Tables for Thermocouples, *Natl. Bur. Std. (U.S.)*, *Cir.* 561, April 1955.
15. West, W. E., and J. W. Westwater: Radiation-Conduction Correction for Temperature Measurements in Hot Gases, *Ind. Eng. Chem.*, vol. 45, pp. 2, 152, 1953.
16. Hennecke, D. K., and E. M. Sparrow: Local Heat Sink on a Convectively Cooled Surface, Application to Temperature Measurement Error, *Int. J. Heat and Mass Trans.*, vol. 13, p. 287, February 1970.
17. Shepard, C. E.. and I. Warshawsky: Electrical Techniques for Compensation of Thermal Time Lag of Thermocouples and Resistance Thermometer Elements, *NACA Tech. Note* 2703, May 1952.
18. Holman, J. P.: "Thermodynamics," 3d ed., McGraw-Hill Book Company, New York, 1980.
19. Powell, R. L.: "Thermocouple Thermometry," in E.R.G. Eckert and R. Goldstein: "Measurements in Heat Transfer," 2d ed., McGraw-Hill Book Company, New York, 1976.
20. Schultz, D. L., and T. V. Jones: Heat Transfer Measurements in Short Duration Hypersonic Facilities, *AGARDograph* 165, 1973.
21. Cooper, T. E., J. F. Meyer, and R. J. Field: Liquid Crystal Thermography and Its Application to the Study of Convective Heat Transfer. *ASME Paper 75-HT-15*, August 1975.
22. Cooper, T. E., and J. P. Groff: Thermal Mapping via Liquid Crystals of the Temperature Field Near a Heated Surgical Probe, *J. Heat Trans.*, vol. 95, p. 250, 1973.
23. Powell, R. L., W. J. Hall, and C. H. Hyink: Thermocouple Reference Tables Based on IPTS-68, *Natl. Bur. Stand. (U.S.)*, *Monograph* 125, 1974.
24. Wobschall, D.: "Circuit Design for Electronic Instrumentation," McGraw-Hill Book Company, New York, 1979.

CHAPTER
NINE

THERMAL- AND TRANSPORT-PROPERTY MEASUREMENTS

9-1 INTRODUCTION

Several types of thermal properties are essential for energy-balance calculations in heat-transfer applications. Values of these properties for a variety of substances and materials are already available in tabular form in various handbooks; however, for new materials, which appear regularly, it is important that the engineer be familiar with some basic methods of measuring these properties.

Most thermal-property measurements involve a determination of heat flow and temperature. We have already discussed several types of temperature-measuring devices in Chap. 8 and shall have occasion to refer to them from time to time in discussing thermal-property measurements. Heat flow is usually measured by making an energy balance on the device under consideration. For example, a metal plate might be heated with an electric heater and the plate immersed in a tank of water during this heating process. The convection heat loss from the plate could thus be determined by making a measurement of the electric power dissipated in the heater. As another example, consider the heating of water by passing it through a heated pipe. The convection heat transfer from the pipe wall to the water may be determined by measuring the mass flow rate of water and the inlet and exit water temperatures to the heated section of pipe. The energy gained by the water is therefore the heat transfer from the pipe, provided that the outside surface of the pipe is insulated so that no losses are incurred. The techniques for measurement of heat transfer by thermal radiation are discussed in Chap. 12.

Thermal conductivity may be classified as a transport property since it is indicative of the energy transport in a fluid or solid. In gases and liquids the transport of energy takes place by molecular motion, while in solids transport of energy by free electrons and lattice vibration is important. Fluid viscosity is also classified as a transport property because it is dependent on the momentum transport that results from molecular motion in the fluid. Mass diffusion is similarly classified as a transport process because it results from molecular movement. The diffusion coefficient is the transport property in this case. In this chapter we shall consider some simple methods for measurement of these transport properties.

The measurement of heat flow falls under the general subject of calorimetry. In this chapter we shall discuss some simple calorimetric determinations that may be performed. The broad subject of thermodynamic-property measurement is beyond the scope of our discussion.

9-2 THERMAL-CONDUCTIVITY MEASUREMENTS

Thermal conductivity is defined by the Fourier equation

$$q_x = -kA\frac{\partial T}{\partial x} \tag{9-1}$$

where
q_x = heat-transfer rate, Btu/h or W
A = area through which the heat is transferred, ft^2 or m^2
$\partial T/\partial x$ = temperature gradient in the direction of the heat transfer, °F/ft or °C/m
k = thermal conductivity, Btu/h · ft · °F or W/m · °C

Experimental determinations of thermal conductivity are based on this relationship. Consider the thin slab of material shown in Fig. 9-1. If the heat-transfer rate through the material, the material thickness, and the difference in temperature are measured, then the thermal conductivity may be calculated from

$$k = \frac{q\,\Delta x}{A(T_1 - T_2)} \tag{9-2}$$

Figure 9-1 Simple thermal-conductivity measurement.

Figure 9-2 Schematic of guarded hot-plate apparatus for measurement of thermal conductivity.

In an experimental setup the heat may be supplied to one side of the slab by an electric heater and removed from the other side by a cooled plate. The temperatures on each side of the slab may be measured with thermocouples or thermistors, whichever is more appropriate.

We must note that the standard English and SI units for thermal conductivity given above are not always the ones employed in practice. Frequently, the thickness of material x is expressed in inches, while the area is expressed in square feet. The units for thermal conductivity appropriate to such dimensions then become $Btu \cdot in/h \cdot ft^2 \cdot °F$.

The main problem of the above method for determining thermal conductivity is that heat may escape from the edges of the slab, or if the edges are covered with insulation, a two-dimensional temperature profile may result, which can cause an error in the determination. This problem may be alleviated by the installation of guard heaters, as shown in Fig. 9-2. In this arrangement the heater is placed in the center and a slab of the specimen is placed on each side of the heater plate. A coolant is circulated through the device to remove the energy, and

Figure 9-3 Apparatus for measurement of thermal conductivity of metals.

thermocouples are installed at appropriate places to measure the temperatures. A guard heater surrounds the main heater, and its temperature is maintained at that of the main heater. This prevents heat transfer out from the edges of the specimen and thus maintains a one-dimensional heat transfer through the material whose thermal conductivity is to be determined. The guarded hot plate, as it is called, is widely used for determining the thermal conductivity of non-metals, i.e., solids of rather low thermal conductivity. For materials of high thermal conductivity a small temperature difference would be encountered which would require much more precise temperature measurement.

A very simple method for the measurement of thermal conductivities of metals is depicted in Fig. 9-3. A metal rod A of known thermal conductivity is connected to a rod of the metal B whose thermal conductivity is to be measured. A heat source and heat sink are connected to the ends of the composite rod, and the assembly is surrounded by insulating material to minimize heat loss to the surroundings and to ensure one-dimensional heat flow through the rod. Thermocouples are attached, or embedded, in both the known and unknown materials as shown. If the temperature gradient through the known material is measured, the heat flow may be determined. This heat flow is subsequently used to calculate the thermal conductivity of the unknown material. Thus,

$$q = -k_A A \left(\frac{dT}{dx}\right)_A = -k_B A \left(\frac{dT}{dx}\right)_B \tag{9-3}$$

The temperatures may be measured at various locations along the unknown and the variation of thermal conductivity with temperature determined from these measurements. Van Dusen and Shelton [6] have used this method for the determination of thermal conductivities of metals up to 600°C.

Example 9-1 A guarded hot-plate apparatus is used to measure the thermal conductivity of a metal having $k = 50$ Btu/h·ft·°F. The thickness of the specimen is 0.125 ± 0.002 in, and the heat flux may be measured within 1 percent. Calculate the accuracy necessary on the ΔT measurement in order to ensure an overall uncertainty in the measurement of k of 5 percent. If one of the plate temperatures is nominally 300°F, calculate the nominal value of the other plate temperature and the tolerable uncertainty in each temperature measurement, assuming a nominal heat flux of 20,000 Btu/h·ft².

SOLUTION We use Eq. (9-2) to estimate the uncertainty in conjunction with Eq. (3-2).

$$k = \frac{(q/A)\, \Delta x}{\Delta T} \tag{a}$$

Equation (3-2) may be written as

$$\frac{w_R}{R} = \left[\left(\frac{\partial R}{\partial x_1}\right)^2 \left(\frac{w_1}{R}\right)^2 + \left(\frac{\partial R}{\partial x_2}\right)^2 \left(\frac{w_2}{R}\right)^2 + \left(\frac{\partial R}{\partial x_3}\right)^2 \left(\frac{w_3}{R}\right)^2\right]^{1/2} \tag{b}$$

We have

$$\frac{w_k}{k} = 0.05$$

$$\frac{\partial k}{\partial (q/A)} = \frac{\Delta x}{\Delta T} \qquad \frac{w_{q/A}}{k} = \frac{0.01 \; q/A(\Delta T)}{(q/A) \, \Delta x} = \frac{0.01 \, \Delta T}{\Delta x}$$

$$\frac{\partial k}{\partial (\Delta x)} = \frac{q/A}{\Delta T} \qquad \frac{w_{\Delta x}}{k} = \frac{(0.002) \, \Delta T}{(q/A)(0.125)}$$

$$\frac{\partial k}{\partial (\Delta T)} = -\frac{(q/A) \, \Delta x}{(\Delta T)^2} \qquad \frac{w_{\Delta T}}{k} = \frac{w_{\Delta T} \, \Delta T}{(q/A) \, \Delta x}$$

Inserting these expressions in Eq. (b) gives

$$0.05 = \left[(0.01)^2 + (0.002/0.125)^2 + \left(\frac{w_{\Delta T}}{\Delta T} \right)^2 \right]^{1/2}$$

or $w_{\Delta T}/\Delta T = 0.0146 = 1.46$ percent. We now calculate the nominal value of ΔT as

$$\Delta T = \frac{(q/A) \, \Delta x}{k} = \frac{(2 \times 10^4)(0.125)}{(12)(50)} = 4.1667°F$$

Since $\Delta T = T_1 - T_2$, the nominal value of T_1 is

$$T_1 = 300 + 4.1667 = 304.1667°F$$

The tolerable uncertainty in each temperature measurement is calculated from

$$\frac{w_{\Delta T}}{\Delta T} = \left\{ \left[\frac{\partial(\Delta T)}{\partial T_1} \frac{w_{T_1}}{\Delta T} \right]^2 + \left[\frac{\partial(\Delta T)}{\partial T_2} \frac{w_{T_2}}{\Delta T} \right]^2 \right\}^{1/2}$$

$$w_{\Delta T} = (w_{T_1}^2 + w_{T_2}^2)^{1/2}$$

and we assume the uncertainties are the same for both temperatures, so that

$$w_T = \frac{1}{\sqrt{2}} w_{\Delta T} = 0.707 w_{\Delta T}$$

$$w_T = (0.707)(0.0146)(4.1667) = \pm 0.043°F$$

We thus see that a very accurate measurement of temperature is necessary in order to give only a 5 percent accuracy in the determination of k.

Example 9-2 An apparatus like that in Fig. 9-3 is used to measure the thermal conductivity of the metal of Example 9-1. The same heat flux is used with a bar 3.0 ± 0.005 in long. Determine the accuracy necessary for the determination of ΔT and the tolerable uncertainty in the temperature measurements. Assume the same conditions as in Example 9-1.

SOLUTION For a 5 percent uncertainty in k the uncertainty in ΔT is given as

$$0.05 = \left[(0.01)^2 + \left(\frac{0.005}{3.0} \right)^2 + \left(\frac{w_{\Delta T}}{\Delta T} \right)^2 \right]^{1/2}$$

$$\frac{w_{\Delta T}}{\Delta T} = 0.049 = 4.9\%$$

The nominal value of ΔT is given as

$$\Delta T = \frac{(2 \times 10^4)(3)}{(12)(50)} = 100°F$$

The allowable uncertainty in the determination of the temperatures is

$$w_T = \frac{1}{\sqrt{2}} \, w_{\Delta T} = (0.707)(0.049)(100) = \pm 3.46°F$$

This value is very easy to obtain. Even if the heat flux were reduced to 2×10^3 Btu/h · ft^2, the nominal temperature difference would be 10°F and the allowable uncertainty would be $\pm 0.346°F$. This is still a value that may be attained with careful experimental techniques.

9-3 THERMAL CONDUCTIVITY OF LIQUIDS AND GASES

Kaye and Higgins [3] have used a guarded hot-plate method for determining thermal conductivity of liquids. Their apparatus is shown in Fig. 9-4. The diameter of the plates is 5 cm, and the thickness of the liquid layer is approximately

Figure 9-4 Guarded hot-plate apparatus for measurement of thermal conductivity of liquids, according to Ref. [3].

0.05 cm. This layer must be sufficiently thin that convection currents are minimized. An annular arrangement, as shown in Fig. 9-5, may also be used for the determination of liquid thermal conductivities. Again, the thickness of the liquid layer must be thin enough to minimize thermal-convection currents.

A concentric-cylinder arrangement may also be used for the measurement of the thermal conductivity of gases. Keyes and Sandell [4] have used such a device for the measurement of thermal conductivity of water vapor, oxygen, nitrogen, and other gases. The inner and outer cylinders were both constructed of silver with a length of 5 in and an outside diameter of $1\frac{1}{2}$ in. The gap space for the gas was 0.025 in. Vines [7] has utilized such a device for the measurement of high-temperature-gas thermal conductivities. A schematic of his apparatus is shown in Fig. 9-6. The emitter serves as the heat source, while the heat stations on either end act as guard heaters. The emitter has an outside diameter of 6 mm and a length of 50 mm, while the receiver has an inside diameter of 10 mm and a length of 125 mm with a wall thickness of 1 mm. For most of the tests it was possible to maintain a temperature difference of 5 to 10°C between the emitter and the receiver. The heat-transfer rate is measured by determining the electric-power input to the emitter, while thermocouples installed on the surface of the emitter and receiver determine the temperature difference.

For the concentric-cylinder device the relation that is used to calculate the thermal conductivity is

$$k = \frac{q \ln (r_2/r_1)}{2\pi L(T_1 - T_2)} \tag{9-4}$$

where q = heat-transfer rate
r_2, r_1 = outside and inside radii of the annular space containing the fluid, respectively
T_2, T_1 = temperatures measured at these radii

Figure 9-5 Concentric-cylinder method for measurements of thermal conductivity of liquids.

Lava spacing piece

Heat station

Emitter

Thermocouples

Emitter winding

Receiver

Heat station

Emitter leads
(current and potential)

Leads to heat
station winding

Figure 9-6 Apparatus for determining thermal conductivity of gases at high temperatures, according to Vines [7].

An investigation by Leidenfrost [8] proposes an apparatus for measuring thermal conductivities of gases and liquids from −180 to +500°C. Complete experimental data are not available, but accuracies of 0.1 percent are claimed for the device.

Thermal conductivities of several materials are given in the Appendix, and additional information is given in Ref. [20].

9-4 MEASUREMENT OF VISCOSITY

The defining equation for *dynamic* or *absolute viscosity* is

$$\tau = \mu \, \frac{du}{dy} \tag{9-5}$$

where τ = shear stress between fluid layers in laminar flow
    μ = dynamic viscosity
    du/dy = normal velocity gradient as indicated in Fig. 9-7

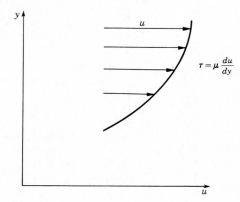

Figure 9-7 Diagram indicating relation of viscosity gradient and fluid shear.

The *kinematic* viscosity is defined by

$$v = \frac{\mu}{\rho}$$

where ρ is the fluid density. A very large number of units are employed for viscosity in both the English and metric systems. Some of the more common units are summarized below along with appropriate conversion factors:

Dynamic viscosity:

$1 \, \text{lbf} \cdot \text{s/ft}^2 = 47.8803 \, \text{N} \cdot \text{s/m}^2$
$= 478.803 \, \text{poise (P)}$
$= 1.158 \times 10^5 \, \text{lbm/h} \cdot \text{ft}$
$= 47.8803 \, \text{kg/m} \cdot \text{s}$
$1 \, \text{N} \cdot \text{s/m}^2 = 2.08854 \times 10^{-2} \, \text{lbf} \cdot \text{s/ft}^2$
$= 10 \, \text{P} = 1000 \, \text{centipoise (cP)}$
$= 2419 \, \text{lbm/h} \cdot \text{ft}$
$= 1 \, \text{kg/m} \cdot \text{s}$
$1 \, \text{P} = 100 \, \text{cP} = 1 \, \text{dyn} \cdot \text{s/cm}^2$
$= 0.1 \, \text{N} \cdot \text{s/m}^2$
$= 2.08854 \times 10^{-3} \, \text{lbf} \cdot \text{s/ft}^2$
$= 241.9 \, \text{lbm/h} \cdot \text{ft}$
$= 0.1 \, \text{kg/m} \cdot \text{s}$

Kinematic viscosity:

$1 \, \text{ft}^2/\text{s} = 9.2903 \times 10^{-2} \, \text{m}^2/\text{s}$
$= 929.03 \, \text{stokes (St)} = 929.03 \, \text{cm}^2/\text{s}$
$1 \, \text{m}^2/\text{s} = 10.7639 \, \text{ft}^2/\text{s}$
$= 10^4 \, \text{St}$
$1 \, \text{St} = 1 \, \text{cm}^2/\text{s} = 100 \, \text{centistokes (cSt)} = 10^{-4} \, \text{m}^2/\text{s}$
$= 10.7639 \times 10^{-4} \, \text{ft}^2/\text{s}$

Various methods are employed for measurement of the viscosity. The two most common methods are the rotating-concentric-cylinder method and the capillary-flow method. We shall discuss the rotating-cylinder method first.

Consider the parallel plates as shown in Fig. 9-8. One plate is stationary, and the other moves with constant velocity u. The velocity profile for the fluid between the two plates is a straight line, and the velocity gradient is

$$\frac{du}{dy} = \frac{u}{b} \tag{9-6}$$

Figure 9-8 Velocity distribution between large parallel plates.

The system could be used to measure the viscosity by measuring the force required to maintain the moving plate at the constant velocity u. The system is impractical from a construction standpoint, however, and the conventional approach is to approximate the parallel flat-plate situation with the rotating concentric cylinders shown in Fig. 9-9. The inner cylinder is stationary and attached to an appropriate torque-measuring device, while the outer cylinder is driven at a constant angular velocity ω. If the annular space b is sufficiently small in comparison with the radius of the inner cylinder, then the rotating-cylinder arrangement approximates the parallel-plate situation, and the velocity profile in the gap space may be assumed to be linear. Then,

$$\frac{du}{dy} = \frac{r_2 \omega}{b} \tag{9-7}$$

where, now, the y distance is taken in the radial direction and it is assumed that $b \ll r_1$. Now, if the torque T is measured, the fluid shear stress is expressed by

$$\tau = \frac{T}{2\pi r_1^2 L} \tag{9-8}$$

Figure 9-9 Rotating concentric-cylinder apparatus for measurement of viscosity.

where L is the length of the cylinder. The viscosity is determined by combining Eqs. (9-5), (9-7), and (9-8) to give

$$\mu = \frac{Tb}{2\pi r_1^2 r_2 L\omega} \qquad (9\text{-}9)$$

If the concentric-cylinder arrangement is constructed such that the gap space a is small, then the bottom disk will also contribute to the torque and influence the calculation of the viscosity. The torque on the bottom disk is

$$T_d = \frac{\mu\pi\omega}{2a} r_1^4 \qquad (9\text{-}10)$$

where a is the gap spacing, as shown in Fig. 9-9. Combining the torques due to the bottom disk and the annular space gives

$$T = \mu\pi\omega r_1^2 \left(\frac{r_1^2}{2a} + \frac{2Lr_2}{b} \right) \qquad (9\text{-}11)$$

Once the torque, angular velocity, and dimensions are measured, the viscosity may be calculated from Eq. (9-11).

Perhaps the most common method of viscosity measurement consists of a measurement of the pressure drop in laminar flow through a capillary tube. Consider the tube cross section shown in Fig. 9-10. If the Reynolds number defined by

$$\text{Re}_d = \frac{\rho u_m d}{\mu} \qquad (9\text{-}12)$$

is less than 1000, laminar flow will exist in the tube and the familiar parabolic-velocity profile will be experienced as shown. If the fluid is incompressible and the flow is steady, it can be shown that the volume rate of flow Q can be written as

$$Q = \frac{\pi r^4 (p_1 - p_2)}{8\mu L} \qquad (9\text{-}13)$$

Figure 9-10 Laminar flow through a capillary tube.

A viscosity determination may be made by measuring the volume rate of flow and pressure drop for flow in such a tube. To ensure that laminar flow will exist, a small-diameter capillary tube is used; the small diameter reduces the Reynolds number as calculated from Eq. (9-12). In Eq. (9-12) the product ρu_m may be calculated from

$$\rho u_m = \frac{\dot{m}}{\pi r^2} \tag{9-14}$$

where $\dot{m}$ is the mass rate of flow.

When a viscosity measurement is made on a gas, the compressibility of the gas must be taken into account. The resulting expression for the mass flow of the gas under laminar flow conditions in the capillary is

$$\dot{m} = \frac{\pi r^4}{16 \mu R T} (p_1^2 - p_2^2) \tag{9-15}$$

where R is the gas constant for the particular gas.

Care must be taken to ensure that the flow in the capillary is *fully developed*, i.e., that the parabolic-velocity profile has been established. This means that the pressure measurements should be taken far enough downstream from the entrance of the tube to ensure that developed flow conditions persist. It may be expected that the flow will be fully developed when

$$\frac{L}{d} > \frac{\mathrm{Re}_d}{8} \tag{9-16}$$

where L is the distance from the entrance of the tube.

The Saybolt viscosimeter is an industrial device that uses the capillary-tube principle for measurement of viscosities of liquids. A schematic of the device is shown in Fig. 9-11. A cylinder is filled to the top with the liquid and enclosed in a constant-temperature bath to ensure uniformity of temperature during the measurements. The liquid is then allowed to drain from the bottom through the short capillary tube. The time necessary for 60 ml to drain is recorded, and this time is taken as indicative of the viscosity of the liquid. Since the capillary tube is short, a fully developed laminar-velocity profile is not established, and it is necessary to apply a correction to account for the actual profile. If the velocity profile were fully developed, the kinematic viscosity would vary directly with the time for drainage; i.e.,

$$v = \frac{\mu}{\rho} = c_1 t$$

To correct for the nonuniform velocity profile, another term is added, to give

$$v = c_1 t + \frac{c_2}{t}$$

Figure 9-11 Schematic of the Saybolt viscosimeter.

With the constants inserted, the relation is

$$v = \left(0.00237t - \frac{1.93}{t} \right) \times 10^{-3} \text{ ft}^2/\text{s} \tag{9-17}$$

The symbol t designates the drainage time in seconds for 60 ml of liquid, and it is common to express the viscosity in units of *Saybolt seconds* when this method of measurement is used. Equation (9-17) provides a method of converting to more useful units for kinematic viscosity. Viscosities for several fluids are given in the Appendix.

Another technique for viscosity measurements is to observe the decrement in the period of oscillation when spheres or disks are suspended in the fluid with an elastic wire and allowed to vibrate from some initially deflected position. Such systems are particularly applicable to measurements of viscosities at high pressures. Kestin and Leidenfrost [18] describe an oscillating-disk system which has met with good success.

Example 9-3 A Saybolt viscosimeter is used to measure the viscosity of a certain motor oil. The time recorded for drainage of 60 ml is estimated at 183 ± 0.5 s. Calculate the percentage uncertainty in the viscosity.

SOLUTION Equation (3-2) is used in conjunction with Eq. (9-17) to estimate the uncertainty. We have

$$\frac{\partial v}{\partial t} = c_1 - \frac{c_2}{t^2} = \left(0.00237 + \frac{1.93}{t^2}\right) \times 10^{-3}$$

so that the uncertainty in the viscosity is given as

$$w_v = \frac{\partial v}{\partial t} w_t$$

$$= \left(0.00237 + \frac{1.93}{t^2}\right)(10^{-3})(0.5) \text{ ft}^2/\text{s}$$

The nominal value of the viscosity is

$$v = \left(0.00237t - \frac{1.93}{t}\right)10^{-3} \text{ ft}^2/\text{s}$$

At $t = 183$ s

$$w_v = 1.21 \times 10^{-6} \text{ ft}^2/\text{s} \ (1.12 \times 10^{-7} \text{ m}^2/\text{s})$$

$$v = 4.22 \times 10^{-4} \text{ ft}^2/\text{s} \ (3.92 \times 10^{-5} \text{ m}^2/\text{s})$$

so that $w_v/v = 2.87 \times 10^{-3} = 0.287$ percent.

9-5 GAS DIFFUSION

Consider a container of a certain gas (2) as shown in Fig. 9-12. At one end of the container another gas (1) is introduced and allowed to diffuse into gas 2. The diffusion rate of gas 1 at any instant of time is given by Fick's law of diffusion as

$$n_1 = -D_{12} A \frac{\partial \bar{N}_1}{\partial x} \tag{9-18}$$

where n_1 is the molal rate of diffusion, A is the cross-sectional area for diffusion, $\bar{N}_1$ is the molal concentration of component 1 at any point, and D_{12} is defined as

Figure 9-12 Schematic of gas-diffusion process.

the diffusion coefficient. It may be noted that Eq. (9-18) is similar to the Fourier law for heat conduction. When a mass balance is made on an element, the one-dimensional diffusion equation results:

$$\frac{\partial \bar{N}_1}{\partial t} = D_{12} \frac{\partial^2 \bar{N}_1}{\partial x^2} \tag{9-19}$$

A solution to this equation may be used as the basis for an experimental determination of the diffusion coefficient D_{12}.

A simple experimental setup for the measurement of gas-diffusion coefficients is shown in Fig. 9-13. Two sections of glass tubing are connected by a flexible Tygon tubing connection, which may be clamped as shown. The sections of tubing are charged with the two gases whose diffusion characteristics are to be investigated. After the initial charging process, the flexible connection is opened and the two gases are allowed to diffuse into one another. After a period of time the Tygon connection is closed again, the exact time recorded, and the two sections allowed to reach equilibrium concentrations. These equilibrium concentrations are then measured with appropriate analytical equipment. It is only necessary to measure the concentration of one of the gases in both sections, since the concentration of the other gas will be obtained from a mole balance. However, measuring both gives a check and information about the accuracy of the experiment. Some check of this kind is always very desirable, although not always feasible.

A solution of Eq. (9-19) gives these concentrations in terms of the diffusion coefficient and dimensions of the tubes. Assuming that the lower portion is filled with gas 1 and the upper portion with gas 2 initially, the solution is

$$F = \frac{N_{1A} - N_{1B}}{N_{1A} + N_{1B}} = \frac{8}{\pi^2} \sum_{k=0}^{\infty} \frac{1}{(2k+1)^2} \exp\left[\frac{-\pi^2 D_{12} t (2k+1)^2}{4L^2}\right] \tag{9-20}$$

where the subscripts A and B refer to the lower and upper sections, respectively, t is the time the two gases are allowed to diffuse, and L is the length of *each* tube. N_{1A} and N_{1B} represent the number of moles of component 1 in the lower and upper parts of the apparatus, respectively, after time t. For an experimental determination of D_{12} the optimum time to be allowed for the diffusion process is

$$t_{\text{opt}} = \frac{4L^2}{\pi^2 D_{12}} \tag{9-21}$$

assuming that the only experimental errors are those involved in the determination of F. If the diffusion process is allowed to run for t_{opt} or longer, it is found that the first term in the series of Eq. (9-20) is the primary term and that the higher-order terms are negligible. Then,

$$F = 0.2982 \qquad \text{at } t = t_{\text{opt}} \tag{9-22}$$

If the second- and higher-order terms are neglected, it is easy to calculate the

Figure 9-13 Loschmidt apparatus for measurement of diffusion coefficients in gases. (*a*) Apparatus; (*b*) detail of flexible connection.

value of D_{12} once a determination of F and t has been made. Neglecting these higher-order terms gives

$$F = \frac{8}{\pi^2} \exp \frac{-\pi^2 D_{12} t}{4L^2} \tag{9-23}$$

or

$$D_{12} = \frac{-4L^2}{\pi^2 t} \ln \frac{\pi^2 F}{8} \tag{9-24}$$

It is necessary to know only an approximate value for D_{12} in order to calculate the value of t_{opt} from Eq. (9-21). Then, the experiment is conducted using a diffusion time greater than this value. The exact value of D_{12} is then calculated from the experimental data using Eq. (9-24).

The simple experimental setup shown in Fig. 9-13 is called a *Loschmidt apparatus*. In the apparatus the heavier gas is usually contained in the lower section, while the lighter gas is placed in the upper section. Extreme care must be exerted to ensure that the system is free of leaks and that each section contains a pure component at the start of the experiment. Care must also be exerted to ensure that both gases are initially at the same temperature and pressure. The apparatus should be constructed so that the volume of the inlet tubing is negligible in comparison with the volume of the main apparatus. While the diffusion process is in progress, the temperature of the apparatus should be maintained constant. Diffusion coefficients for several gases are given in the Appendix.

Example 9-4 An apparatus like that shown in Fig. 9-13 is constructed so that $L = 50$ cm. CO_2 is placed in the lower tube and air in the upper tube. Both gases are at standard atmospheric pressure and 77°F. The diffusion coefficient for CO_2 in air is 0.164 cm²/s. Calculate the mole fraction of CO_2 in each tube for (1) 5 min and (2) 2 h after the connection between the tubes is opened.

SOLUTION The mole fractions are given by

$$x_{1A} = \frac{N_{1A}}{N_{1_0}} \qquad x_{1B} = \frac{N_{1B}}{N_{1_0}}$$

where N_{1_0} is the initial number of moles in each tube and equal to $N_{1A} + N_{1B}$. Gas 1 is the CO_2 in this case. From Eq. (9-20),

$$F = x_{1A} - x_{1B} \qquad (a)$$

We calculate the optimum time as

$$t_{opt} = \frac{(4)(50)^2}{\pi^2(0.164)} = 6180 \text{ s}$$

The 5-min condition is far short of this, so that the series solution of Eq. (9-20) must be used to find F. Thus, with $t = 300$ s,

$$F = \frac{8}{\pi^2}\left[e^{-300/6180} + \frac{1}{9}e^{-(9)(300)/6180} + \frac{1}{25}e^{-(25)(300)/6180} + \cdots\right]$$

$$= \frac{8}{\pi^2}(0.9526 + 0.0718 + 0.0119 + \cdots)$$

$$= 0.8400$$

flow rate of cooling water $\dot{m}_w$, the condensate temperature T_{c_2}, the entering fuel and air temperatures T_{f_1} and T_{a_1}, and the relative humidity of the inlet air ϕ_1. In addition, an analysis of the products of combustion is made to determine the oxygen, carbon dioxide, and carbon monoxide content. From all these data, mass and energy balances may be made to determine the heating value of the fuel. Let us consider the simple case of the combustion of methane, CH_4. We write the chemical balance equation as

$$aCH_4 + xO_2 + 3.76xN_2 + bH_2O \rightarrow dCO_2 + eCO + fO_2 + gH_2O + 3.76xN_2$$

$$(9\text{-}25)$$

An analysis of the products gives the values of d, e, and f for 100 mol of dry products. The mass balance then becomes

CARBON

$$a = d + e \tag{9-26}$$

HYDROGEN

$$4a + 2b = 2g \tag{9-27}$$

OXYGEN

$$2x + b = 2d + e + 2f + g \tag{9-28}$$

Equation (9-26) permits an immediate determination of a. The relative humidity of the incoming air ϕ_1 gives a relationship for b as

$$\frac{b}{4.76x} = \frac{p_{g_1}\phi_1}{p_1 - \phi_1 p_{g_1}} \tag{9-29}$$

where p_{g_1} is the saturation pressure of water vapor at the inlet air temperature and p_1 is the total pressure of the incoming air–water-vapor mixture. We thus have three unknowns (b, g, and x) and three equations [(9-27) to (9-29)] which may be solved for the unknowns. With the mass balance now complete, an energy balance may be made to determine the chemical energy content of the incoming fuel. We observe that

$$\dot{m}_{f_1} = 16a \qquad \dot{m}_{a_1} = 32x + (28)(3.76x)$$
$$\dot{m}_{v_1} = 18b$$

The water vapor in the products of combustion is calculated from

$$\dot{m}_{v_2} = 18g - \dot{m}_c \tag{9-30}$$

The energy balance becomes

$$\dot{m}_{f_1}(h_{f_1} + E_{f_1}) + \dot{m}_{a_1}h_{a_1} + \dot{m}_{v_1}h_{v_1} - \dot{m}_{p_2}h_{p_2} - \dot{m}_{c_2}h_{c_2} = \dot{m}_w(h_{w_2} - h_{w_1}) \tag{9-31}$$

where E_{f_1} is designated as the chemical energy of the fuel. The energy term for the products $\dot{m}_p h_{p_2}$ is calculated from the energies of the individual constituents. Thus,

$$\dot{m}_p h_{p_2} = \dot{m}_{CO_2} h_{CO_2} + \dot{m}_{CO} h_{CO} + \dot{m}_{O_2} h_{O_2} + \dot{m}_{N_2} h_{N_2} + \dot{m}_{v_2} h_{v_2} \qquad (9\text{-}32)$$

The mass flow rate of these constituents is determined from the balance conditions in Eqs. (9-27) to (9-29).

The purpose of the above discussion is to illustrate the number of experimental measurements that must be made in even a simple calorimetric determination like the one shown and the fact that all these measurements influence the accuracy of the final result. Of course, the calculated value of heating value (or chemical energy content) will be more sensitive to some measurements than

Figure 9-16 Schematic of the nonflow bomb calorimeter.

others. For details on the measurement techniques the reader should consult Ref. [5]. It may be noted that there are standardized procedures available for this particular test which make it unnecessary to perform all the analysis indicated above; nevertheless, this analysis indicates the theory behind the measurement process.

The *bomb calorimeter* is another device that is frequently used for heating-value determinations in solid and liquid fuels. In contrast to the flow calorimeter described above, the measurements are made under constant-volume, nonflow conditions as shown in Fig. 9-16. A measured sample of the fuel is enclosed in a metal container, which is subsequently charged with oxygen at high pressure. The bomb is then placed in a container of water and the fuel ignited through external electric connections. The temperature of the water is measured as a function of time after the ignition process; and from a knowledge of the mass of water in the system, mass and specific heat of the container, and transient heating and cooling curves, the energy release during combustion may be determined. A motor-driven stirrer ensures uniformity in temperature of the water surrounding the bomb. In some circumstances external heating may be supplied to the jacket water to maintain a uniform temperature, while in other instances the jacket may be left empty in order to maintain nearly adiabatic conditions on the inner water container. A compensation for the heat lost to the environment may be made through an analysis of the transient heating and cooling curve. Doolittle [1] gives detailed procedures for use of the bomb calorimeter, and the interested reader should consult this reference for more information.

9-7 CONVECTION HEAT-TRANSFER MEASUREMENTS

Determination of convection heat-transfer coefficients covers a very broad range of experimental activities. In this section we shall illustrate only two simple experimental setups: one for a forced-convection system and one for a free-convection system. The interested reader should consult Ref. [2] for more information.

Consider the experimental setup shown in Fig. 9-17. It is desired to obtain convection heat-transfer coefficients for the flow of water in a smooth tube. Heat is supplied to the tube by electric heating as shown. The tube is usually made of a high-resistance material like stainless steel to reduce the electric current necessary for heating. Thermocouples are spot-welded or cemented to the outside surface of the tube to measure the wall temperature. Either thermocouples or thermometers are inserted in the flow to measure the water temperature at inlet and outlet to the heated section. A voltmeter and ammeter measure the power input to the tube, while some type of flowmeter measures the water-flow rate. The electrically heated tube delivers a constant heat flux to the water (constant $Btu/h \cdot ft^2$ or W/m^2 of tube surface), so that it is reasonable to assume a straight-line variation of water bulk temperature from inlet to outlet. Thus the wall and bulk tem-

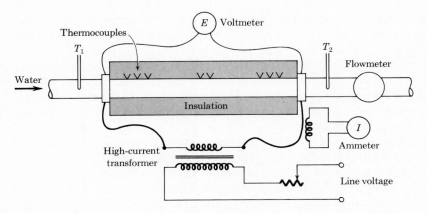

Figure 9-17 Schematic of apparatus for determination of forced-convection heat-transfer coefficients in smooth tubes.

peratures are known along the length of the tube, and the heat-transfer coefficient may be calculated at any axial location from

$$q = hA(T_w - T_B) \qquad (9\text{-}33)$$

where A is the total *inside-heated* surface area of the tube, T_w and T_B are the wall and bulk temperatures at the particular location, and q is the total heat-transfer rate given by

$$q = EI \qquad (9\text{-}34)$$

E and I are the voltage and current impressed on the test section. The heated surface area is

$$A = \pi d_i L \qquad (9\text{-}35)$$

Figure 9-18 Schematic of heated cylinder arrangement for determination of free-convection heat-transfer coefficients.

The outside surface of the tube must be covered with insulation to ensure that all the electric heating is dissipated in the water.

A simple experimental setup for determination of natural-convection heat-transfer coefficients on a horizontal cylinder is shown in Fig. 9-18. The heated horizontal cylinder consists of three sections. The center section is the test element, while the two end sections are guard heaters that eliminate end losses from the center section. All three sections are electrically heated and instrumented with thermocouples, which measure their respective surface temperatures. The electric current to the guard heaters is adjusted so that their temperature is just equal to the temperature of the center section. At this balance condition all the electric heating of the center section is dissipated to the surrounding air or enclosure. The heat loss from the test element is given by

$$q = hA(T_w - T_\infty) + \sigma \epsilon A(T_w^4 - T_s^4) \tag{9-36}$$

where h is the free-convection heat-transfer coefficient, T_w is the test section temperature, ϵ is the surface emissivity, T_∞ is the surrounding air temperature, T_s is the temperature of the surrounding room or enclosure, and A is the surface area, given by

$$A = \pi dL \tag{9-37}$$

The heat-transfer rate q is determined from measurement of the electric-power input to the test section. To determine the value of h from Eq. (9-36), the surface emissivity must be known. In practice, the test-section surface is usually nickel- or chrome-plated so that the radiation loss is small compared with the convection loss.

Example 9-5 A 30×30-cm square metal plate is used for a determination of free-convection heat-transfer coefficients. The plate is placed in a vertical position and exposed to room air at 20°C. The plate is electrically heated to a uniform surface temperature of 50°C, and the heating rate is measured as 15.0 ± 0.2 W. The emissivity of the surface is estimated as 0.07 ± 0.02. Determine the nominal value and uncertainty of the heat-transfer coefficient, assuming that the temperature measurements are exact. Assume that the effective radiation temperature of the surroundings is 20°C.

SOLUTION The nominal value of the heat-transfer coefficient is obtained from Eq. (9-36):

$$\frac{q}{A} = h(T_w - T_\infty) + \sigma \epsilon(T_w^4 - T_s^4) \tag{a}$$

where $T_w = 50°C$ $T_\infty = T_s = 20°C$

$$\epsilon = 0.07 \qquad \frac{q}{A} = \frac{15}{(0.3)^2} = 166.7 \text{ W/m}^2$$

The resulting value of h is

$$h = 5.091 \text{ W/m}^2 \cdot °C$$

From Eq. (a) the explicit relation for h is

$$h = \frac{1}{T_w - T_\infty} \left[\frac{q}{A} - \sigma\epsilon(T_w^4 - T_s^4) \right] \tag{b}$$

The uncertainty may be obtained from this expression using Eq. (3-2). We have

$$\frac{\partial h}{\partial(q/A)} = \frac{1}{T_w - T_\infty} \qquad w_{q/A} = \frac{0.2}{(0.3)^2} = 2.22 \text{ W/m}^2$$

$$\frac{\partial h}{\partial\epsilon} = \frac{-\sigma(T_w^4 - T_s^4)}{T_w - T_\infty} \qquad w_\epsilon = 0.02$$

The uncertainty in h is thus

$$W_h = \left\{ \left(\frac{1}{30}\right)^2 (2.22)^2 + \left[\frac{(5.669 \times 10^{-8})(323^4 - 293^4)}{30} \right]^2 (0.02)^2 \right\}^{1/2}$$

$$= (5.476 \times 10^{-3} + 0.01764)^{1/2}$$

$$= 0.152 \text{ W/m}^2 \cdot {}^\circ\text{C}$$

or

$$\frac{W_h}{h} = 0.0299 = 2.99\%.$$

9-8 HUMIDITY MEASUREMENTS

The water-vapor content of air is an important parameter in many processes. Proper control of critical operations with fabrics, papers, and chemicals frequently hinges upon a suitable control of the humidity of the surrounding environment. We shall discuss four basic techniques for measurement of the water-vapor content of air. First, let us consider some definitions. The *specific humidity*, or *humidity ratio*, is the mass of water vapor per unit mass of dry air. The *dry-bulb temperature* is the temperature of the air–water-vapor mixture as measured by a thermometer exposed to the mixture. The *wet-bulb temperature* is the temperature indicated by a thermometer covered with a wicklike material saturated with liquid after the arrangement has been allowed to reach evaporation equilibrium with the mixture, as indicated in Fig. 9-19. The *dew point* of the mixture is the temperature at which vapor starts to condense when the mixture is cooled at constant pressure.

The *relative humidity* ϕ is defined as the ratio of the actual mass of vapor to the mass of vapor required to produce a saturated mixture at the same temperature. If the vapor behaves like an ideal gas,

$$\phi = \frac{m_v}{m_{\text{sat}}} = \frac{p_v V/R_v T}{p_g V/R_v T} = \frac{p_v}{p_g} \tag{9-38}$$

Vapor
diffusion

Saturated
wick material

Dry bulb

Wet bulb

Figure 9-19 Measurement of dry- and wet-bulb temperatures.

where p_v is the actual partial pressure of the vapor and p_g is the saturation pressure of the vapor at the temperature of the mixture. The specific humidity is

$$\omega = \frac{m_v}{m_a} \qquad (9\text{-}39)$$

which, for ideal-gas behavior, becomes

$$\omega = 0.622 \frac{p_v}{p_a} \qquad (9\text{-}40)$$

where p_a is the partial pressure of the air.

The most fundamental method for measuring humidity is a gravimetric procedure, which is used by the National Bureau of Standards for calibration purposes. A sample of the air–water-vapor mixture is exposed to suitable chemicals until the water is absorbed. The chemicals are then weighed to determine the amount of vapor absorbed. Uncertainties as low as 0.1 percent can be obtained with this method.

There is a definite analytical relationship between the dry-bulb, wet-bulb, and dew-point temperatures of a mixture and its humidity. A determination of any two of these temperatures may be used to calculate the humidity. The classic method used for humidity determination in large open spaces is a measurement of the dry- and wet-bulb temperatures with a sling psychrometer. Both thermometers are rotated with a speed of about 5 m/s and the temperatures recorded. The vapor pressure of the mixture may then be calculated with Carrier's equation:

$$p_v = p_{gw} - \frac{(p - p_{gw})(T_{DB} - T_{WB})}{K_w - T_{WB}} \qquad (9\text{-}41)$$

where p_v = actual vapor pressure

p_{gw} = saturation pressure corresponding to wet-bulb temperature

p = total pressure of mixture

T_{DB} = dry-bulb temperature, °F or °C

T_{WB} = wet-bulb temperature, °F or °C

K_w = 2800 when T is in °F

= 1537.8 when T is in °C

The specific or relative humidity is then calculated from Eq. (9-38) or (9-40).

The sling psychrometer is not particularly suitable to automation or continuous-recording requirements. For such applications a Dunmore-type electrical humidity transducer is used. This transducer consists of a resistance element which is constructed by winding dual noble-metal elements on a plastic form with a carefully controlled spacing between them. The windings are then coated with a lithium-chloride solution, which forms a conducting path between the wires. The resistance of this coating is a very strong function of relative humidity and thus may be used to sense the humidity. An ac bridge circuit, as discussed in Chap. 4, is normally used for sensing the resistance of the transducer. Because of wide nonlinear variation of the resistance of the coating with humidity, a single transducer is normally suitable for use only over a narrow range of humidities. If measurements over a wide range of humidities are desired, multiple sensors can be employed, each with a coating suitable for some segment of the humidity spectrum. In general, a single transducer is not used over much more than a 10 percent relative-humidity range. Accuracies for the device are about 1.5 percent relative humidity and may be used over an operating range of -40 to $+150$°F (-40 to 65°C).

The humidity of a mixture may also be determined with a measurement of dew-point temperature. For this purpose the conventional technique is to cool a highly polished metal surface (mirror) until the first traces of fogging or condensation appear when the surface is in contact with the mixture. The temperature of the surface for the onset of fogging is then the dew point. The sensitivity of the measurement process may be improved by using a photocell to detect a beam of light that is reflected from the surface. The intensity of the reflected beam will decrease at the onset of fogging, and this point may be used to mark the temperature signal from the mirror as the dew point. For continuous monitoring of dew point, the mirror-photocell technique may be modified by connecting the mirror to a low-temperature source like a dry-ice–acetone bath and simultaneously to an electric heating element. An appropriate electronic circuit senses the output of the photocell and causes the electric heater to be actuated when the output drops as a result of fogging. The heater then remains on until the mirror surface is clear and the photocell output rises, thereby reducing the heater current. Basically, the system just described is a control system to maintain the mirror at the dew-point temperature regardless of changes in environment. A thermocouple attached to the mirror may then be used to continuously monitor the dew point. A summary of humidity-measurement techniques is given in Refs. [9] to [11]. The thermodynamics of air–water-vapor mixtures is discussed in Ref. [12].

9-9 HEAT-FLUX METERS

There are many applications where a direct measurement of heat flux is desired. In this section we shall discuss a few devices for such measurements. The first type of heat-flux meter we shall consider is the slug sensor shown in Fig. 9-20. A slug of high thermal conductivity is installed in the wall as shown with a surrounding layer of insulation. When the slug is subjected to a heat flux at the surface, its temperature will rise and give an indication of the magnitude of the heat flux. Assuming that the temperature of the slug remains uniform at any instant of time, the transient-energy balance becomes

$$\frac{q}{A} = \frac{mc}{A} \frac{dT_s}{d\tau} + U(T_s - T_w) \tag{9-42}$$

where q/A is the imposed heat flux, m is the mass of the slug, c is the specific heat of the slug, U is a coefficient expressing the conduction loss to the surrounding wall, T_s is the temperature of the slug, and T_w is the surrounding wall temperature. It is clear that a solution of Eq. (9-42) will give the slug temperature as a function of the heat flux and the properties of the slug. Thus, a measurement of T_s may be taken as an indication of the heat flux.

Some problems are encountered in using slug sensors. First, for very high heat fluxes the temperature of the slug may not remain uniform and a more cumbersome analysis must be employed to calculate the heat flux from the temperature measurements. Second, the presence of the sensor alters the temperature profile in the wall; in particular, the temperature at the sensor location is higher than it might have been otherwise (for slug conductivity greater than wall conductivity), thereby influencing the net heat flux at the point of measurement, whether it occurs by convection, radiation, or a combination of both. Third, the

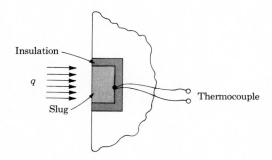

Figure 9-20 Slug-type heat-flux meter.

slug sensor is clearly adapted only to transient measurements. As soon as the wall reaches equilibrium, the temperature of the slug will no longer be indicative of the heat flux. Additional information on the response of slug-type sensors is available in Refs. [13], [14], and [19]. Figures 8-35 and 8-36 give the approximate response times to be expected from semi-infinite slabs subjected to various heat fluxes. Signal levels from thermocouples are also indicated, and these values may be employed for preliminary design studies.

The Gardon [15] heat-flux meter is depicted in Fig. 9-21. A thin constantan disk is mounted to a copper heat sink, as shown, with the heat sink installed in the wall material. A small copper wire is then attached to the center of the disk and, along with a wire connected to the copper sink, forms the two wires of a copper-constantan thermocouple to measure the temperature difference between the center of the disk and the sides. A heat-flux incident on the disk is absorbed and conducted radially outward, thereby creating the temperature difference that is sensed by the thermocouple. The only losses in the gage are the radiation losses from the back side of the disk to the copper heat sink, but by careful calibration this may be taken into account with little difficulty. Heat fluxes over a range of 5×10^4 to 10^6 Btu/h·ft² (0.15 to 3.2 MW/m²) may be measured with the device. Time constants of the order of 0.1 s are experienced. Increased sensitivity may be obtained by using a copper disk along with a positively doped bismuth-telluride center connection [17].

A very versatile heat-flux meter has been described by Hager [16] and is marketed by the RdF Corp., Hudson, N.H. The meter operates on the principle of measuring the temperature drop across a thin insulating material of about 0.0003 in (7.6 μm) thickness. The temperature drop is directly proportional to the heat conducted through the material in accordance with Fourier's law [Eq. (9-1)]. The unique feature of these meters is the butt-bonding process, which allows a distortion-free joining of copper-constantan on both sides of the material. The detailed construction is indicated in Fig. 9-22, with a number of junctions connected in series to produce a thermopile. As many as 40 junctions may be produced in this manner within a space of about 1 in². The junctions are covered with a protective coating, but the total thickness of the heat-flux meter is never more than 0.012 in (0.3 mm). The maximum heat flux that may be accom-

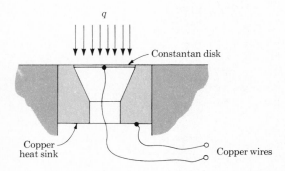

Figure 9-21 Gardon foil-type heat-flux meter, according to Ref. [15].

Figure 9-22 Detail of heat-flux meter operating on conduction principle. (*Courtesy of RdF Corp., Hudson, N.H.*)

Figure 9-23 Photographs of assembled microfoil heat-flux meters. (*Courtesy of RdF Corp., Hudson, N.H.*)

modated with the meter is about 2×10^5 Btu/h·ft² (0.63 MW/m²) at a maximum operating temperature of 500°F (260°C). Sensitivity ranges from 0.07 to 27 mV·ft²·s/Btu (0.062 to 23.8 mV·cm²/W), depending on the number of junctions employed. Time constants range between 0.02 and 0.4 s. Figure 9-23 shows some photographs of typical meter configurations.

9-10 pH MEASUREMENT

The pH of a solution is a measure of the hydrogen ion concentration C_H and is defined by

$$\text{pH} = \log C_H \qquad (9\text{-}43)$$

A typical pH meter is shown in Fig. 9-24. A special glass electrode is used which is permeable only to hydrogen ions. The very thin bottom therefore acts as a membrane. A reference electrode of calomel is also placed in the solution. Reversible electrodes are attached to both the glass and reference electrode, and hydrochloric acid (1 N) is contained in the sealed glass electrode.

The net potential across the membrane is given by

$$\Delta v_i = -2.30 \frac{\mathscr{R}T}{\mathscr{F}} \log \frac{C_H}{C_R} \qquad (9\text{-}44)$$

where $\mathscr{R}$ = universal gas constant, 8314 J/kg·mol·K
T = absolute temperature, K
$\mathscr{F}$ = Faraday's constant, 9.647×10^7 C/kg·mol
C_H = hydrogen ion concentration in solution
C_R = concentration in glass electrode
($C_R = 1.0$ for 1 N HCl)

Figure 9-24 Schematic of pH meter.

Δv_i

Inserting the constants and making use of the definition of pH gives

$$\Delta v_i = -1.98 \times 10^{-4} T \text{ (pH units)} \qquad (9\text{-}45)$$

Because the resistance of the glass membrane is very high (on the order of 100 MΩ) a dc amplifier with high input impedance must be used. Fig. 4-32j is an example of an operational-amplifier configuration which might be employed. Calibration is achieved by inserting the electrodes in a solution of known pH and adjusting the output meter for a proper reading.

REVIEW QUESTIONS

9-1 Write the defining equation for thermal conductivity.

9-2 What is a guarded hot-plate apparatus? Under what conditions is it applied?

9-3 How are thermal conductivities of metals measured?

9-4 What are some of the considerations that must be applied for gas and liquid thermal conductivity that are not important in thermal-conductivity measurements for solids?

9-5 Write the defining equation for viscosity.

9-6 What is the operating principle of the Saybolt viscosimeter?

9-7 Write the defining equation for diffusion coefficient.

9-8 How does a Loschmidt apparatus function?

9-9 Describe the flow calorimeter.

9-10 Describe the bomb calorimeter.

9-11 Write the defining equation for the heat-transfer coefficient.

9-12 How does a sling psychrometer measure relative humidity?

9-13 Describe a technique for measuring dew point.

9-14 What is the operating principle of the slug-type heat-flux meter?

9-15 What is a Gardon heat-flux meter?

PROBLEMS

9-1 The following data are taken from Ref. [6] on an apparatus like that shown in Fig. 9-3. The measurements were made on a lead sample whose thermal conductivity is taken to be 0.352 W/cm·°C at 0°C.

Thermocouple positions	Temperature, °C	Distance between adjacent thermocouples, cm
1	277.3	3.14
2	231.6	3.14
3	186.5	3.14
4	143.0	3.15
5	100.3	
6	78.6	3.16
7	37.3	

Using these data, construct a graph of the thermal conductivity of lead versus temperature.

9-2 The apparatus of Fig. 9-3 is used for measurement of the thermal conductivity of an unknown metal. The lengths of the unknown and known bars are the same, and five thermocouples are equally spaced on each bar. From the temperature and heat flux measurement the thermal conductivity of the unknown is to be established as a function of temperature. Design the apparatus to measure the thermal conductivity of a material, with $k \sim 100$ Btu/h·ft·°F. Consider the influence of thermocouple spacing, dimensions of the bars, heat flux measurement, and temperature measurement on accuracy. Take the thermal conductivity of the known material as exactly 20 Btu/h·ft·°F.

9-3 A concentric-cylinder device like that in Fig. 9-5 is used for a measurement of the thermal conductivity of water at 35°C. Using the dimensions given in the text, discuss the influence of the heat flux and temperature measurements on the accuracy of the thermal-conductivity determination.

9-4 The viscosity of water is 1.65 lbm/h·ft at 100°F. A small capillary tube 50 ± 0.01 ft long is to be used to check this value. Calculate the maximum allowable flow rate for an inside tube diameter of 0.100 ± 0.001 in. Calculate the pressure drop for this flow rate, and estimate the allowable uncertainty in the pressure drop in order that the uncertainty in the viscosity does not exceed 5 percent. Assume that the flow rate is measured within ± 0.01 lbm/h.

9-5 The rotating concentric-cylinder apparatus of Fig. 9-9 is used to measure the viscosity of water at 100°F. The inner cylinder radius is 1.50 ± 0.001 in, and its length is 4.00 ± 0.002 in. The outer cylinder is rotated at a speed of 1800 ± 2 rpm. The outer cylinder radius is 1.60 ± 0.001 in, and the gap spacing at the bottom of the cylinder is 0.500 ± 0.005 in. Calculate the nominal torque which will be measured, and estimate the allowable uncertainty in this measurement in order that the overall uncertainty in the viscosity does not exceed 5 percent. Repeat the calculation for glycerine at 100°F.

9-6 The tolerance limits on the radii of the inner and outer cylinders of Fig. 9-9 exert a strong influence on the accuracy of the viscosity measurement. Establish a relation expressing the percent uncertainty in the viscosity measurement as a function of the uncertainty in these radius measurements. Assume that the torque, angular velocity, and bottom-spacing measurements are exact for this calculation. Assume that the uncertainty (tolerance) is the same for both cylinders.

9-7 The viscosity of a gas is to be determined by measuring the flow rate and pressure drop through a capillary tube. The experimental variables which are measured are:

Inlet pressure p_1
Pressure differential $\Delta p = p_1 - p_2$
Temperature T
Mass flow rate $\dot{m}$
Tube diameter $d = 2r$

Derive an expression for the percent uncertainty in the viscosity measurement in terms of the uncertainties in these five primary measurements. In view of the information in Chaps. 5 to 8, discuss the relative influence of each of the measurements, i.e., which primary measurement will probably have the most effect on the final estimated uncertainty in the viscosity.

9-8 The viscosity of an oil having a density of 51.9 lbm/ft^3 is measured as 200 Saybolt seconds. Determine the dynamic viscosity in units of centipoise and pounds-mass per hour-foot.

9-9 A Loschmidt apparatus is used to measure the diffusion coefficient for benzene in air. At 25°C the diffusion coefficient is 0.088 cm^2/s for a pressure of 1 standard atmosphere. Both tubes have a length of 60 cm. Calculate the mole fraction of benzene in each tube 10 min and 3 h after the connection between the tubes is opened.

9-10 A Loschmidt apparatus with $L = 50$ cm is used to measure the diffusion coefficient for CO_2 in air. After a time of 6.52 min the mole fraction of CO_2 in the upper tube is 0.0912. Calculate the value of the diffusion coefficient.

9-11 For $t = t_{opt}$, calculate the relative magnitudes of the first four terms in the series expansion of Eq. (9-20). Repeat for $t = \frac{1}{2} t_{opt}$ and $t = 2t_{opt}$.

9-12 Obtain a simplified expression for the heating value of methane as measured with a flow calorimeter with the following conditions:

$$T_{a_1} = T_{f_1} = T_{p_2} = T_{c_2} \qquad \phi_1 = 0$$

The incoming gaseous fuel is saturated with water vapor.

9-13 Design an apparatus like that shown in Fig. 9-17 to measure forced-convection heat-transfer coefficients for water at about 200°F in a range of Reynolds numbers from 50,000 to 100,000 based on tube diameter; i.e.,

$$\text{Re}_d = \frac{\rho u_m d}{\mu}$$

where u_m is the mean flow velocity in the tube. Be sure to specify the accuracy and range of all instruments required and estimate the uncertainty in the calculated values of the heat-transfer coefficients.

9-14 Design an apparatus like that shown in Fig. 9-18 to measure free-convection heat-transfer coefficients for air at atmospheric pressure and ambient temperature of 20°C. The apparatus is to produce data over a range of Grashof numbers from 10^5 to 10^7 where

$$\text{Grashof number} = \text{Gr} = \frac{\rho^2 g \beta (T_w - T_\infty) d^3}{\mu^2}$$

where β = volume coefficient of expansion for air
$\quad\quad d$ = tube diameter
$\quad\quad T_w$ = tube surface temperature
$\quad\quad T_\infty$ = ambient air temperature
The properties of the air are evaluated at the film temperature defined by (see Table 8-5)

$$T_f = \frac{T_w + T_\infty}{2}$$

The Grashof number is dimensionless when a consistent set of units is used. Be sure to specify the accuracy and range of all instruments used and estimate the uncertainty in the calculated values of the heat-transfer coefficient.

9-15 A copper sphere 2.5 cm in diameter is used as a slug-type heat-flux meter. A thermocouple is attached to the interior and the transient temperature recorded when the sphere is suddenly exposed to a convection environment. Plot the time response of the system for sudden exposure to an environment with $h = 570$ W/m$^2 \cdot$ °C. Assume the sphere is initially at 95°C and the environment temperature remains constant at 35°C. What is the time constant of this system?

9-16 A heat-flux meter is to be constructed of a thin layer of glasslike substance having a thermal conductivity of 0.4 Btu/h $\cdot$ ft $\cdot$ °F. The temperature drop across the layer will be measured with a thermopile consisting of 10 copper-constantan junctions. The glass thickness is 0.0008 in. Calculate the voltage output of the thermopile when a heat flux of 10^5 Btu/h $\cdot$ ft^2 is imposed on the layer.

9-17 The dry- and wet-bulb temperatures for a quantity of air at 14.7 psia are measured as 95 and 75°F, respectively. Calculate the relative humidity and the dew point of the air–water-vapor mixture.

9-18 An air–water-vapor mixture is contained at a total pressure of 206 kPa. The dry-bulb temperature is 32°C. What will be the wet-bulb temperature if the relative humidity is 50 percent?

9-19 The thermal conductivity of an insulating material is to be determined using a guarded hot-plate apparatus. A 30 × 30-cm sample is used, and the approximate thermal conductivity is about 1.7 W/m $\cdot$ °C. A differential-thermocouple arrangement is used to measure the temperature drop across the sample with an uncertainty of ± 0.3°C. A heat input of 5 kW $\pm$ 1 percent is supplied, with a resulting temperature drop of 55°C. The sample thickness is 2.0 mm. What is the thermal conductivity and its uncertainty, assuming no heat losses and that the sample dimensions are known exactly?

9-20 Wet bulb and dry bulb thermometers of air at 1 atm read 20°C and 35°C, respectively, within ±0.5°C. Calculate the relative humidity and dew point and their uncertainties.

9-21 Devise an apparatus for measuring the thermal conductivity of window glass. Specify heating requirements and temperature measurement techniques. Estimate uncertainty in the measurement.

9-22 Devise a capillary tube arrangement for measuring the viscosity of air at 20°C. Specify dimensions, flow rates, and temperature and pressure measurement requirements and techniques.

9-23 The dew point of air at 40°C is measured at 25°C ± 0.5°C. Calculate the relative humidity and wet bulb temperatures and their uncertainties.

REFERENCES

1. Doolittle, J. S.: "Mechanical Engineering Laboratory," McGraw-Hill Book Company, New York, 1957.
2. Jakob, M.: "Heat Transfer," vol. 1, John Wiley & Sons, Inc., New York, 1949.
3. Kaye, G. W. C., and W. F. Higgins: The Thermal Conductivities of Certain Liquids, *Proc. Roy. Soc. London*, vol. 117, no. 459, 1928.
4. Keyes, F. G., and D. J. Sandell: New Measurements of Heat Conductivity of Steam and Nitrogen, *Trans. ASME*, vol. 72, p. 768, 1950.
5. Shoop, C. F., and G. L. Tuve: "Mechanical Engineering Practice," 5th ed., McGraw-Hill Book Company, New York, 1956.
6. Van Dusen, M. S., and S. M. Shelton: Apparatus for Measuring Thermal Conductivity of Metals up to 600°C, *J. Res. Natl. Bur. Std.*, vol. 12, no. 429, 1934.
7. Vines, R. G.: Measurements of Thermal Conductivities of Gases at High Temperatures. *Trans. ASME*, vol. 82C, p. 48, 1960.
8. Leidenfrost, W.: An Attempt to Measure the Thermal Conductivity of Liquids, Gases and Vapors with a High Degree of Accuracy over Wide Ranges of Temperature and Pressure, *Intern. J. Heat and Mass Transfer*, vol. 7, no. 4, pp. 447–476, 1964.
9. Fraade, D. J.: Measuring Moisture in Gases, *Instr. Control Systems*, p. 100, April 1963.
10. Moisture and Humidity, *Instr. Control Systems*, p. 121, October 1964.
11. Amdur, E. J.: Humidity Sensors, *Instr. Control Systems*, p. 93, June 1963.
12. Holman, J. P.: "Thermodynamics," 3d ed., McGraw-Hill Book Company, New York, 1980.
13. Kirchhoff, R. H.: Calorimetric Heating-Rate Probe for Maximum Response Time Interval, *AIAA J.*, p. 966, May 1964.
14. Hornbaker, D. R., and D. L. Rall: Thermal Perturbations Caused by Heat Flux Transducers and Their Effect on the Accuracy of Heating Rate Measurements, *ISA Trans.*, p. 100, April 1963.
15. Gardon, R.: An Instrument for the Direct Measurement of Intense Thermal Radiation, *Rev. Sci. Instr.*, p. 366, May 1953.
16. Hager, N. E., Jr.: Thin Foil Heat Meter, *Rev. Sci. Instr.*, November 1965.
17. Carey, R. M.: "Experimental Test of a Heat Flow Meter," M.S. thesis, Southern Methodist University, Dallas, Tex., July 1963.
18. Kestin, J., and W. Leidenfrost: An Absolute Determination of the Viscosity of Eleven Gases Over a Range of Pressure, *Physica*, vol. 25, p. 1033, 1959.
19. Schultz, D. L., and T. V. Jones: Heat Transfer Measurements in Short-Duration Hypersonic Facilities, *AGARDograph 165*, 1973.
20. Leidenfrost, W.: Measurement of Thermophysical Properties in "Measurements in Heat Transfer," pp. 475–517, McGraw-Hill Book Company, New York, 1976.

FORCE, TORQUE, AND STRAIN
MEASUREMENTS

10-1 INTRODUCTION

Some types of force measurements applicable to pressure-sensing devices have been discussed in Chap. 6. We now wish to consider other methods for measuring forces and torques and relate these methods to basic strain measurements and experimental stress analysis.

Force is represented mathematically as a vector with a point of application. Physically, it is a directed push or pull. According to Newton's second law of motion, as written for a particle of constant mass, force is proportional to the product of mass and acceleration. Thus,

$$F = \frac{1}{g_c} ma \qquad (10\text{-}1)$$

where the $1/g_c$ term is the proportionality constant. When force is expressed in lbf, mass in lbm, and acceleration in ft/s^2, g_c has the value of 32.1739 lbm $\cdot$ ft/lbf $\cdot$ s^2 and is numerically equal to the acceleration of gravity at sea level. The weight of a body is the force exerted on the body by the acceleration of gravity at sea level, so that

$$F = W = \frac{mg}{g_c} \qquad (10\text{-}2)$$

At sea level, the weight in lbf units is numerically equal to the mass in lbm units. In Sec. 2-5 we discussed the relationships between different unit systems. In the SI

Figure 10-1 Vector representation for torque.

system the newton is the unit of force and the kilogram is the standard unit of mass. Thus, the weight of a body should be expressed in newtons in the SI system. Regretfully, there will be users of metric units who will express force in kilograms-*force* (kilopond), and the reader should be aware of this possible practice.

Torque is represented as a moment vector formed by the cross product of a force and radius vector. In Fig. 10-1 the torque that the force F exerts about the point 0 is given mathematically as

$$\mathbf{T} = \mathbf{r} \times \mathbf{F} = |r||F| \sin \theta \qquad (10\text{-}3)$$

In the SI system torque is expressed in units of newton-meters, while in the English system the common unit is foot-pounds-force.

Mass represents a quantity of matter and is an inertial property related to force through Eq. (10-1). As such, mass is not a physical property that is measured directly; it is determined through a force measurement or by comparison with mass standards. Even with a comparison of two masses, as with a simple balance, a force equivalency is still involved.

In this chapter we shall be concerned with force measurements as related to mechanical systems. Forces arising from electrical and magnetic effects are not considered.

10-2 MASS BALANCE MEASUREMENTS

Consider the schematic of an analytical balance as shown in Fig. 10-2. The balance arm rotates about the fulcrum at point 0 (usually a knife edge) and is shown in an unbalanced position as indicated by the angle ϕ. Point G represents the center of gravity of the arm, and d_G is the distance from 0 to this point. W_B is the weight of the balance arm and pointer. When $W_1 = W_2$, ϕ will be zero and the weight of the balance arms will not influence the measurements. The sensitivity of the balance is a measure of the angular displacement ϕ per unit unbalance in the two weights W_1 and W_2. Expressed analytically this relation is

$$S = \frac{\phi}{W_1 - W_2} = \frac{\phi}{\Delta W} \qquad (10\text{-}4)$$

Figure 10-2 Schematic of analytical balance.

We now wish to determine the functional dependence of this sensitivity on the physical size and mass of the balance. According to Fig. 10-2 the moment equation for equilibrium is

$$W_1(L \cos \phi - d_B \sin \phi) = W_2(L \cos \phi + d_B \sin \phi) + W_B d_G \sin \phi \quad (10\text{-}5)$$

For small deflection angles $\sin \phi \approx \phi$ and $\cos \phi \sim 1.0$, so that Eq. (10-5) becomes

$$W_1(L - d_B \phi) = W_2(L + d_B \phi) + W_B d_G \phi$$

or

$$\frac{\phi}{\Delta W} = \frac{L}{(W_1 + W_2)d_B + W_B d_G} \quad (10\text{-}6)$$

Near equilibrium, $W_1 \approx W_2$, and S becomes

$$S = \frac{\phi}{\Delta W} = \frac{L}{W_B d_G + 2W d_B} \quad (10\text{-}7)$$

where, now, we have used the single symbol W to designate the loading on the balance. If the balance arm is constructed so that $d_B = 0$, we obtain

$$S = \frac{L}{W_B d_G} \quad (10\text{-}8)$$

an expression that is independent of the loading of the balance. Precision balances are available that have an accuracy of about 1 part in 10^8. For precision instruments optical methods are usually employed to sense the deflection angle and determine the equilibrium position.

Example 10-1 A balance is constructed with the following dimensions:

$$W_B = 50 \text{ g}$$

$$d_G = 0.3 \text{ cm}$$

$$L = 21.2 \text{ cm}$$

$$d_B = 0.01 \text{ cm}$$

The pointer scale is graduated so that readings may be taken within one-quarter of a degree. Estimate the uncertainty due to sensitivity in determining the weight of a mass of 1000 g. It may be noted that other uncertainties exist in the measurement, such as frictional effects, errors in reading, etc.

SOLUTION From Eq. (10-7) we have

$$S = \frac{\phi}{\Delta W} = \frac{L}{2W d_B + W_B d_G}$$

$$= \frac{21.2}{(2)(1000)(0.01) + (50)(0.3)} = 0.606 \text{ rad/g}$$

For an uncertainty in ϕ of 0.25°, the uncertainty in W is then given by

$$\Delta W = \frac{\phi}{S} = \frac{0.25\pi}{(180)(0.606)} = 0.0072 \text{ g}$$

or about 1 part in 100,000.

When a balance is used for mass determinations, errors may result if corrections are not made for buoyancy forces of the air surrounding the samples. Typically, an unknown mass is balanced with a group of standard brass weights. The forces that the instrument senses are not the weight forces of the unknown mass and brass weights but the weight forces *less* the buoyancy force on each mass. If the measurement is conducted in a vacuum or the unknown brass and mass weights have equal volume, the buoyancy forces will cancel out and there will be no error. Otherwise, the error may be corrected with the following analysis. The two forces exerted on the balance arms will be

$$W_1 = (\rho_u - \rho_a)V_u \tag{10-9}$$

$$W_2 = (\rho_s - \rho_a)V_s \tag{10-10}$$

where ρ_u = density of the unknown
ρ_s = density of the standard weights
ρ_a = density of the surrounding air
V_u, V_s = volumes of the unknown and standard weights, respectively

The true weights of the unknown and standard weights are

$$W_u = \rho_u V_u \tag{10-11}$$

$$W_s = \rho_s V_s \tag{10-12}$$

At balance $W_1 = W_2$, and there results

$$W_u = W_s\left(1 + \frac{\rho_a \rho_s - \rho_u}{\rho_s \rho_u - \rho_a}\right) \tag{10-13}$$

Example 10-2 A quantity of a plastic material having a density of about 1280 kg/m³ is weighed on a standard equal-arm balance. Balance conditions are achieved with brass weights totaling 152 g in a room where the ambient air is at 20°C and 1 atm. The density of the brass weights may be taken as 8490 kg/m³. Calculate the true weight of the plastic material and the percent error that would result if the balance reading were taken without correction.

SOLUTION The true weight of the brass weights is 152 g. The density of the air may be calculated as

$$\rho_a = \frac{p}{RT} = \frac{1.0132 \times 10^5}{(287)(293)} = 1.2049 \text{ kg/m}^3$$

The true weight of the plastic is then calculated from Eq. (10-13) as

$$W_u = 152\left[1 + \frac{(1.2049)(8490 - 1280)}{(8490)(1280 - 1.2049)}\right] = 152.122 \text{ g}$$

The error is $(0.122/152) \times 100$ or 0.08 percent.

10-3 ELASTIC ELEMENTS FOR FORCE MEASUREMENTS

Elastic elements are frequently employed to furnish an indication of the magnitude of an applied force through a displacement measurement. The simple spring is an example of this type of force-displacement transducer. In this case the force is given by

$$F = ky \tag{10-14}$$

where k is the spring constant and y is the displacement from the equilibrium position. For the simple bar shown in Fig. 10-3 the force is given by

$$F = \frac{AE}{L} y \tag{10-15}$$

y

F

Figure 10-3 Simple elastic element.

where A = cross-sectional area
L = length
E = Young's modulus for the bar material

The deflection of the cantilever beam shown in Fig. 10-4 is related to the loading force by

$$F = \frac{3EI}{L^3}\, y \tag{10-16}$$

where I is the moment of inertia of the beam about the centroidal axis in the direction of the deflection. Any one of the three devices mentioned above is suitable for use as a force transducer provided that accurate means are available for indicating the displacements. The differential transformer (Sec. 4-21), for example, may be useful for measurement of these displacements, as well as capacitance and piezoelectric transducers (Secs. 4-22 and 4-23).

Another elastic device frequently employed for force measurements is the thin ring shown in Fig. 10-5. The force-deflection relation for this type of elastic element is

$$F = \frac{16}{\pi/2 - 4/\pi}\, \frac{EI}{d^3}\, y \tag{10-17}$$

where d is the outside ring diameter and I is the moment of inertia about the centroidal axis of the ring section. The *proving ring* is a ring transducer that employs a sensitive micrometer for the deflection measurement, as shown in Fig. 10-6. To obtain a precise measurement, one edge of the micrometer is

F

L

y

Figure 10-4 Cantilever elastic element.

Figure 10-5 Thin-ring elastic element.

Figure 10-6 Proving ring.

mounted on a vibrating-reed device R, which is plucked to obtain a vibratory motion. The micrometer contact is then moved forward until a noticeable damping of the vibration is observed. Deflection measurements may be made within ± 0.00002 in (0.5 μm) with this method. The proving ring is widely used as a calibration standard for large tensile-testing machines.

Example 10-3 A small cantilever beam is constructed of spring steel having $E = 28.3 \times 10^6$ psi. The beam is 0.186 in wide and 0.035 in thick, with a length of 1.00 ± 0.001 in. An LVDT is used for the displacement-sensing device, and it is estimated that the uncertainty in the displacement measurement is ± 0.001 in. The uncertainties in the bar dimensions are ± 0.0003 in. Calculate the indicated force and uncertainty due to dimension tolerances when $y = 0.100$ in.

SOLUTION The moment of inertia is calculated with

$$I = \frac{bh^3}{12} \tag{a}$$

where b is the width and h is the thickness. Thus,

$$I = \frac{(0.186)(0.035)^3}{12} = 6.64 \times 10^{-7} \text{ in}^4$$

Equation (10-16) may be rewritten as

$$F = \frac{3Ebh^3}{12L^3} \, y \tag{b}$$

Using Eq. (3-2) the uncertainty in the force measurement can be written

$$\frac{w_F}{F} = \left[\left(\frac{w_b}{b} \right)^2 + 9\left(\frac{w_h}{h} \right)^2 + 9\left(\frac{w_L}{L} \right)^2 + \left(\frac{w_y}{y} \right)^2 \right]^{1/2} \tag{c}$$

The nominal value of the force becomes

$$F = \frac{(3)(28.3 \times 10^6)(6.64 \times 10^{-7})}{(1.00)^3} \, 0.100$$

$$= 5.64 \text{ lbf (25.09 N)}$$

From Eq. (c) the uncertainty is

$$\frac{w_F}{F} = \left[\left(\frac{0.0003}{0.186} \right)^2 + 9\left(\frac{0.0003}{0.035} \right)^2 + 9\left(\frac{0.001}{1.00} \right)^2 + \left(\frac{0.001}{0.100} \right)^2 \right]^{1/2}$$

$$= 0.0278 \text{ or } 2.78\%$$

In this instance 1.0 percent uncertainty would be present even if the dimensions of the beam were known exactly.

The surface strain (deformation) in elastic elements like those discussed above is, of course, a measure of the deflection from the no-load condition. This surface strain may be measured very readily by the electrical-resistance strain gage to be discussed in subsequent paragraphs. The output of the strain gage may thus be taken as an indication of the impressed force. The main problem with the use of these gages for the force-measurement applications is that a moment may be impressed on the elastic element because of eccentric loading. This would result in an alternation of the basic strain distribution as measured by the strain gage. There are means for compensating for this effect through installation of multiple gages, properly interconnected to cancel out the deformation resulting from the impressed moment. The interested reader should consult Refs. [4] and [5] for a summary of these methods.

10-4 TORQUE MEASUREMENTS

Torque, or moment, may be measured by observing the angular deformation of a bar or hollow cylinder, as shown in Fig. 10-7. The moment is given by

$$M = \frac{\pi G(r_o^4 - r_i^4)}{2L} \phi \qquad (10\text{-}18)$$

where G = shear modulus of elasticity
   r_i = inside radius
   r_o = outside radius
   L = length of the cylinder
   ϕ = angular deflection

Strain gages attached at 45° angles as shown will indicate strains of

$$\epsilon_{45^\circ} = \pm \frac{Mr_0}{\pi G(r_o^4 - r_i^4)} \qquad (10\text{-}19)$$

Either the deflection or the strain measurement may be taken as an indication of the applied moment. Multiple strain gages may be installed and connected so

Figure 10-7 Hollow cylinder as an elastic element for torque measurement.

that any deformation due to axial or transverse load is canceled out in the final readout circuit.

A rather old device for the measurement of torque and dissipation of power from machines is the Prony brake.† A schematic diagram is shown in Fig. 10-8. Wooden blocks are mounted on a flexible band or rope, which is connected to the arm. Some arrangement is provided to tighten the rope to increase the frictional resistance between the blocks and the rotating flywheel of the machine. The torque exerted on the Prony brake is given by

$$T = FL \qquad (10\text{-}20)$$

The force F may be measured by conventional platform scales or other methods discussed in the previous paragraphs.

The power dissipated in the brake is calculated from

$$P = \frac{2\pi TN}{33,000} \qquad \text{hp} \qquad (10\text{-}21)$$

where the torque is in foot-pounds-force and N is the rotational speed in revolutions per minute.

Various other types of brakes are employed for power measurements on mechanical equipment. The *water brake*, for example, dissipates the output energy through fluid friction between a paddle wheel mounted inside a stationary chamber filled with water. The chamber is freely mounted on bearings so that the torque transmitted to it can be measured through an appropriate moment arm similar to that used with the Prony brake.

The dc cradled dynamometer is perhaps the most widely used device for power and torque measurements on internal-combustion engines, pumps, small steam turbines, and other mechanical equipment. The basic arrangement of this device is shown in Fig. 10-9. A dc motor generator is mounted on bearings as shown, with a moment arm extending from the body of the motor to a force-measurement device, which is usually a pendulum scale. When the device is connected to a power-producing machine, it acts as a dc generator whose output may be varied by dissipating the power in resistance racks. The torque impressed on the dynamometer is measured with the moment arm and the output power

---

† Named for G. C. F. M. Riche, Baron de Prony (1755–1839), French hydraulic engineer.

Figure 10-8 Schematic of a Prony brake.

Figure 10-9 Schematic of a cradled dynamometer.

calculated with Eq. (10-21). The dynamometer may also be used as an electric motor to drive some power-absorbing device like a pump. In this case, it furnishes a means for measurement of torque and power input to the machine. Commercial dynamometers are equipped with controls for precise variation of the load and speed of the machine and are available with power ratings as high as 3700 kW (5000 hp).

10-5 STRESS AND STRAIN

Stress analysis involves a determination of the stress distribution in materials of various shapes and under different loading conditions. Experimental stress analysis is performed by measuring the deformation of the piece under load and inferring from this measurement the local stress which prevails. The measurement of deformation is only one facet of the overall problem, and the analytical work that must be applied to the experimental data in order to determine the local stresses is of equal importance. Our concern in the following sections is with the methods that may be employed for deformation measurements. Some simple analyses of these measurements will be given to illustrate the reasoning necessary to obtain local stress values. For more detailed considerations the reader should consult Refs. [2] and [5] and the periodical publication [8].

   Consider the bar shown in Fig. 10-10 subjected to the axial load T. Under no-load conditions the length of the bar is L and the diameter is D. The cross-sectional area of the bar is designated by A. If the load is applied such that the stress does not exceed the elastic limit of the material, the axial strain is given by

$$\epsilon = \frac{T/A}{E} = \frac{\sigma_a}{E} \tag{10-22}$$

where σ_a is the axial stress and E is Young's modulus for the material. The unit axial strain ϵ_a is defined by the relation

$$\epsilon_a = \frac{dL}{L} \tag{10-23}$$

i.e., it is the axial deformation per unit length. Resulting from the deformation in the axial direction is a corresponding deformation in the cross-sectional area of

Figure 10-10 Simple bar in axial strain.

the bar. The change in area is evidenced by a change in the diameter or, more specifically, by a change in the transverse dimension. The ratio of the unit strain in the transverse direction to the unit strain in the axial direction is defined as *Poisson's ratio* and must be determined experimentally for various materials.

$$\mu = -\frac{\epsilon_t}{\epsilon_a} = -\frac{dD/D}{dL/L} \tag{10-24}$$

A typical value for Poisson's ratio for many materials is 0.3. If the material is in the plastic state, the volume remains constant with the change in strain, so that

$$dV = L\,dA + A\,dL = 0$$

or

$$\frac{dA}{A} = -\frac{dL}{L}$$

Expressed in terms of the diameter, this relation is

$$2\frac{dD}{D} = -\frac{dL}{L}$$

so that $\mu = 0.5$ under these conditions.

10-6 STRAIN MEASUREMENTS

Let us first consider some basic definitions. Any strain measurement must be made over a finite length of the workpiece. The smaller this length, the more nearly the measurement will approximate the unit strain at a point. The length over which the average strain measurement is taken is called the *base length*. The *deformation sensitivity* is defined as the minimum deformation that can be indicated by the appropriate gage. *Strain sensitivity* is the minimum deformation that can be indicated by the gage per unit base length.

A simple method of strain measurement is to place some type of grid marking on the surface of the workpiece under zero-load conditions and then measure the deformation of this grid when the specimen is subjected to a load. The grid may be scribed on the surface, drawn with a fine ink pen, or photoetched. Rubber threads have also been used to mark the grid. The sensitivity of the grid method depends on the accuracy with which the displacement of the grid lines may be measured. A micrometer microscope is frequently employed for such measurements. An alternative method is to photograph the grid before and after the deformation and make the measurements on the developed photograph. Photographic paper can have appreciable shrinkage, so that glass photographic plates are preferred for such measurements. The gird may also be drawn on a rubber model of the specimen and the local strain for the model related to that which would be present in the actual workpiece. Grid methods are usually applicable to materials and processes having appreciable deformation under load. These methods might be applicable to a study of the strains encountered in sheet-metal-forming processes. The grid could be installed on a flat sheet of metal before it is formed. The deformation of the grid after forming gives the designer an indication of the local stresses induced in the material during the forming process.

Brittle coatings offer a convenient means for measuring the local stress in a material. The specimen or workpiece is coated with a special substance having very brittle properties. When the specimen is subjected to a load, small cracks appear in the coating. These cracks appear when the state of tensile stress in the workpiece reaches a certain value and thus may be taken as a direct indication of this local stress. The brittle coatings are valuable for obtaining an overall picture of the stress distribution over the surface of the specimen. They are particularly useful for determination of stresses at stress concentration points that are too small or inconveniently located for installation of electrical resistance or other types of strain gages. In some instances, stress data obtained from brittle-coating tests may be used to plan more precise strain measurements with resistance strain gages. A popular brittle coating is manufactured by the Magnaflux Corporation under the trademark of Stresscoat. Strain sensitivities of this coating range from 400 to 2000 μin/in. The theory and application of brittle coatings are discussed very completely by Durelli et al. [3], and the interested reader should consult this reference for additional information. We may mention, however, that residual stresses in the coating, differential thermal expansion between the coating and the workpiece, and curing procedure can influence the stress properties of the coating significantly.

10-7 ELECTRICAL-RESISTANCE STRAIN GAGES

The electrical-resistance strain gage is the most widely used device for strain measurement. Its operation is based on the principle that the electrical resistance of a conductor changes when it is subjected to mechanical deformation. Typically, an electric conductor is bonded to the specimen with an insulating cement under no-load conditions. A load is then applied, which produces a deformation in both the specimen and the resistance element. This deformation is indicated through a measurement of the change in resistance of the element and a calculation procedure which is described below.

Let us now develop the basic relations for the resistance strain gage. The resistance of the conductor is

$$R = \rho \frac{L}{A} \tag{10-25}$$

where L = length

A = cross-sectional area

ρ = resistivity of the material

Differentiating Eq. (10-25), we have

$$\frac{dR}{R} = \frac{d\rho}{\rho} + \frac{dL}{L} - \frac{dA}{A} \tag{10-26}$$

The area may be related to the square of some transverse dimension, such as the diameter of the resistance wire. Designating this dimension by D, we have

$$\frac{dA}{A} = 2 \frac{dD}{D} \tag{10-27}$$

Introducing the definition of the axial strain and Poisson's ratio from Eqs. (10-23) and (10-24), we have

$$\frac{dR}{R} = \epsilon_a(1 + 2\mu) + \frac{d\rho}{\rho} \tag{10-28}$$

The gage factor F is defined by

$$F = \frac{dR/R}{\epsilon_a} \tag{10-29}$$

so that

$$F = 1 + 2\mu + \frac{1}{\epsilon_a} \frac{d\rho}{\rho} \tag{10-30}$$

We may thus express the local strain in terms of the gage factor, the resistance of the gage, and the change in resistance with the strain:

$$\epsilon = \frac{1}{F} \frac{\Delta R}{R} \tag{10-31}$$

The value of the gage factor and the resistance are usually specified by the manufacturer so that the user need only measure the value of ΔR in order to determine the local strain. For most gages the value of F is constant over a rather wide range of strains. It is worthwhile, however, to examine the influence of various physical properties of the resistance material on the value of F. If the resistivity of the material does not vary with the strain, we have, from Eq. (10-30),

$$F = 1 + 2\mu \qquad (10\text{-}32)$$

Taking a typical value of μ as 0.3, we would obtain $F = 1.6$. In this case, the change in resistance of the material results solely from the change in physical dimensions. If the resistivity decreases with strain, the value of F will be less than 1.6. When the resistivity increases with strain, F will be greater than 1.6. Gage factors for various materials have been observed from -140 to $+175$. If the resistance material is strained to the point that it is operating in the plastic region, $\mu = 0.5$ and the resistivity remains essentially constant. Under these conditions the gage factor approaches a value of 2. For most commercial strain gages the gage factor is the same for both compressive and tensile strains. A high gage factor is desirable in practice because a larger change in resistance ΔR is produced for a given strain input, thereby necessitating less sensitive readout circuitry.

Three common types of resistance strain gages are shown in Fig. 10-11. The bonded-wire gage employs wire sizes varying between 0.0005 and 0.001 in (12 and 25 μm). The foil gage usually employs a foil less than 0.001 in thick and is available in a wide variety of configurations which may be adapted to different stress-measurement situations. Because of this flexibility it is the most commonly used gage.

The semiconductor gage employs a silicon base material that is strain-sensitive and has the advantage that very large values of F may be obtained ($F \sim 100$). The material is usually produced in brittle wafers having a thickness of about 0.01 in (0.25 mm). Semiconductor gages also have very high temperature coefficients of resistance. Table 10-1 presents a summary of characteristics of several strain-gage materials. Most commercial gages employ constantan (or other alloys) or isoelastic for materials of construction. Reference [7] furnishes additional information.

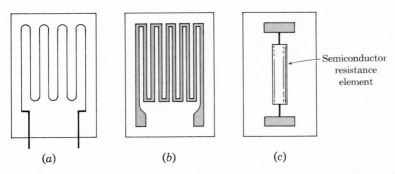

(a)   *(b)*   *(c)*

Figure 10-11 Three types of resistance strain gages. (*a*) Wire gage; (*b*) foil gage; (*c*) semiconductor gage.

Table 10-1 Characteristics of some resistance strain-gage materials

Material	Trade name	Approx. gage factor, F	Approx. resistivity at 20°C, $\mu\Omega \cdot$ cm	Temp. Coeff. of resistance, $°C^{-1} \times 10^6$	Remarks
55% Cu, 45% Ni	Advance, Constantan, Copel	2.0	49	11	F constant over wide range of strain; low-temperature use below 360°C
4% Ni, 12% Mn, 84% Cu	Manganin	0.47	44	20	Same
80% Ni, 20% Cu	Nichrome V	2.0	108	400	Suitable for high temperature use to 800°C
36% Ni, 8% Cr, 0.5% Mo, 55.5% Fe	Isoelastic	3.5	110	450	Used for low temperatures to 300°C
67% Ni, 33% Cu	Monel	1.9	40	1900	Useful to 750°C
74% Ni, 20% Cr, 3% Al, 3% Fe	Karma	2.4	125	20	Useful to 750°C
95% Pt, 5% Ir	—	5.0	24	1250	Useful to 1000°C
Silicon semiconductor	—	−100 to +150	10^9	90,000	Brittle but has high gage factor; not suitable for large strain measurements

Wire and foil gages may be manufactured in various ways, but the important point is that the resistance element must be securely bonded to its mounting. It is essential that the bond between the resistance element and the cement joining it to the workpiece be stronger than the resistance wire itself. In this way, the strength of the resistance element is the smaller, and hence the deformation of the entire gage is governed by the deformation of the resistance element. Most wire strain gages employ either a nitrocellulose cement or a phenolic resin for the bonding agent with a thin paper backing to maintain the wire configuration. Such gages may be used up to 150°C (300°F). A Bakelite mounting is usually employed for temperatures up to 260°C (500°F). Foil gages are manufactured by an etching process similar to that used with printed circuit boards and use base materials of paper, Bakelite, and epoxy film. Epoxy cements are also employed for both wire and foil gages.

When strain gages are mounted on a specimen, two notes of caution should be observed: (1) The surface must be absolutely clean. Cleaning with an emery cloth followed by acetone is usually satisfactory. (2) Sufficient time must be allowed for the cement to dry and harden completely. Even though the cement is dry around the edge of the gage, it may still be wet under the gage. If possible, 24 h should be allowed for drying at room temperature. Drying time may be reduced for higher temperatures.

For low-temperature applications (-100 to $+100$°C) Duco cement (nitrocellulose) is normally employed with paper-covered gages and Eastman 910 (cyanoacrylate) with foil gages mounted on epoxy. These cements are discussed in Ref. [2] along with rather detailed instructions for mounting the various types of gages. The interested reader should consult this reference and the literature of various manufacturers of strain gages for more information.

Problems associated with strain-gage installations generally fall into three categories: (1) temperature effects, (2) moisture effects, and (3) wiring problems. It is assumed that the gage is properly mounted. Temperature problems arise because of differential thermal expansion between the resistance element and the material to which it is bonded. Semiconductor gages offer the advantage that they have a lower expansion coefficient than either wire or foil gages. In addition to the expansion problem, there is a change in resistance of the gage with temperature, which must be adequately compensated for. We shall see how this compensation is performed in a subsequent section. Moisture absorption by the paper and cement can change the electrical resistance between the gage and the ground potential and thus affect the output-resistance readings. Methods of moisture-proofing are discussed in Refs. [1] and [2]. Wiring problems are those situations that arise because of faulty connections between the gage-resistance element and the external readout circuit. These problems may develop from poorly soldered connections or from inflexible wiring, which may pull the gage loose from the test specimen or break the gage altogether. Proper wiring practices are discussed in Refs. [2] and [6].

Electrical-resistance strain gages cannot be easily calibrated because once they are attached to a calibration workpiece, removal cannot be made without destroying the gage. In practice, then, the gage factor is taken as the value specified by the manufacturer and a semicalibration effected by checking the bridge measurement and readout system.

In addition to the general references on strain gages given at the end of this chapter, excellent technical manuals are available from manufacturers. Balwin-Lima Hamilton and Micro-Measurements Group are two such companies.

10-8 MEASUREMENT OF RESISTANCE STRAIN-GAGE OUTPUTS

Consider the bridge circuit of Fig. 4-25. The voltage present at the detector is given by Eq. (4-26) as

$$E_g = E_D = E\left(\frac{R_1}{R_1 + R_4} - \frac{R_2}{R_2 + R_3}\right) \tag{10-33}$$

If the bridge is in balance, $E_D = 0$. Let us suppose that the strain gage represents R_1 in this circuit and a high-impedance readout device is employed so that the bridge operates as a voltage-sensitive deflection circuit. We assume that the bridge is balanced at zero-strain conditions and that a strain on the gage of ϵ results in a change in resistance ΔR_1. R_1 will be used to represent the resistance of the gage at zero-strain conditions. We immediately obtain a voltage due to the strain as

$$\frac{\Delta E_D}{E} = \frac{R_1 + \Delta R_1}{R_1 + \Delta R_1 + R_4} - \frac{R_2}{R_2 + R_3} \tag{10-34}$$

Solving for the resistance change gives

$$\frac{\Delta R_1}{R_1} = \frac{(R_4/R_1)[\Delta E_D/E + R_2/(R_2 + R_3)]}{1 - \Delta E_D/E - R_2/(R_2 + R_3)} - 1 \tag{10-35}$$

Equation (10-35) expresses the resistance changes as a function of the voltage unbalance at the detector ΔE_D.

The bridge circuit may also be used as a current-sensitive device. Combining Eqs. (4-24) to (4-26) gives

$$i_g = \frac{E(R_1 R_3 - R_2 R_4)}{R_1 R_2 R_4 + R_1 R_3 R_4 + R_1 R_2 R_3 + R_2 R_3 R_4 + R_g(R_1 + R_4)(R_2 + R_3)} \tag{10-36}$$

Again, we assume the bridge to be balanced under zero-strain conditions and take R_1 as the resistance of the gage under these conditions. Thus,

$$R_1 R_3 = R_2 R_4 \tag{10-37}$$

The galvanometer current ΔI_g is that value which results from a change in resistance ΔR_1 from the balanced condition. It may be shown that the denominator of Eq. (10-36) is not very sensitive to small changes in R_1 and hence is very nearly a constant which we shall designate as C. Thus,

$$\Delta I_g = \frac{E}{C}\left[(R_1 + \Delta R_1)R_3 - R_2 R_4\right] \qquad (10\text{-}38)$$

Applying the balance conditions from Eq. (10-37) gives

$$\Delta I_g = \frac{E}{C} R_3 \Delta R_1 \qquad (10\text{-}39)$$

Introducing the gage factor from Eq. (10-31),

$$\Delta I_g = \frac{E}{C} R_3 R_1 F\epsilon = \text{const} \times \epsilon \qquad (10\text{-}40)$$

Thus, the deflection current may be taken as a direct indication of the strain imposed on the gage.

We have already mentioned a bridge amplifier circuit in connection with electric resistance thermometers (Fig. 8-11). This same arrangement can be employed for use with strain-gage bridge circuits.

10-9 TEMPERATURE COMPENSATION

It is generally not possible to calculate corrections for temperature effects in strain gages. Consequently, compensation is made directly by means of the experimental setup. Such a compensation arrangement is shown in Fig. 10-12.

Figure 10-12 Temperature-compensation arrangement for electra-resistance strain gages.

Gage 1 is installed on the test specimen, while gage 2 is installed on a like piece of material that remains unstrained throughout the tests but at the same temperature as the test piece. Any changes in the resistance of gage 1 due to temperature are thus canceled out by similar changes in the resistance of gage 2, and the bridge circuit detects an unbalanced condition resulting only from the strain imposed on gage 1. Of course, care must be exerted to ensure that both gages are installed in exactly the same manner on their respective workpieces.

Example 10-4 A resistance strain gage with $R = 120 \ \Omega$ and $F = 2.0$ is placed in an equal-arm bridge in which all resistances are equal to 120 Ω. The power voltage is 4.0 V. Calculate the detector current in microamperes per micro-inch of strain. The galvanometer resistance is 100 Ω.

SOLUTION The denominator of Eq. (10-36) is calculated as

$$C = (4)(120)^3 + (100)(240)^2 = 1.267 \times 10^7$$

We calculate the current sensitivity from Eq. (10-40) as

$$\frac{\Delta I_g}{\epsilon} = \frac{E}{C} R_3 R_1 F = \frac{(4.0)(120)^2(2.0)}{1.267 \times 10^7} = 9.08 \times 10^{-3} \ \text{A/in}$$

$$= 9.08 \times 10^{-3} \ \mu\text{A}/\mu\text{in} \ (0.357 \ \text{mA/mm})$$

Example 10-5 For the gage and bridge in Example 10-4, calculate the voltage indication for a strain of 1.0 μin/in when a high-impedance detector is used.

SOLUTION We have $R_1 = R_2 = R_3 = R_4 = 120 \ \Omega$. Under these conditions Eq. (10-34) becomes

$$\frac{\Delta E_D}{E} = \frac{1 + (\Delta R_1/R_1)}{2 + (\Delta R_1/R_1)} - \frac{1}{2} = \frac{\Delta R_1/R_1}{4 + 2(\Delta R_1/R_1)}$$

The resistance change is calculated from Eq. (10-31)

$$\frac{\Delta R_1}{R_1} = F\epsilon = 2.0 \times 10^{-6}$$

so that

$$\frac{\Delta E_D}{E} = \frac{2.0 \times 10^{-6}}{4 + 4 \times 10^{-6}} = 0.5 \times 10^{-6}$$

and $\Delta E_D = (4.0)(0.5 \times 10^{-6}) = 2.0 \ \mu\text{V}$

10-10 STRAIN-GAGE ROSETTES

The installation of a strain gage on a bar specimen like that shown in Fig. 10-10 is a useful application of the gage, but it is quite restricted. The strain that is measured in such a situation is a principal strain, since we assumed that the bar is operating under only a tensile load. Obviously, a more general measurement problem will involve strains in more than one direction, and the orientation of the principal stress axes will not be known. It would, of course, be fortunate if the strain gages were installed on the specimen so that they were oriented exactly with the principal stress axes. We now consider the methods that may be used to calculate the principal stresses and strains in a material from three strain-gage measurements. The arrangements for strain gages in such applications are called *rosettes*. We shall give only the final relations that are used for calculation purposes. The interested reader should consult Refs. [2] and [5] for the derivations of these equations.

Consider the rectangular rosette shown in Fig. 10-13. The three strain gages are oriented as shown, and the three strains measured by these gages are ϵ_1, ϵ_2, and ϵ_3. The principal strains for this situation are

$$\epsilon_{max}, \epsilon_{min} = \frac{\epsilon_1 + \epsilon_3}{2} \pm \frac{1}{\sqrt{2}} [(\epsilon_1 - \epsilon_2)^2 + (\epsilon_2 - \epsilon_3)^2]^{1/2} \qquad (10\text{-}41)$$

The principal stresses are

$$\sigma_{max}, \sigma_{min} = \frac{E(\epsilon_1 + \epsilon_3)}{2(1 - \mu)} \pm \frac{E}{\sqrt{2}(1 + \mu)} [(\epsilon_1 - \epsilon_2)^2 + (\epsilon_2 - \epsilon_3)^2]^{1/2} \quad (10\text{-}42)$$

The maximum shear stress is designated by τ_{max} and calculated from

$$\tau_{max} = \frac{E}{\sqrt{2}(1 + \mu)} [(\epsilon_1 - \epsilon_2)^2 + (\epsilon_2 - \epsilon_3)^2]^{1/2} \qquad (10\text{-}43)$$

The principal stress axis is located with the angle θ according to

$$\tan 2\theta = \frac{2\epsilon_2 - \epsilon_1 - \epsilon_3}{\epsilon_1 - \epsilon_3} \qquad (10\text{-}44)$$

Figure 10-13 Rectangular strain-gage rosette.

This is the axis at which the maximum stress σ_{max} occurs. A problem arises with the determination of the quadrant for θ, since there will be two values obtained as solutions to Eq. (10-44). The angle θ will lie in the first quadrant $(0 < \theta < \pi/2)$ if

$$\epsilon_2 > \frac{\epsilon_1 + \epsilon_3}{2} \qquad (10\text{-}45)$$

and in the second quadrant if ϵ_2 is less than this value.

Another type of strain-gage rosette in common use is the delta rosette shown in Fig. 10-14. The principal strains in this instance are given by

$$\epsilon_{max}, \epsilon_{min} = \frac{\epsilon_1 + \epsilon_2 + \epsilon_3}{3} \pm \frac{\sqrt{2}}{3} [(\epsilon_1 - \epsilon_2)^2 + (\epsilon_2 - \epsilon_3)^2 + (\epsilon_3 - \epsilon_1)^2]^{1/2}$$

$$(10\text{-}46)$$

The principal stresses are

$$\sigma_{max}, \sigma_{min} = \frac{E(\epsilon_1 + \epsilon_2 + \epsilon_3)}{3(1 - \mu)} \pm \frac{\sqrt{2}\,E}{3(1 + \mu)} [(\epsilon_1 - \epsilon_2)^2 + (\epsilon_2 - \epsilon_3)^2$$

$$+ (\epsilon_3 - \epsilon_1)^2]^{1/2} \qquad (10\text{-}47)$$

The maximum shear stress is calculated from

$$\tau_{max} = \frac{\sqrt{2}\,E}{3(1 + \mu)} [(\epsilon_1 - \epsilon_2)^2 + (\epsilon_2 - \epsilon_3)^2 + (\epsilon_3 - \epsilon_1)^2]^{1/2} \qquad (10\text{-}48)$$

The principal stress axis is located according to

$$\tan 2\theta = \frac{\sqrt{3}(\epsilon_3 - \epsilon_2)}{2\epsilon_1 - \epsilon_2 - \epsilon_3} \qquad (10\text{-}49)$$

The angle θ will be in the first quadrant when $\epsilon_3 > \epsilon_2$ and in the second quadrant when $\epsilon_2 > \epsilon_3$.

It is worthwhile to mention that resistance strain gages may be sensitive to transverse as well as axial strains. The resistance change produced by a transverse strain, however, is usually less than 2 or 3 percent of the change produced by an axial strain. For this reason, it may be neglected in many applications. If the transverse strain is to be considered, the above rosette formulas must be modified accordingly. We shall not present this modification, but refer the reader to Refs. [1] and [5] for more information.

Figure 10-14 Delta strain-gage rosette.

Example 10-6 A rectangular rosette is mounted on a steel plate having $E = 29 \times 10^6$ psi and $\mu = 0.3$. The three strains are measured as

$$\epsilon_1 = + 500 \ \mu in/in$$

$$\epsilon_2 = +400 \ \mu in/in$$

$$\epsilon_3 = - 100 \ \mu in/in$$

Calculate the principal strains and stresses and the maximum shear stress. Locate the axis of principal stress.

SOLUTION As an intermediate step we calculate the quantities

$$A = \frac{\epsilon_3 + \epsilon_1}{2} = 200 \ \mu in/in$$

$$B = [(\epsilon_1 - \epsilon_2)^2 + (\epsilon_2 - \epsilon_3)^2]^{1/2} = 510 \ \mu in/in$$

Then,

$$\epsilon_{max} = A + \frac{1}{\sqrt{2}} B = 561 \ \mu in/in$$

$$\epsilon_{min} = A - \frac{1}{\sqrt{2}} B = - 161 \ \mu in/in$$

$$\sigma_{max} = \frac{EA}{1 - \mu} + \frac{EB}{\sqrt{2}(1 + \mu)} = \frac{(29 \times 10^6)(200 \times 10^{-6})}{1 - 0.3}$$

$$+ \frac{(29 \times 10^6)(510 \times 10^{-6})}{\sqrt{2}(1 + 0.3)} = 8280 + 8050 = 16,330 \ psi \ (112.6 \ MN/m^2)$$

$$\sigma_{min} = 8280 - 8050 = 230 \ psi \ (1.59 \ kN/m^2)$$

$$\tau_{max} = \frac{EB}{\sqrt{2}(1 + \mu)} = 8050 \ psi \ (55.5 \ MN/m^2)$$

We also have

$$\tan 2\theta = \frac{2\epsilon_2 - \epsilon_1 - \epsilon_3}{\epsilon_1 - \epsilon_3} = \frac{(2)(400) - (500) - (-100)}{(500) - (-100)} = 0.667$$

$$2\theta = 33.7 \ or \ 213.7°$$

$$\theta = 16.8 \ or \ 106.8°$$

We choose the first-quadrant angle ($\theta = 16.8°$) in accordance with Eq. (10-45).

Figure 10-15 Schematic of unbonded resistance strain gage.

10-11 THE UNBONDED RESISTANCE STRAIN GAGE

The bonded electrical-resistance strain gage discussed above is the most widely used device for strain measurements. An alternative resistance gage is the unbonded type shown in Fig. 10-15. A springloaded mechanism holds the two plates in a close position while the fine-wire filaments are stretched around the mounting pins as shown. The mounting pins must be rigid and also serve as electrical insulators. When plate A moves relative to B, a strain is imposed on these filaments, which may be detected through a measurement of the change in resistance. The allowable displacement of commercial gages is of the order of ± 0.0015 in (0.038 mm), and the wire diameter is usually less than 0.001 in (0.025 mm). The I^2R heating in the unbonded gage can be a problem because the wires have no ready means for heat dissipation other than convection to the surrounding air. The principle of the unbonded gage has been applied to acceleration and diaphragm pressure transducers with good success.

REVIEW QUESTIONS

10-1 Why must a correction be applied for the specific gravity of a sample when a mass balance is used?

10-2 How are elastic elements used for force or torque measurements?

10-3 What is a proving ring?

10-4 What is a Prony brake?

10-5 What does a dynamometer measure?

10-6 Distinguish between stress and strain.

10-7 What is Poisson's ratio?

10-8 What is deformation sensitivity?

10-9 What is strain sensitivity?

10-10 How may a brittle coating be used for strain measurements?

10-11 Define the gage factor for strain gages.

10-12 What is the main advantage of a semiconductor strain gage?

10-13 How are measurements performed with a resistance strain gage?

10-14 How may temperature compensation be performed on resistance strain gages?

10-15 What is meant by a strain-gage rosette? How is it used?

PROBLEMS

10-1 Many balances use brass weights ($\rho_s = 8490$ kg/m^3) in atmospheric air ($\rho_a = 1.2$ kg/m^3). Plot the percent error in the balance determination as a function of ρ_a/ρ_s.

10-2 Using the conditions given in Example 10-1, calculate the angular sensitivity required if the uncertainty in the weight determination is to be of the order of 1 part in 10^8.

10-3 An LVDT is used to measure the axial deformation of the simple bar of Fig. 10-3. The uncertainty in this measurement is ± 0.025 mm. A steel bar is to be used to measure a force of 1100 N with an uncertainty of 1 percent. Determine a suitable diameter and length of rod for this measurement, assuming the rod dimensions will be known exactly.

10-4 The diaphragm of Fig. 6-10c is to be used as a force-measurement device with the load applied at the center boss. Steel is to be used as the diaphragm material, and the maximum displacement is not to exceed one-third the thickness. A load of 4500 N is to be measured with an uncertainty of 1 percent using an LVDT with an uncertainty of ± 2.5 μm. Determine suitable diaphragm dimensions for this measurement.

10-5 A farmer decides to build a crude weighing device to weigh bags of grain up to 150 lbf. For this purpose he employs a section of 1-in steel pipe (1.315 in OD, 0.957 in ID) as a cantilever beam. He intends to measure the deflection of the beam with a metal carpenter's scale having graduations of $\frac{1}{16}$ in. On the basis of these data what length of pipe would you recommend for the farmer's application?

10-6 A proving ring is constructed of steel ($E = 30 \times 10^6$ psi) having a cross section with 0.250-in thickness and 1.000-in depth. The overall diameter of the ring is 6.000 in. A micrometer is used for the deflection measurement, and its uncertainty is ± 0.0001 in. Assuming that the dimensions of the ring are exact, calculate the applied load when the uncertainty in this load is 1.0 percent. Calculate the percent uncertainty in the load when $y = 0.01$ in and the uncertainties in the dimensions are

$$w_h = w_b = 0.0002 \text{ in}$$

$$w_d = 0.01 \text{ in}$$

where b and h are the thickness and depth of the ring, respectively.

10-7 A hollow steel cylinder ($G = 12 \times 10^6$ psi) is subjected to a moment such that ϕ is $1.50°$. The dimensions of the cylinder are

$$r_i = 0.500 \pm 0.0003 \text{ in}$$

$$r_o = 0.625 \pm 0.0003 \text{ in}$$

$$L = 5.000 \pm 0.001 \text{ in}$$

The uncertainty in the angular deflection is $\pm 0.05°$. Calculate the nominal value of the impressed moment and its uncertainty. Also calculate the $45°$ strains.

10-8 In a bridge circuit like that of Fig. 4-25, $R_2 = R_3 = 100$ Ω. The galvanometer resistance is 50 Ω. The strain-gage resistance at zero strain is 120 Ω, and the value of R_4 is adjusted to bring the bridge into balance at zero-strain conditions. The gage factor is 2.0. Calculate the galvanometer current when $\epsilon = 400$ μm/m. Take the battery voltage as 4.0 V.

10-9 Calculate the voltage output of the bridge in Prob. 10-8, assuming a detector of very high impedance.

10-10 Show that Eq. (10-41) reduces to

$$\epsilon_{max} = \epsilon_1$$

$$\epsilon_{min} = \epsilon_3$$

when $\theta = 0$. Obtain relations for the principal stresses and maximum shear stress in this instance.

10-11 A delta rosette is placed on a steel plate and indicates the following strains:

$$\epsilon_1 = 400 \ \mu m/m$$

$$\epsilon_2 = 84 \ \mu m/m$$

$$\epsilon_3 = -250 \ \mu m/m$$

Calculate the principal strains and stresses, the maximum shear stress, and the orientation angle for the principal axes.

10-12 Obtain simplified relations for the delta rosette under the conditions that $\epsilon_1 = \epsilon_{max}$; that is, $\theta = 0$.

10-13 Suppose a rectangular rosette is used to measure the same stresses as in Prob. 10-11. The bottom leg of this rosette is placed in the same location as the bottom arm of the delta rosette. Calculate the strains that would be indicated by the rectangular rosette under these conditions.

10-14 Calculate the percent uncertainty in the maximum and minimum stresses resulting from uncertainties of 2 percent in the strains as measured with a rectangular rosette. For this calculation assume E and μ are known exactly.

10-15 Rework Prob. 10-14 for the delta rosette.

10-16 A rectangular rosette is placed on a steel plate and indicates the following strains:

$$\epsilon_1 = 563 \ \mu m/m$$

$$\epsilon_2 = -155 \ \mu m/m$$

$$\epsilon_3 = -480 \ \mu m/m$$

Calculate the principal strains and stresses, the maximum shear stress, and the orientation angle for the principal axes.

10-17 A strip of steel sheet, $1.6 \times 50 \times 500$ mm, is available for use as a force-measuring elastic element. The strip is to be used by cementing strain gages to its flat surfaces and measuring the deformation under load. The strain gages have a maximum strain of 2000 $\mu m/m$ and a gage factor of 1.90. A battery is available with a voltage of 4.0 V, and the readout voltmeter has a high impedance and an accuracy of $\pm 1 \ \mu V$. The nominal resistance of the strain gages is 120 Ω. Calculate the force range for which the measurement system may be applicable.

10-18 A bridge circuit is used to measure the output of a strain gage having a resistance of 110 Ω under zero conditions. $R_2 = R_3 = 100 \ \Omega$, and the galvanometer resistance is 75 Ω. The gage factor is 2.0. Calculate the galvanometer current when $\epsilon = 300 \ \mu m/m$ for a power voltage of 6.0 V. Also calculate the voltage output of the bridge for a very-high-impedance dectector.

10-19 An electrical-resistance strain gage records a strain of 400 $\mu m/m$ on a steel tension member. What is the axial stress?

10-20 A hollow steel cylinder is used as a torque-sensing element. The dimensions are: $r_i = 2.5$ cm, $r_o = 3.2$ cm, $L = 15$ cm. Calculate the angular deflection for an applied moment of 22.6 N·m. Calculate the strain that would be indicated by gages attached as shown in Fig. 10-7.

10-21 A certain automotive engine is reported to produce a torque of 540 N·m at 3000 rpm. What power dynamometer would be necessary to test this engine? Express in both horsepower and kilowatts.

10-22 A stainless-steel hollow cylinder has dimensions of 2.5 cm ID and 2.16 cm OD. The length of the cylinder is 25 cm. This cylinder is to be used for axial and torque force measurements by instrumenting it with appropriate strain gages. Specify the types, number, and location for the gages and the bridge circuit and readout equipment required. Also specify the range of variables suitable for measurement by the device and estimate the uncertainties which may be experienced.

10-23 Something similar to the Prony brake of Fig. 10-8 is to be designed for measurement of a small internal combustion engine producing a power output of 15 hp. Copper blocks are used instead of wood to dissipate the heat. Present a design for this device with appropriate dimensions and force measurement system described. Estimate the uncertainty which might be expected in the measurement.

10-24 A bridge circuit is used to measure a strain gage output. Under zero strain conditions the gage resistance is 120 Ω. $R_2 = R_3 = 110$ Ω and the galvanometer resistance is 70 Ω. The gage factor is 1.8. Calculate the galvanometer current when $\epsilon = 350$ μm/m for a power supply voltage of 4.0 V. Also calculate the voltage output for the bridge when used with a high impedance detector. If the gage is installed on steel, what stress does this measurement represent?

REFERENCES

1. Dean, M. (ed.): "Semiconductor and Conventional Strain Gages," Academic Press, Inc., New York, 1962.
2. Dove, R. C., and P. H. Adams: "Experimental Stress Analysis and Motion Measurement," Charles E. Merrill Books, Inc., Columbus, Ohio, 1964.
3. Durelli, A. J., E. A. Phillips, and C. H. Tsao: "Introduction to the Theoretical and Experimental Analysis of Stress and Strain," McGraw-Hill Book Company, New York, 1958.
4. Harris, C. M., and C. E. Crede (eds.): "Basic Theory and Measurements," vol. 1, "Shock and Vibration Handbook," McGraw-Hill Book Company, New York, 1961.
5. Hetenyi, M. (ed.): "Handbook of Experimental Stress Analysis," John Wiley & Sons, Inc., New York, 1950.
6. Perry, C. C., and H. R. Lissner: "The Strain Gage Primer," 2d ed., McGraw-Hill Book Company, New York, 1962.
7. Stein, Peter K.: Material Considerations for Strain Gages, *Inst. & Control Systems*, vol. 37, no. 10, pp. 132–135, October 1964.
8. *Proc. Soc. Exptl. Stress Anal.*, published by the Society.
9. Dally, J. W., and W. F. Riley: "Experimental Stress Analysis," McGraw-Hill Book Company, New York, 1965.
10. Juvinall, R. C.: "Engineering Considerations of Stress, Strain, and Strength," McGraw-Hill Book Company, New York, 1967.
11. Beckwith, T. G., and N. L. Buck: "Mechanical Measurements," 2d ed., Addison-Wesley Publishing Company, Inc., Reading, Mass., 1969.

ELEVEN

MOTION AND VIBRATION MEASUREMENT

11-1 INTRODUCTION

Measurements of motion and vibration parameters are important in many applications. The experimental quantity that is desired may be velocity, acceleration, or vibration amplitude. These quantities may be useful in predicting the fatigue failure of a particular part or machine or may play an important role in analyses which are used to reduce structure vibration or noise level.

The central problem in any type of motion or vibration measurement concerns a determination of the appropriate quantities in reference to some specified state, i.e., velocity, displacement, or acceleration in reference to ground. Ideally, one would like to have a motion or vibration transducer which connects to the body in motion and furnishes an output signal proportional to the vibrational input. The ideal transducer should be independent of its location; i.e., it should function equally well whether it is connected to a vibrating structure on the ground, in an aircraft, or in a space vehicle. In this chapter we shall see some of the compromises which are necessary to approach such behavior.

Sound may be classified as a vibratory phenomenon, and we shall indicate some of the important parameters necessary for specification of sound level. The measurement and analysis of sound levels are very specialized subjects which are becoming increasingly important in modern building and equipment design [8].

11-2 TWO SIMPLE VIBRATION INSTRUMENTS

Consider the simple wedge shown in Fig. 11-1 which is attached to a vibrating wall. When the wall is at rest, the wedge appears as in Fig. 11-1a; when the wall is in motion, the wedge appears as in Fig. 11-1b. An observation is made of the

(a)   (b)

Figure 11-1 Simple wedge as a device for amplitude-displacement measurements. (a) At rest; (b) in motion.

distance x. At this distance the thickness of the wedge is equal to the double amplitude of the motion. In terms of x the amplitude is given by

$$a = x \tan \frac{\theta}{2} \qquad (11\text{-}1)$$

where θ is the total included angle of the wedge. The wedge-measurement device is necessarily limited to rather large-amplitude motions, usually for $a > 0.8$ mm.

A simple vibration-frequency-measurement device is shown in Fig. 11-2. The small cantilever beam mounted on the block is placed against the vibrating surface, and some appropriate method is provided for varying the beam length. When the beam length is properly adjusted so that its natural frequency is equal to the frequency of the vibrating surface, the resonance condition shown in Fig. 11-2b will result. The natural frequency of such a beam is given by

$$\omega_n = 11.0 \sqrt{\frac{EI}{mL^4}} \qquad (11\text{-}2)$$

where ω_n = natural frequency, Hz
        E = Young's modulus for the beam material, psi or N/m^2
        I = moment of inertia of the beam about the centroidal axis
            in the direction of deflection, in^4 or m^4
        m = mass per unit length, lbm/in, or kg/m
        L = beam length, in or m

(a)   (b)

Figure 11-2 Cantilever beam used as frequency-measurement device. (a) No vibration or vibration other than ω_n; (b) vibration at ω_n.

Example 11-1 A $\frac{1}{16}$-in-diameter spring-steel rod is to be used for a vibration-frequency measurement as shown in Fig. 11-2. The length of the rod may be varied between 1 and 4 in. The density of this material is 489 lbm/ft^3, and the modulus of elasticity is 28.3 × 10^6 psi. Calculate the range of frequencies that may be measured with this device and the allowable uncertainty in L at $L = 4$ in in order that the uncertainty in the frequency is not greater than 1 percent. Assume the material properties are known exactly.

SOLUTION We have

$$E = 28.3 \times 10^6 \text{ psi}$$

$$I = \frac{\pi r^4}{4} = \frac{\pi (1/32)^4}{4} = 7.49 \times 10^{-7} \text{ in}^4$$

$$m = \rho \pi r^2 = \frac{\pi (489)(1/32)^2}{(144)} \frac{1}{12} = 8.69 \times 10^{-4} \text{ lbm/in}$$

At $x = 1$ in,

$$\omega_n = 11.0 \left[\frac{(28.3 \times 10^6)(7.49 \times 10^{-7})}{(8.69 \times 10^{-4})(1)^4} \right]^{1/2} = 1718 \text{ Hz}$$

At $x = 4$ in,

$$\omega_n = 107.5 \text{ Hz}$$

We use Eq. (3-2) to determine the allowable uncertainty in the length measurement in terms of the uncertainty in the frequency measurement.

$$\frac{\partial \omega_n}{\partial L} = 11.0 \sqrt{\frac{EI}{m}} \left(\frac{-2}{L^3} \right)$$

$$w_{\omega_n} = \left[\left(\frac{\partial \omega_n}{\partial L} w_L \right)^2 \right]^{1/2}$$

so that

$$w_L = \frac{w_{\omega_n} L^3}{22.0 \sqrt{EI/m}}$$

At $L = 4$ in,

$$w_{\omega_n} = (0.01)(107.5) = 1.075 \text{ Hz}$$

$$w_L = \frac{(1.075)(4)^3}{(22.0)(156.2)} = 0.02 \text{ in}$$

11-3 PRINCIPLES OF THE SEISMIC INSTRUMENT

The seismic instrument is a device that has the functional form of the system shown in Fig. 2-3. A schematic of a typical instrument is shown in Fig. 11-3. The mass is connected through the parallel spring and damper arrangement to the housing frame. This frame is then connected to the vibration source whose characteristics are to be measured. The mass tends to remain fixed in its spatial position, so that the vibrational motion is registered as a relative displacement between the mass and the housing frame. This displacement is then sensed and indicated by an appropriate transducer, as shown in the schematic diagram. Of course, the seismic mass does not remain absolutely steady, but for selected frequency ranges it may afford a satisfactory reference position.

The seismic instrument may be used for either displacement or acceleration measurements by proper selection of mass, spring, and damper combinations. In general, a large mass and soft spring are desirable for vibrational displacement measurements, while a relatively small mass and stiff spring are used for acceleration indications. This will be apparent from the theoretical discussion that follows.

Figure 2-3 is reproduced as Fig. 11-4 to show the mechanical system to be analyzed. Using Newton's second law of motion, we have

$$m \frac{d^2 x_2}{dt^2} + c \frac{dx_2}{dt} + k x_2 = c \frac{dx_1}{dt} + k x_1 \tag{11-3}$$

where it is assumed that the damping force is proportional to velocity. We assume that a harmonic vibratory motion is impressed on the instrument such that

$$x_1 = x_0 \cos \omega_1 t \tag{11-4}$$

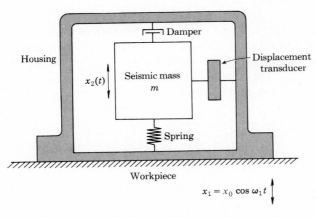

Housing

Damper

Seismic mass
m

$x_2(t)$

Spring

Displacement transducer

Workpiece

$x_1 = x_0 \cos \omega_1 t$

Figure 11-3 Schematic of typical seismic instrument.

$x_1(t) = x_0 \cos \omega_1 t$

$x_2(t)$

Figure 11-4 Mechanical system for a seismic instrument.

and wish to obtain an expression for the relative displacement $x_2 - x_1$ in terms of this impressed motion. The relative displacement is that which is detected by the transducer shown in Fig. 11-3. Rewriting Eq. (11-3) and substituting Eq. (11-4) gives

$$\frac{d^2x_2}{dt^2} + \frac{c}{m}\frac{dx_2}{dt} + \frac{k}{m} x_2 = x_0 \left(\frac{k}{m} \cos \omega_1 t - \frac{c}{m} \omega_1 \sin \omega_1 t \right) \qquad (11\text{-}5)$$

The solution to Eq. (11-5) is

$$x_2 - x_1 = e^{-(c/2m)t}(A \cos \omega t + B \sin \omega t) + \frac{mx_0 \omega_1^2 \cos (\omega_1 t - \phi)}{[(k - m\omega_1^2)^2 + c^2\omega_1^2]^{1/2}} \qquad (11\text{-}6)$$

where the frequency is given by

$$\omega = \left[\frac{k}{m} - \left(\frac{c}{2m} \right)^2 \right]^{1/2} \qquad \text{for } \frac{c}{c_c} < 1.0 \qquad (11\text{-}7)$$

and the phase angle by

$$\phi = \tan^{-1} \frac{c\omega_1}{k - m\omega_1^2} \qquad (11\text{-}8)$$

A and B are constants of integration determined from the initial or boundary conditions.

Note that Eq. (11-6) is composed of two terms: (1) the transient term involving the exponential function and (2) the steady-state term. This means that after the initial transient has died out a steady-state harmonic motion is established in accordance with the second term. The frequency of this steady-state motion is the same as that of the impressed motion, and its amplitude is

$$(x_2 - x_1)_0 = \frac{x_0(\omega_1/\omega_n)^2}{\{[1 - (\omega_1/\omega_n)^2]^2 + [2(c/c_c)(\omega_1/\omega_n)]^2\}^{1/2}} \qquad (11\text{-}9)$$

where the natural frequency ω_n and critical damping coefficient c_c are given by

$$\omega_n = \sqrt{\frac{k}{m}} \qquad (11\text{-}10)$$

$$c_c = 2\sqrt{mk} \qquad (11\text{-}11)$$

The phase angle may also be written

$$\phi = \tan^{-1} \frac{2(c/c_c)(\omega_1/\omega_n)}{1 - (\omega_1/\omega_n)^2} \tag{11-12}$$

A plot of Eq. (11-9) is given in Fig. 11-5. It may be seen that the output amplitude is very nearly equal to the input amplitude when $c/c_c = 0.7$ and $\omega_1/\omega_n > 2$. For low values of the damping ratio the amplitude may become quite large. The output becomes essentially a linear function of input at high-frequency ratios. Thus, a seismic-vibration pickup for measurement of displacement amplitude should be utilized for measurement of frequencies substantially higher than its natural frequency. The instrument constants c/c_c and ω_n should be known or obtained from calibration. The anticipated accuracy of measurement may then be calculated for various frequencies.

The acceleration amplitude of the input vibration is

$$a_0 = \left(\frac{d^2 x_1}{dt^2} \right)_0 = \omega_1^2 x_0 \tag{11-13}$$

We may thus use the measured output of the instrument as a measure of acceleration. There are problems associated with this application, however. In Eq. (11-9) the bracketed term is the one that governs the linearity of the acceleration response, since ω_n will be fixed for a given instrument. In Fig. 11-6 we have a plot of

$$\frac{(x_2 - x_1)_0 \omega_n^2}{a_0} \qquad \text{versus} \qquad \frac{\omega_1}{\omega_n}$$

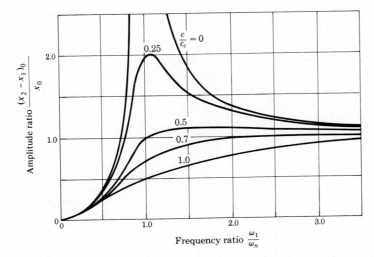

Figure 11-5 Displacement response of a seismic instrument as given by Eq. (11-9).

Figure 11-6 Acceleration response of a seismic instrument as given Eq. (11-13).

which indicates the nondimensionalized acceleration response. Thus, by measuring $(x_2 - x_1)_0$, we can calculate the input acceleration a_0. Generally unsatisfactory performance is observed at frequency ratios above 0.4. Thus, for acceleration measurements we wish to operate at frequencies much lower than the natural frequency, in contrast to the desirable region of operation for amplitude measurements. From the standpoint of instrument construction this means that we wish to have a low natural frequency (soft spring, large mass) for amplitude measurements and a high natural frequency (stiff spring, small mass) for acceleration measurements in order to be able to operate over a wide range of frequencies and still enjoy linear response.

The seismic instrument may also be used for vibration-velocity measurements by employing a variable-reluctance magnetic pickup as the sensing transducer. The output of such a pickup will be proportional to the relative velocity amplitude, i.e., the quantity

$$\left[\frac{d}{dt}(x_2 - x_1)\right]_0$$

From the above discussion it may be seen that the seismic instrument is a very versatile device that may be used for measurement of a variety of vibration parameters. For this reason, many commercial vibration and acceleration pickups operate on the seismic-instrument principle. Calibration methods for the seismic instrument are discussed in Refs. [1], [4], [6], and [7].

The transient response of the seismic instrument is governed partially by the exponential decay term in Eq. (11-6). The time constant for this term could be taken as

$$T = \frac{2m}{c} \tag{11-14}$$

or, in terms of the natural frequency and critical damping ratio,

$$T = \frac{1}{\omega_n(c/c_c)} \tag{11-15}$$

The specific transient response of the seismic-instrument system is also a function of the type of input signal, i.e., whether it is a step function, harmonic function, ramp function, etc. The linearity of a vibration transducer is thus influenced by the frequency-ratio requirements that are necessary to give linear response as indicated by Eqs. (11-6) and (11-9). The design of a transducer for particular response characteristics must involve a compromise between these two effects, combined with a consideration of the sensitivity of the displacement-sensing transducer and its transient response characteristics. The response of the seismic instrument to different types of inputs is discussed by Dove and Adams [4].

Example 11-2 Using Eq. (11-9) and $c/c_c = 0.7$, calculate the value of ω_1/ω_n such that $(x_2 - x_1)_0/x_0 = 0.99$; that is, the error is 1 percent.

SOLUTION We have

$$0.99 = \frac{(\omega_1/\omega_n)^2}{\{[1 - (\omega_1/\omega_n)^2]^2 + [2(0.7)(\omega_1/\omega_n)]^2\}^{1/2}}$$

Rearranging this equation gives the quadratic relation

$$\left[\left(\frac{1}{0.99}\right)^2 - 1\right]\left(\frac{\omega_1}{\omega_n}\right)^4 + 0.04\left(\frac{\omega_1}{\omega_n}\right)^2 - 1 = 0$$

which yields

$$\frac{\omega_1}{\omega_n} = 2.47$$

Example 11-3 A small seismic instrument is to be used for measurement of linear acceleration. It has $\omega_n = 100$ rad/s and a displacement-sensing transducer which detects a maximum of ± 0.1 in. The uncertainty in the displacement measurement is ± 0.001 in. Calculate the maximum acceleration that may be measured with this instrument and the uncertainty in the measurement, assuming ω_n is known exactly.

SOLUTION From Fig. 11-6 with $\omega_1/\omega_n = 0$, we have

$$a_0 = \omega_n^2 (x_2 - x_1)_0$$
$$= 100^2(0.1/12) = 83.3 \text{ ft/s}^2 \ (25.39 \text{ m/s}^2)$$

The uncertainty is calculated from

$$w_a = \left[\left(\frac{\partial a}{\partial x} w_x \right)^2 \right]^{1/2} = \omega_n^2 w_x = (100)^2(0.001/12) = 0.833 \text{ ft/s}^2$$

or an uncertainty of 1 percent. It may be noted that larger uncertainties would be present when smaller accelerations were measured.

Example 11-4 Calculate the frequency ratio for which the error in acceleration measurement is 1 percent, with $c/c_c = 0.7$; that is, $[(x_2 - x_1)_0 \, w_n^2]/a_0 = 0.99$.

SOLUTION We have

$$0.99 = \frac{1}{\{[1 - (\omega_1/\omega_n)^2]^2 + [2(0.7)(\omega_1/\omega_n)^2]^2\}^{1/2}}$$

This gives the quadratic relation

$$\left(\frac{\omega_1}{\omega_n} \right)^4 - 0.04 \left(\frac{\omega_1}{\omega_n} \right)^2 + 1 - (1/0.99)^2 = 0$$

which yields

$$\frac{\omega_1}{\omega_n} = 0.306$$

Example 11-5 Calculate the time required for the exponential transient term in Eq. (11-6) to decrease by 99 percent with $\omega_n = 100$ rad/s and $c/c_c = 0.7$. Compare this time with the period of an acceptable frequency ω_1 for displacement and acceleration amplitude measurements as calculated from the 99 percent relationships of Examples 11-2 and 11-4.

SOLUTION We have

$$e^{-(c/2m)t} = 0.01 \quad \text{and} \quad \frac{ct}{2m} = 4.61$$

$$4.61 = \omega_n \left(\frac{c}{c_c} \right) t$$

$$t = 0.0658 \text{ s}$$

From Example 11-2, $\omega_1/\omega_n = 2.47$ and the corresponding period for ω_1 in the displacement measurement is

$$P = \frac{2\pi}{\omega_1} = \frac{2\pi}{247} = 0.0257 \text{ s}$$

From Example 11-4, $\omega_1/\omega_n = 0.306$, and the corresponding period for ω_1 in the acceleration measurement is

$$P = \frac{2\pi}{\omega_1} = \frac{2\pi}{30.6} = 0.205 \text{ s}$$

No specific conclusions may be drawn from this comparison because the transient response depends on the manner in which the instrument is subjected to a change.

11-4 PRACTICAL CONSIDERATIONS FOR SEISMIC INSTRUMENTS

The previous paragraphs have shown the basic response characteristics of seismic instruments to an impressed harmonic vibrational motion. Let us now consider some of the ways that these instruments might be constructed in practice.

In Fig. 11-7 a seismic instrument is illustrated that uses a voltage-divider potentiometer for sensing the relative displacement between the frame and the seismic mass. To provide the damping for the system, the case of the instrument might be filled with a viscous liquid which would interact continuously with the frame and the mass. Because of the relatively large mass of the potentiometer, such systems have rather low natural frequencies (less than about 100 Hz) and, as such, are limited to acceleration measurements at frequencies less than about 50 Hz. Such acceleration transducers may weigh a pound or more.

The linear variable differential transformer (LVDT) described in Sec. 4-21 offers another convenient means for measurement of the relative displacement between the seismic mass and accelerometer housing. Such devices have somewhat higher natural frequencies than potentiometer devices (270 to 300 Hz) but

Figure 11-7 Seismic instrument using a voltage-divider potentiometer for sensing relative displacement.

are still restricted to applications with low frequency-response requirements. The differential transformer, however, has a much lower resistance to motion than the potentiometer and is capable of much better resolution. In addition, the seismic accelerometer using an LVDT can be considerably lighter in construction than one with a potentiometer. Table 4-4 gives a comparison of the characteristics of the LVDT and potentiometer.

The electrical-resistance strain gage discussed in Chap. 10 may also be used for a displacement-sensing device in a seismic instrument. Consider the schematic in Fig. 11-8. The seismic mass is mounted on a cantilever beam. On each side of the beam a resistance strain gage is mounted to sense the strain in the beam resulting from the vibrational displacement of the mass. Damping for the system is provided by the viscous liquid, which fills the housing. The outputs of the strain gages are connected to an appropriate bridge circuit, which is used to indicate the relative displacement between the mass and housing frame. The natural frequencies of such systems are fairly low and roughly comparable to those for the LVDT systems. The low natural frequencies result from the fact that the cantilever beam must be sufficiently large to accommodate the mounting of the resistance strain gages.

For high-frequency measurements the seismic instrument frequently employs a piezoelectric transducer (Sec. 4-23), as shown in Fig. 11-9. The natural frequency of such instruments may be as high as 100 kHz, and the entire instrument may be quite small and light in weight. A total weight of 1 oz for a piezoelectric accelerometer is not uncommon. Piezoelectric devices have rather high electric outputs but are not generally suitable for measurements at frequencies below about 10 Hz. Electrical impedance matching between the transducer and readout circuitry is usually a critical matter requiring careful design considerations.

Seismic vibration instruments may be influenced by temperature effects. Devices employing variable-resistance displacement sensors will require a correction to account for resistance changes due to temperature. The damping of the instrument is affected by changes in viscosity of the fluid surrounding the seismic mass. Silicone oils are frequently employed in such applications, and their viscosity is strongly dependent on temperature. It would not be uncommon for such oils to experience a change in viscosity by a factor of 2 for a temperature change of only 50°F (28°C). The viscosity-temperature effect could be eliminated, of

Cantilever beam

Housing filled with damping fluid

m $x_2(t)$

Strain gages

$x_1(t)$

Figure 11-8 Seismic instrument utilizing an electrical-resistance strain gage for sensing relative displacement.

Figure 11-9 Seismic instrument utilizing a piezoelectric transducer.

course, if a fluid with a small change in viscosity with temperature were used. Unfortunately, fluids which have this desirable behavior, viz., gases, have low viscosities as well, so that they are generally unsuitable for use as damping agents in the seismic instrument. One way of alleviating the viscosity-temperature problem is to install an electrical-resistance heater in the fluid to maintain the temperature at a constant value regardless of the surrounding temperature.

Obviously, vibration measurements are not always performed on large, massive pieces of equipment. It is frequently of interest to measure the vibration of a thin plate or other small, low-mass structures. The question immediately arises about the effect of the presence of the vibration instrument in such measurements. If the mass of the instrument is not small in comparison with the mass of the workpiece on which it is to be installed, the vibrational characteristics may be altered appreciably. In such cases a correct interpretation of the output of the vibrational instrument may only be obtained by analyses beyond the scope of our discussion, and the interested reader should consult Ref. [6] for more information about this subject. A simple experimental technique, however, will tell the engineer whether the presence of the instrument has altered the vibrational characteristics of the structure under test. First, a measurement is made with the vibration instrument in place. Then, an additional mass is mounted on the structure along with the instrument and a second measurement taken under the same conditions as the first. If there is no appreciable difference between the two measurements, it may be assumed that the presence of the vibrometer has not altered the vibrational characteristics. If there is an appreciable difference in the measurements, an analysis will be necessary to determine the "free" vibrational characteristics in terms of the experimental measurements.

11-5 SOUND MEASUREMENTS

Sound waves are a vibratory phenomenon and for this reason are appropriately discussed in this chapter. They might also be discussed in the chapter on pressure measurements because acoustic effects are usually measured in terms of harmonic pressure fluctuations that they produce in a liquid or gaseous medium. They are also characterized by an energy flux per unit area and per unit time as the acoustic wave moves through the medium. A mathematical description of different acoustic waves will be given later in this section.

It is standard practice in acoustic measurements to relate sound intensity and sound pressure to certain reference values I_0 and p_0, which correspond to the intensity and mean pressure fluctuations of the faintest audible sound at a frequency of 1000 Hz. These reference levels are:

$$I_0 = 10^{-16} \text{ W/cm}^2 \tag{11-16}$$

$$p_0 = 2 \times 10^{-4} \text{ dyn/cm}^2$$

$$= 2 \times 10^{-5} \text{ N/m}^2 \ (0.0002 \ \mu\text{bar}) \tag{11-17}$$

$$= 2.9 \times 10^{-9} \text{ psi}$$

Intensity and pressure levels are measured in decibels. Thus,

$$\text{Intensity level (dB)} = 10 \log \frac{I}{I_0} \tag{11-18}$$

$$\text{Pressure level (dB)} = 20 \log \frac{p}{p_0} \tag{11-19}$$

When pressure fluctuations and particle displacements are in phase, such as in a plane acoustic wave, these levels are equal, $10 \log I/I_0 = 20 \log p/p_0$.

The magnitudes of the particle velocity and pressure fluctuations created by a sound wave are small. For example, a plane sound wave having an intensity of 140 dB generates a maximum oscillation velocity of approximately 2.4 ft/s (0.73 m/s) and a root mean square pressure fluctuation of about 0.029 psi (200 Pa). A

Figure 11-10 Calculation aid for adding or subtracting sound levels measured in decibels.

sound intensity of 90 dB is considered the maximum permissible level for extended human exposure.

In many circumstances we shall be interested in the sound intensity that results from several sound sources. This calculation could, of course, be performed with Eqs. (11-18) and (11-19), but the chart in Fig. 11-10 is frequently used to advantage in such computations. The following examples illustrate the use of the figure.

Example 11-6 Calculate the total sound intensity resulting from two sound sources at 60 and 70 dB.

SOLUTION The numerical difference between the two levels being added is 10 dB. Consulting Fig. 11-10, we find that the numerical difference between the total and larger level is about 0.4 dB. Therefore, the total sound intensity is 70 + 0.4 = 70.4 dB.

Example 11-7 Calculate the total sound intensity resulting from three sound sources at 50, 55, and 60 dB.

SOLUTION We first combine the two lower levels to find their equivalent intensity. This intensity is then combined with the highest level to obtain the total sound level. Figure 11-10 is used for the calculations.

50- and 55-dB levels:

Difference = 5 dB
Amount to be added to larger level = 1.2 dB
Equivalent level = 55 + 1.2 = 56.2 dB

56.2- and 60-dB combination:

Difference = 3.8 dB
Amount to be added to larger level = 1.5 dB
Total sound level = 60 + 1.5 = 61.5 dB

The human ear responds differently to different sound frequencies and to certain groups of frequencies, which are called "noise." The acoustical engineer is responsible for designing rooms or systems in such a way that unwanted frequencies or noise is filtered out while clarity of transmission of certain other sounds is still retained. An auditorium, for example, should be designed so that very little outside noise is transmitted through the walls, while the inside structure and fixtures should allow clear transmission of audio performances on the stage. The walls should have good sound insulation, while the inside surfaces should be properly designed so that unwanted reverberations are eliminated without causing undue attenuation of sound from the stage. Almost all acoustical design problems involve the exercise of considerable judgment based on experience as well as a knowledge of some basic sound-level measurements.

Sound-level measurements are performed with some type of microphone, which may be considered a type of seismic vibration instrument. The electric output of the microphone is proportional to sound-pressure level, which may be used to calculate the sound intensity according to Eq. (11-18). Appropriate power amplifiers and readout meters or recorders indicate the sound level. In general, the microphone must be calibrated in a test facility with a source of known frequency and intensity. Even with careful calibration, accuracies of better than ± 1 dB may not be expected in sound-pressure-level measurements. The characteristics of some typical microphones are given in Table 11-1.

A typical practical application of sound-level measurements may call for an analysis of the noise spectrum in a certain sound source. For this purpose a noise analyzer with bandpass-filter circuits is used. Several sound-pressure-level measurements are taken to determine the sound intensity in various wavelength bands. Based on experience, the acoustical engineer may then take steps to attenuate whatever frequencies he or she feels are necessary to achieve the design objective. For many commercial noise and vibration meters calibration may be achieved with a simple whistle or tuning fork in a quiet room. Sound-level readings will generally follow the inverse square law with respect to distance.

Table 11-1 Characteristics of microphones according to Ref. [9]

Mechanism	Frequency range	Approximate open-circuit sensitivity, dB, below $1 \text{ V/dyn} \cdot \text{cm}^2$	Approximate impedance, Ω	Application and remarks
Carbon microphone	Up to 5 kHz	-40	100	Telephones
Electrodynamic	Up to 20 kHz	-85	10	Field measurements, communications, etc.
Electrostatic	Up to 50 kHz	-50	500,000	Precision measurements, standards
Piezoelectric:				
Rochelle	Audio region ultrasonic	-50	100,000	Temperature-dependent Hygroscopic
ADP crystal		-50		Use in underwater sound, solids, etc.
Quartz	Mainly ultrasonic	-90 to -100	High	
Barium titanate		-90	Low	High-intensity work in air; water; measurements with small probes
Magnetostrictive	Mainly ultrasonic	-100	Low	Underwater sound
Ribbon microphone	Audio region	-100	1	Directive

Example 11-8 A certain microphone has an open-circuit sensitivity of -80 dB referenced to 1 V for a sound-pressure excitation of 1 dyn/cm². Calculate the voltage output when exposed to a sound field of (a) 100 dB and (b) 50 dB.

SOLUTION We first calculate the sound pressure from Eq. (11-19). We have, for the 100-dB level,

$$100 = 20 \log \frac{p}{2 \times 10^{-4}}$$

$$p = 20 \text{ dyn/cm}^2$$

The reference voltage for an excitation of 1 dyn/cm² is calculated as

$$-80 \text{ dB} = 20 \log \frac{E}{1.0}$$

$$E = 10^{-4} \text{ V}$$

Because we assume the output voltage varies in a linear manner with the impressed sound field, the output voltage at 100-dB sound level is

$$E = 20 \times 10^{-4} = 2 \text{ mV}$$

For a sound level of 50 dB the sound pressure is obtained from

$$50 \text{ dB} = 20 \log \frac{p}{2 \times 10^{-4}}$$

$$p = 6.24 \times 10^{-2} \text{ dyn/cm}^2$$

The output voltage is therefore

$$E = (6.24 \times 10^{-2})(10^{-4}) = 6.24 \times 10^{-6} \text{ V}$$

A good summary of acoustical principles and materials for use in sound-attenuation applications is given in Refs. [2], [3], and [8]. The following paragraphs present a summary discussion of some of the types of acoustical phenomena that are encountered in practice.

Traveling Plane Waves

A plane acoustic wave may be described in terms of the displacement of a particle at a distance x from the wave source and at a time t. The analytical relation expressing the particle displacement as a function of distance and time for a harmonic wave is

$$\xi = \xi_m \cos \frac{2\pi}{\lambda} (ct - x) \tag{11-20}$$

where ξ is the particle displacement, ξ_m is the particle-displacement amplitude, c represents the sound velocity, and λ is the wavelength. An observer located at some particular distance x_1 from a plane wave source will experience a periodic variation of the particle displacement and velocity. Another observer located at a distance x_2 from the source will experience exactly the same relative periodic motion; however, the absolute time at which the maximum and minimum points in the motion occur will depend on the distance $(x_2 - x_1)$ and the wave velocity c. The amplitude of the oscillatory motion will decrease with the distance from the wave source owing to viscous dissipation in the fluid. The plane wave described by Eq. (11-20) is called a traveling wave because the particle displacement is dependent on time and the distance from the wave source. Traveling waves may be described in a similar manner for cylindrical wave propagation.

The fluctuations in pressure due to the passage of a plane sound wave may be described in terms of the amplitude of the particle displacement through the relation

$$p = -\beta \frac{\partial \xi}{\partial x} = -\beta \frac{2\pi}{\lambda} \xi_m \sin \frac{2\pi}{\lambda} (ct - x) \tag{11-21}$$

where β is the adiabatic bulk modulus of the fluid defined by

$$\beta = -V \frac{dp}{dV} \tag{11-22}$$

The intensity of a sound wave is defined as the flux of energy per unit time and per unit area. Several equivalent expressions for the intensity of a plane wave are given in Eq. (11-23).

$$\begin{aligned}
I &= \tfrac{1}{2}\rho_a c (\dot{\xi}_m)^2 \\
&= \tfrac{1}{2}\rho_a c \omega^2 \xi_m^2 \\
&= \rho_a c (\dot{\xi}_{rms})^2 \\
&= p_{rms} \xi_{rms} \\
&= \frac{p_{rms}^2}{\rho_a c}
\end{aligned} \tag{11-23}$$

where ρ_a = air density
ω = frequency
$\dot{\xi}_m$ = velocity amplitude of the wave
$\dot{\xi}_{rms}$ = rms value of the particle velocity
p_{rms} = rms pressure fluctuation

It is worthwhile to note that the intensity expressions containing the pressure terms do not apply for the case of cylindrical or spherical waves unless the radius of curvature of these waves is large enough that the wave front may be approximated by a plane.

Standing Waves

Standing waves, like traveling waves, are described in terms of periodic particle displacements and the corresponding periodic particle velocities. For the case of traveling waves the amplitude of the particle displacement is the same regardless of the position x provided that viscous dissipation effects are neglected. For standing waves the amplitude of the particle displacement will follow a periodic variation with the distance x from the sound source. At even quarter-wavelengths the amplitude will be zero, and at odd quarter-wavelengths the particle-displacement amplitude will take on its maximum value. Consequently, the term " standing wave" is derived from the fact that the amplitude of the particle displacement has a periodic variation which is independent of time. Thus, we have the idea that the wave " stands " in a certain position. The particle motion in a standing wave may be represented by the relation

$$\xi = 2\xi_m \sin\left(\frac{2\pi x}{\lambda}\right) \cos\left(\frac{2\pi ct}{\lambda} + \text{const}\right) \tag{11-24}$$

Constant-Pressure Sound Field

Constant-sound-pressure waves are distinctly different from either traveling waves or standing waves and are encountered in reverberant chambers whose walls reflect the major portion of the wave energy striking them. For the ideal case, the energy of the sound source is distributed uniformly throughout the reverberant room so that the fluid will experience a uniform compression and expansion in all directions.

Traveling waves and standing waves represent a vector field; i.e., they depend on the direction in which the wave is propagating. Reverberant waves, for the ideal case, are independent of either distance or direction from the wave source; hence, they represent a scalar field. There is evidently no simple correlation between the pressure fluctuations and the particle displacements in the free space of a reverberant room.

Psychoacoustic Factors

The sound-pressure scale for typical sound sources is indicated in Fig. 11-11, ranging from the threshold of hearing to the intensive sound of a rocket engine. We have already noted that the human ear responds differently to different frequencies; therefore, in order to interpret the response of the ear to a particular sound source we must know the distribution of the sound over the entire frequency spectrum. The nonlinear response of the ear has been determined, resulting in the Fletcher-Munson [11] curves for equal loudness shown in Fig. 11-12. According to these curves, the ear is much more sensitive to mid-frequencies than to low or high frequencies. Based on the above concepts, it is possible to define the *loudness level* as a relative measure of sound strengths

	180	Rocket engine
	160	Supersonic boom
	140	Threshold of pain
Hydraulic press (3 ft)	130	
Large pneumatic riveter (4 ft)		
		Boiler shop (maximum level)
Pneumatic chipper (5 ft)		
	120	
Overhead jet aircraft, four-engine (500 ft)		
		Jet-engine test-control room
Unmuffled motorcycle		
	110	Construction noise (compressors and
Chipping hammer (3 ft)		hammers) (10 ft)
		Woodworking shop
Annealing furnace (4 ft)		
		Loud power mower
Rock and roll band	100	
Subway train (20 ft)		
Heavy trucks (20 ft)		Inside subway car
Train whistles (500 ft)		Food blender
	90	
10-hp outboard (50 ft)		
		Inside sedan in city traffic
Small trucks accelerating (30 ft)		
		Heavy traffic (25 to 50 ft)
	80	Office with tabulating machines
Light trucks in city (20 ft)		
Automobiles (20 ft)		
	70	
Dishwashers		Average traffic (100 ft)
		Accounting office
Conversational speech (3 ft)		
	60	
	50	Private business office
		Light traffic (100 ft)
		Average residence
	40	
	30	
		Broadcasting studio (music)
	20	
	10	
	0	

Figure 11-11 Typical overall sound levels; decibels RE 0.0002 μbar, the minimum audible sound level at 1000 Hz.

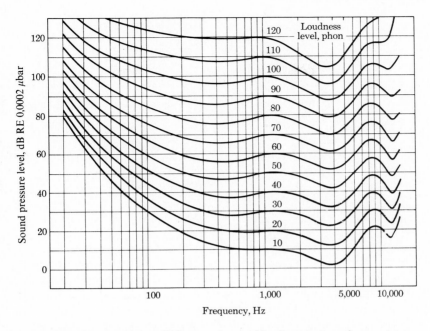

Figure 11-12 Fletcher-Munson [11] curves for equal-loudness contours for pure tones.

judged by an average human listener. The unit of loudness level is the *phon*, which is defined so that the loudness in phons is numerically equal to the sound-pressure level at a frequency of 1000 Hz. The loudness levels are noted on the curves of Fig. 11-12.

Measurement Systems

The typical sound-measurement system is shown in Fig. 11-13. An appropriate microphone is used to sense the sound wave. Before measurements begin, an acoustic calibrator is placed over the microphone to effect a calibration. The

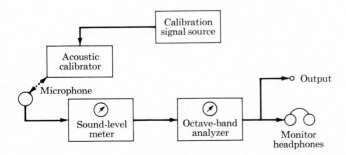

Figure 11-13 Schematic of instrument arrangement for sound measurements.

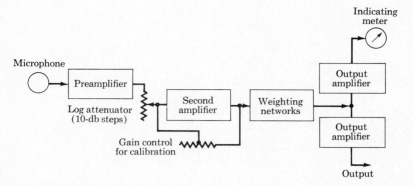

Figure 11-14 Schematic of sound-level meter.

calibrator consists of a special cavity for the particular microphone and a transducer that will produce a known sound-pressure level inside the cavity for given driving voltage inputs. The response of the microphone to the calibration signals is noted on the sound-level meter and thereby furnishes reference levels for use in comparing the signals produced by the unknown sound field.

In practice, the microphone may be mounted directly on the sound-level meter or separated with a connecting cable in order to accommodate some particular measurement objective. The sound-level meter incorporates the functions shown in Fig. 11-14. An input preamplifier amplifies the signal which feeds the attenuator switch, which may be varied in 10-dB steps. The signal is then amplified further, and three standardized weighting networks modify the signal in

Figure 11-15 Standardized weighting networks for sound-level meters for random sound incidence upon the microphone. (*American Standard Specification for General Purpose Sound-Level Meters, SI-4-1961.*)

Table 11-2 Standard octave-band-analyzer filters

Lowpass to 75 Hz
75–150 Hz
150–300 Hz
300–600 Hz
600–1200 Hz
1200–2400 Hz
2400–4800 Hz
Highpass from 4800 Hz

accordance with approximations to the standard loudness contours. The frequency-response characteristics of these networks are shown in Fig. 11-15. The A scale approximates the 40-phon contour and is used for levels below 55 dB. The B scale approximates the 70-phon contour and is used for levels between 55 and 85 dB. Finally, the C scale is used above 85 dB and provides a fairly flat response curve. The output from the weighting networks is either observed on the self-contained indicating meter or recorded with some type of external device such as an oscillograph or tape recorder.

The entire sound spectrum is of importance because of the influence of different frequencies on human response. For this reason the output of the sound-level meter is usually fed to an octave-band analyzer, which is an instrument containing amplifiers, standard filter circuits, and an indicating-meter circuit to measure the signal strength in each filter passband. The standard octave pass-bands are given in Table 11-2. Facilities are provided to measure the output signal of the octave-band analyzer with standard recording equipment. In general, a large variety of instruments is available for measurement and analysis of sound level, and new products appear on the market regularly. The reader is urged to consult well-disposed manufacturers' representatives for information on the instruments that are currently available and that may be most suitable to a particular application.

Acoustical Properties of Materials

One of the important objectives of acoustic measurements is a specification of the sound-absorption properties of materials. With such information available, the acoustical engineer may then choose the proper material to accomplish environmental objectives in noise control or architectural design. There are many complicating factors to be considered in the selection of sound-insulation materials, and we shall discuss only two parameters that are of common interest. For further information the reader should consult Refs. [3] and [10].

The sound-absorption coefficient $\bar{\alpha}$ is defined by

$$\bar{\alpha} = \frac{\text{sound energy absorbed}}{\text{sound energy incident upon surface}}$$

Figure 11-16 Increase in sound absorption resulting from a doubling in thickness of a typical homogeneous porous sound-absorbing material, according to Ref. [3].

The absorption coefficient is strongly frequency-dependent and influenced by such factors as porosity of the surface, thickness of material, rigidity of material, etc. Figure 11-16 illustrates the frequency dependence for a typical porous sound-absorbing material. The material is clearly most effective in absorbing sound fields at mid- and high frequencies. The absorption characteristics of some different types of materials are shown in Fig. 11-17. The more rigid wood panels are substantially less effective as absorbers than the porous material. In general, rigid materials are poor sound absorbers because they have the ability to vibrate and reemit sound. A highly damped, nonrigid material like the porous sheet or, to a lesser degree, the assembly represented by curve B in Fig. 11-17 is more effective in absorbing an incident sound wave. A thin sheet of steel mounted in an undamped fashion would be a very poor absorber because of the ease with which it may be set into vibration.

Figure 11-17 Absorption coefficients for (a) 1-in-thick porous acoustical material; (b) two sheets of 1/8-in-thick plywood, separated by tufts of randomly glued felt causing 1/8-in spacing between sheets; (c) one sheet of 3/8-in-thick randomly braced plywood, according to Ref. [3].

Noise-Reduction Coefficient

Since the sound-absorption characteristics of materials are so strongly frequency-dependent, some average parameter must be selected which represents the material behavior over the entire frequency spectrum. One could choose an integrated average of the absorption coefficient, but the accepted practice is to define a noise-reduction coefficient (NRC) as

$$\text{NRC} = \sum \frac{\begin{array}{c} \text{sound-absorption coefficients of} \\ \text{materials at 250, 500, 1000, and 2000 Hz} \end{array}}{4}$$

Thus the NRC may be determined by measuring the absorption coefficients at the four test frequencies and taking their arithmetic average. It should be noted that this definition does not take into account the absorption characteristics at a low frequency, say 125 Hz. For this reason, comparison of materials on the basis of NRC alone may be misleading if the absorption characteristics at low frequencies are of particular interest. Certain inconsistencies are present in published data of noise-reduction coefficients because of the influence of sample mounting and test procedures on the final reported measurements. For further information the interested reader should consult Refs. [3] and [12].

REVIEW QUESTIONS

11-1 Describe the basic concept of the seismic instrument.

11-2 Under what conditions is a seismic instrument suitable for (a) amplitude measurements and (b) acceleration measurements?

11-3 Describe some alternative constructions of a seismic instrument. When would each construction be used?

11-4 When would a piezoelectric transducer be used to advantage in a seismic instrument?

11-5 What are the reference-energy flux and pressure levels for sound fields?

11-6 What is the maximum permissible sound level for human exposure?

11-7 What is meant by the open-circuit sensitivity of a microphone?

11-8 Distinguish between traveling, standing, and constant-pressure sound fields.

11-9 Why is an octave-band analyzer used in sound measurements?

11-10 What is meant by loudness level?

11-11 Is the ear more or less sensitive to low frequencies than to midfrequencies (that is, 1000 Hz)?

11-12 How is a sound-level meter calibrated?

11-13 Define the sound-absorption coefficient.

11-14 Define the noise-reduction coefficient (NRC).

PROBLEMS

11-1 The vibrating wedge shown in Fig. 11-1 is used for an amplitude measurement. The length of the wedge is 15 cm ± 0.5 mm, and the thickness is 2.5 cm ± 0.2 mm. The x distance is measured as $x = 5.6$ cm ± 1.3 mm. Calculate the vibration amplitude and its uncertainty in percent.

11-2 A small cantilever vibrometer is available for measurement of vibration frequency, but the specification sheet is lost, so that the properties of the device are not known. The instrument is calibrated by placing it on a large compressor in the laboratory which is rotating at 300 ± 2.0 rpm. The measured length for resonance conditions is 5.6 cm $\pm$ 0.2 mm. Calculate the frequency that the instrument will indicate when $L = 10$ cm $\pm$ 0.5 mm. Also calculate the uncertainty in the measurement at this length.

11-3 Consider the cantilever beam of Example 11-1. Suppose the uncertainties in the material properties and dimensions are

$$w_E = 2\%$$

$$w_r = 0.0005 \text{ in}$$

$$w_\rho = 5 \text{ lbm/ft}^3$$

$$w_L = 0.01 \text{ in}$$

Calculate the resulting uncertainty in ω_n at $L = 1.0$ and $L = 4.0$ in.

11-4 A cantilever beam is to be used for the measurement of frequencies between 100 and 1000 Hz. The beam is to be made from a rod of spring steel having the properties given in Example 11-1. The maximum length of the rod is to be 12.5 cm, and the uncertainties in material dimensions and properties are those given in Prob. 11-3. Calculate the nominal diameter of the rod and the uncertainty in the frequency at 100 and 1000 Hz.

11-5 A seismic accelerometer is to be used to measure linear acceleration over a range from 30 to 300 m/s^2. The natural frequency of the instrument is 200 Hz, and this value may vary by ± 2 Hz owing to temperature fluctuations. Calculate the allowable uncertainty in the relative displacement measurement in order to ensure an uncertainty of no more than 5 percent in the acceleration measurement.

11-6 A seismic instrument is to be used to measure velocity. Show that the input velocity amplitude is

$$v_0 = \left(\frac{dx_1}{dt}\right)_0 = \omega_1 x_0$$

Subsequently show that this velocity amplitude may be expressed in terms of the steady-state relative displacement as

$$(x_2 - x_1)_0 = \frac{v_0(\omega_1/\omega_n)}{\omega_n\{[1 - (\omega_1/\omega_n)^2]^2 + [2(c/c_c)(\omega_1/\omega_n)]^2\}^{1/2}}$$

11-7 Calculate the value of the time constant for the instrument in Prob. 11-5 if $c/c_c = 0.65$.

11-8 Plot the error in acceleration measurement of a seismic instrument for $c/c_c = 0.7$ versus frequency ratio, i.e.,

$$1 - \frac{(x_2 - x_1)_0 \, \omega_n^2}{a_0} \qquad \text{versus} \qquad \frac{\omega_1}{\omega_n}$$

11-9 A large seismic instrument is constructed so that $m = 45$ kg and $c/c_c = 0.707$. A spring with $k = 2.9$ kN/m is used so that the instrument will be relatively insensitive to low-frequency signals for displacement measurements and relatively insensitive to high-frequency signals for acceleration measurements. Calculate the value of linear acceleration that will produce a relative displacement of 2.5 mm on the instrument. Calculate the value of ω_1/ω_n such that

$$\frac{(x_2 - x_1)_0}{x_0} = 0.99$$

11-10 A seismic accelerometer is to be designed so that the time constant calculated from Eq. (11-15) is equal to the period of the maximum acceptable frequency for a 1 percent error in measurement; that is, $(x_2 - x_1)_0/a_0 = 0.99$. Plot ω_n versus c/c_c in accordance with this condition.

11-11 Calculate the phase angle ϕ for the conditions of Example 11-2.

11-12 Calculate the phase angle ϕ for the conditions of Example 11-3.

11-13 Calculate the energy flux and rms pressure variation corresponding to a sound intensity of 100 dB.

11-14 The estimated accuracy of a sound-intensity measurement is usually expressed in decibels. Calculate the uncertainties in intensity in watts per square centimeter corresponding to ± 1, ± 2, and ± 3 dB. Express these as percentage values at sound levels of 50, 80, and 120 dB.

11-15 A standing sound wave is created in room air (20°C, 1 atm) with a frequency of 1000 Hz. The peak sound intensity occurring in the wave is 120 dB. Plot the sound intensity and pressure fluctuations as functions of x for one wavelength. Over what fractional portion of a wavelength could the intensity be considered constant, consistent with an estimated uncertainty of ± 1 dB in the intensity measurement?

11-16 Calculate the equivalent sound intensity of a combination of two, three, and four 50-dB sources.

11-17 Calculate the total sound intensity of a combination of 63- and 73-dB sources.

11-18 Four sound sources of 62, 58, 65, and 70 dB are added together. Calculate the total sound intensity in two different ways showing that the same result is obtained.

11-19 A microphone has an open-circuit sensitivity of -55 dB. Calculate the voltage output for sound levels of 60 and 80 dB.

11-20 Two microphones having sensitivities of -55 and -70 dB are available. What amplification factor must be applied to the signal of the least sensitive unit in order to make its output the same as the other microphone?

11-21 Calculate the noise-reduction coefficient (NRC) for the materials shown in Figs. 11-16 and 11-17.

11-22 Calculate the attenuation in decibels for the 2-in porous material of Fig. 11-16 within each octave band of Table 11-2. For this calculation choose some average absorption coefficient for each range of interest. Assume that all energy not absorbed is transmitted through the material.

11-23 A 1000-Hz tone is generated with an intensity of 70 dB. What must be the intensities of tones at 50, 100, and 10,000 Hz in order for them to be detected with the same loudness by the ear?

11-24 Suppose that the playback of a certain musical passage in a high-fidelity system produces an intensity of 80 dB at 1000 Hz when the reproduction is judged to be most authentic to the original source. For various reasons, a person may wish to play the music at a lower level. When the 1000-Hz tones are reduced to 45 dB, how much must the 60- and 100-Hz tones be boosted in order to retain the same sense of authenticity to the original source?

11-25 Compare the sound-level-meter weighting-characteristic curves with the loudness curves of 40, 70, and 90 phon, corresponding to the A, B, and C scales.

11-26 You are asked to design a cantilever device which will have a natural frequency fo 440 Hz to be used in music applications. Present a suitable design and specify the uncertainty which might be expected in the natural frequency.

11-27 Four sound sources of 63, 78, 89, and 92 dB are to be added together. Calculate the total sound intensity in two different ways showing that the same result is obtained.

11-28 A seismic instrument is to be designed to measure frequencies between 100 and 400 Hz for obtaining both displacement and acceleration measurements. Specify some mass-spring-damper arrangements to accomplish this and estimate the uncertainty.

11-29 A condenser microphone has an open circuit sensitivity of -40 dB. Calculate the voltage output for sound levels of 100 and 120 dB.

11-30 Calculate the attenuation in decibels for material B in Fig. 11-17 within each octave band of Table 11-2.

REFERENCES

1. Beckwith, T. G., and N. L. Buck: "Mechanical Measurements," 2d ed., Addison-Wesley Publishing Company, Inc., Reading, Mass., 1969.
2. Beranek, L. L.: "Acoustics," McGraw-Hill Book Company, New York, 1954.
3. ———: "Noise Reduction," McGraw-Hill Book Company, New York, 1960.
4. Dove, R. C., and P. H. Adams: "Experimental Stress Analysis and Motion Measurement," Charles E. Merrill Books, Inc., Columbus, Ohio, 1964.
5. Gross, E. E.: Noise Measuring and Sound Control, *Refrig. Eng.*, vol. 66, p. 49, 1957.
6. Harris, C. M., and C. M. Crede (eds.): "Basic Theory and Measurements," vol. 1, "Shock and Vibration Handbook," McGraw-Hill Book Company, New York, 1961.
7. Hetenyi, M.: "Handbook of Experimental Stress Analysis," John Wiley & Sons, Inc., New York, 1950.
8. "ASHRAE Guide and Data Book," Amer. Soc. Heating, Refrig., and Air Cond., 1963.
9. Condon, E. U. (ed.): "Handbook of Physics," McGraw-Hill Book Company, New York, 1958.
10. Harris, C. M. (ed.): "Handbook of Noise Control," McGraw-Hill Book Company, New York, 1957.
11. Fletcher, H., and W. A. Munson: Loudness, Its Definition, Measurement, and Calculation, *J. Acous. Soc. Am.*, vol. 5, pp. 82–108, 1933.
12. Broch, J. T.: Acoustic Noise Measurements, 2d ed., Bruel and Kjaer Instrument Company, 1971.

TWELVE

THERMAL- AND NUCLEAR-RADIATION MEASUREMENTS

12-1 INTRODUCTION

In Chap. 8 we mentioned the importance of thermal-radiation measurements in temperature determinations. In this chapter we shall examine the physical principles and operating characteristics of some of the more important thermal-radiation detectors and indicate their range of applicability. For a rather complete survey of the field of infrared engineering and thermal-radiation detectors, the reader should consult the references at the end of this chapter.

Radioactivity measurements are a broad field of activity performed by engineers and are becoming increasingly important in many applications. In this chapter we shall discuss some of the measurements that find wide application and indicate the devices used for performing these measurements. Our discussion is, of necessity, a cursory one, and the interested reader should consult the book by Price [10] for detailed information on detection of nuclear radiation.

12-2 DETECTION OF THERMAL RADIATION

The measurement of thermal radiation is basically a measurement of radiant energy flux. The detection of this energy flux may be accomplished through a measurement of the temperature of a thin metal strip exposed to the radiation. The strip is usually blackened to absorb most of the radiation incident upon it and is constructed as thin as possible to minimize the heat capacity and thereby bring about the most desirable transient characteristics. A schematic of the

q_{cond}

q_{conv} **Figure 12-1** Schematic of the general thermal-radiation detector.

general thermal-radiation detector is given in Fig. 12-1. The temperature attained by the element is not only a function of the radiant energy absorbed, but is also dependent on the convection losses to the surroundings and conduction to the mounting fixtures. Convection losses from the element may be reduced by enclosing the detector in an evacuated system with an appropriate window for transmission of the radiation. The infrared-transmission characteristics of several substances employed as window materials are given in Fig. 12-2 according to Ref. [1]. Conduction losses may be reduced with suitable insulating materials.

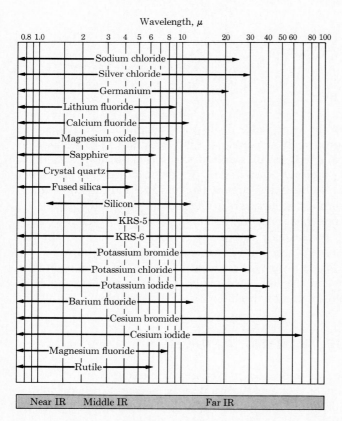

Figure 12-2 Infrared transmission characteristics of several optical materials, according to Ref. [1] as arranged by Hackforth [7]. Sample thickness = 2 mm, long-wavelength cutoff at 10 percent transmission.

Either thermocouples or thermopiles may be used for detecting the temperature of the blackened radiation-sensitive element. Thermopiles offer the advantage that they produce a higher voltage output. Many ingenious methods of construction have been devised for such thermopiles, a commercial example of which is given in Fig. 12-3. The expanded view of the thermopile shows the blackened junction pairs surrounded by an annular ring of mica that serves as both electrical and thermal insulation. The thermopile registers the difference in temperature between the hot junctions and the ambient temperature surrounding

Figure 12-3 A commercial radiometer utilizing a thermopile sensor. (*a*) Cutaway of instrument; (*b*) detail of thermopile sensor. (*Courtesy of Honeywell, Inc.*)

Detector with silver
chloride window

Top view
(window removed)

Detector with
KRS-5 window

Figure 12-4 A commercial thermistor radiation detector. (*Courtesy of Barnes Engineering Co., Stamford, Conn.*)

the detector. The lens at the front of the device focuses the radiation on the thermopile junctions. Special circuitries are used to provide for compensation of ambient temperatures between 50 and 250°F (10 to 121°C).

Thermal radiation may also be sensed by a metal bolometer, which consists of a thin strip of blackened metal foil such as platinum. The temperature of the foil is indicated through its change in resistance with temperature. An appropriate bridge circuit is used to measure the resistance.

Thermistors are widely used as thermal-radiation detectors, and a cutaway drawing of a commercial detection device is given in Fig. 12-4. Two thermistors are enclosed in the detector case, which is covered by an appropriate glass window having satisfactory transmission characteristics. One thermistor element is exposed to the incoming radiation that is to be measured, while the other element is shielded from this radiation. The shielded element is connected in the circuit so that it furnishes a continuous compensation for the temperature of the detector enclosure. A schematic of a commercial radiometer using thermistor

Figure 12-5 Schematic of a commercial radiometer utilizing a thermistor detector. (*Courtesy of Barnes Engineering Co., Stamford, Conn.*)

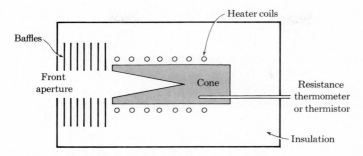

Figure 12-6 Typical construction of a blackbody source.

detectors is given in Fig. 12-5. The incoming radiation is focused by the mirror system onto the thermistor detector. A motor-driven chopper periodically interrupts the radiation so that an alternating signal is produced which may be amplified more easily. The amplified signal is subsequently rectified to produce an output voltage proportional to the radiant flux incident on the thermistor detector. In this system mirrors may be adjusted so that the optical system can be focused on an area as small as 1 mm². The movable mirror and focusing lamp are used for this purpose. The system is exceedingly sensitive and may even be used for detection of radiation from sources near room temperatures.

Thermal-radiation detectors are calibrated directly by obtaining the output as a function of the known radiation from a blackbody source at various temperatures. A typical blackbody source is constructed as shown in Fig. 12-6. The conical cavity is constructed of some high-conductivity material such as aluminum or copper, and the inside surface is blackened. An electric heater maintains the cavity at a desired temperature, which is indicated and controlled through a sensitive electric-resistance thermometer or thermistor. Baffles near the opening prevent stray radiation from the surroundings from influencing the radiation output of the cavity. The detailed construction of blackbody standards is discussed by Marcus [9]. It may be noted that cavities like the one discussed above may give effective emissivities within 1 percent of blackbody conditions.

A comparison of the approximate transient and energy-threshold characteristics of three types of thermal-radiation receivers is given in Table 12-1 accord-

Table 12-1 Characteristics of thermal receivers†

Type	Response time	Energy threshold, W
Thermocouple	0.2 s	3.0×10^{-6}
Thermistor bolometer	3.0 ms	7.2×10^{-8}
Metal bolometer	4.0 ms	3.3×10^{-8}

† According to Ref. [2].

ing to Ref. [2]. The energy-threshold values are the minimum energy which may adequately be detected by such a device.

Photoconductive-radiation detectors have been discussed in Sec. 4-25 and the response characteristics of a number of practical sensors noted. The lead-sulfide cell is widely used for detection of radiation in the 1- to 4-μm range.

Example 12-1 A certain thermistor radiation bolometer detector has characteristics such that the minimum input power necessary to produce a signal-to-noise ratio of unity is 10^{-8} W. Suppose such a detector is located 1.0 m from a blackbody radiation source and is perfectly black. Calculate the temperature of the source necessary to produce a signal-to-noise ratio of 10, assuming that the detector and surroundings are maintained at 20°C. The source is a black sphere, 5.0 cm in diameter, and the detector area is 1 mm².

SOLUTION The net energy absorbed by the detector is

$$q = \sigma A(T_s^4 - 293^4)(2.5/100)^2$$

where A is the detector area. For a signal-to-noise ratio of 10, the input power must be $10 \times 10^{-8} = 10^{-7}$ W $= q$. Thus,

$$T_s^4 - 293^4 = \frac{(10^{-7})(100/2.5)^2}{(5.669 \times 10^{-8})(1 \times 10^{-6})} = 2.82 \times 10^9$$

$$T_s = 317.7 \text{ K} = 44.7°\text{C}$$

For a signal-to-noise ratio of 100, the corresponding value of T_s would be 161°C. From the results of this example we see why optical systems like those shown in Figs. 12-3 and 12-5 are used to collect radiation over a fairly large area and focus it on the small surface area of the detector.

12-3 MEASUREMENT OF EMISSIVITY

An apparatus for the measurement of total normal emissivity has been described by Snyder, Gier, and Dunkle [11]. The apparatus uses a thermopile radiometer and is constructed as shown in Fig. 12-7. An electric heater is used to maintain the temperature of the sample, while thermocouples embedded in the sample furnish an indication of its temperature. A detailed drawing of the thermopile receiver is shown in Fig. 12-7b. It is constructed of 160 junctions of silver constantan mounted in a cylindrical housing which is blackened on the inside. Two blackened aluminum-foil strips are attached to the junctions. The rear shield has a narrow slot which allows the hot-junction strip to be exposed to the radiant flux from the sample while the cold-junction strip is exposed to the temperature of the thermopile enclosure. The temperature difference between the hot and cold junctions is thus taken as an indication of the radiant energy flux, which, in turn, is related to the emissivity of the sample. The conical shield on the front of the

Figure 12-7 Apparatus for measurement of normal emissivity, according to Ref. [11]. (a) Schematic; (b) detail of thermopile construction.

device ensures a proper concentration of energy from the sample. According to Ref. [11], the device furnishes essentially a linear output of 0.0404 mV per W/m² radiant flux. If the temperatures of the hot and cold junctions are assumed to be essentially the same as the inside temperature of the radiometer (this apparently is a good assumption because of the small total energy absorbed in the radiometer) and the sample is assumed gray, the total normal emissivity of the sample ϵ_s is

$$\epsilon_s = \frac{7.84E}{F_{ts}\,\sigma(T_s^4 - T_R^4)} \tag{12-1}$$

where E = voltage output of the thermopile, mV
   T_s = sample temperature, °R
   T_R = radiometer temperature, °R

The view factor F_{ts} is calculated from

$$F_{ts} = \frac{r_1^2}{r_1^2 + L^2} \tag{12-2}$$

In order for measurement to be valid the temperature of the radiometer must be maintained at very nearly that of the surroundings.

Glaser and Blau [6] have presented a method for measurement of spectral emissivities, and the interested reader should consult their paper for additional information on spectral characteristics of surfaces.

Example 12-2 A measurement of total emissivity is made with the apparatus shown in Fig. 12-7. $L = 14.25$ in and $r_1 = 1.50$ in. The output is 0.823 ± 0.005 mV, $T_s = 703 \pm 1.0°F$, and $T_R = 70 \pm 0.5°F$. Calculate the nominal value of the emissivity, and estimate the uncertainty in the measurement.

SOLUTION We first calculate the view factor F_{ts} as

$$F_{ts} = \frac{(1.50)^2}{(1.50)^2 + (14.25)^2} = 0.011$$

The emissivity is then calculated with the use of Eq. (12-1).

$$\epsilon_s = \frac{(7.84)(0.823)}{(0.011)(0.1714 \times 10^{-8})(1163^4 - 530^4)}$$

$$= 0.195$$

We use Eq. (3-2) to calculate the uncertainty. The appropriate parameters are

$$\frac{\partial \epsilon}{\partial E} = \frac{7.84}{F_{ts}\,\sigma(T_s^4 - T_R^4)} = 0.237 \text{ mV}^{-1}$$

$$w_E = 0.005 \text{ mV}$$

$$\frac{\partial \epsilon}{\partial T_s} = \frac{-(7.84)(4)T_s^3\,E}{F_{ts}\,\sigma(T_s^4 - T_R^4)^2} = -7.0 \times 10^{-4} \text{ }°R^{-1}$$

$$w_{T_s} = 1.0°R$$

$$\frac{\partial \epsilon}{\partial T_R} = \frac{(7.84)(4)T_R^3\,E}{F_{ts}\,\sigma(T_s^4 - T_R^4)^2} = 0.663 \times 10^{-5} \text{ }°R^{-1}$$

$$w_{T_R} = 0.5°R$$

and the uncertainty in the emissivity is calculated as

$$w_\epsilon = \{[(0.237)(0.005)]^2 + [(-7.0 \times 10^{-4})(1.0)]^2 + [(0.663 \times 10^{-5})(0.5)]^2\}^{1/2}$$

$$= 0.00137 \text{ or } 0.7\%$$

If the uncertainty in the source temperature had been $\pm 5°F$, the resulting uncertainty in the emissivity would be 0.0037 or 1.9 percent.

12-4 REFLECTIVITY AND TRANSMISSIVITY MEASUREMENTS

In Chap. 8 we described the nature of thermal radiation and the properties which describe surface characteristics. Emissivity is one important property, and Sec. 12-3 has described one method for its measurement. Reflectivity and transmissivity are other important surface properties, and we shall now describe a technique for their measurement through the use of an integrating sphere. For completeness we shall describe the technique for determining monochromatic properties, i.e., properties over a range of thermal-radiation wavelengths.

Let us assume that a suitable *monochromator* is available to produce monochromatic radiation. This radiation might be produced by some kind of prism arrangement or through the use of narrowband optical filters. We wish to determine the fraction of some incident radiation that is either reflected or transmitted by a surface. For this purpose the integrating sphere of Fig. 12-8 is employed. The monochromatic radiation is chopped to produce an ac source to a detector

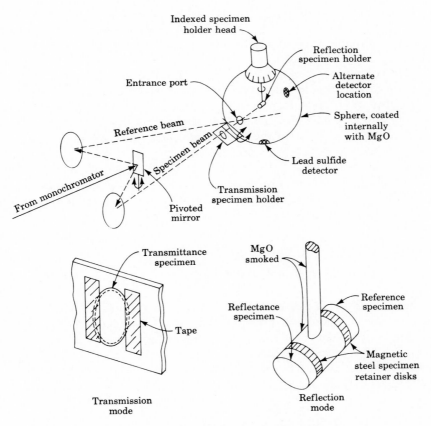

Figure 12-8 Integrating-sphere reflectometer and specimen mounting details.

that may be easily amplified. Consider first the measurment of reflectivity. The pivoted mirror permits a divergence of the monochromatic beam either directly to the specimen inside the sphere or to some location on the inner surface of the sphere. The inside of the sphere is coated with a thick layer (about 2 mm) of magnesium oxide to produce a very high and diffuse reflectance of about 0.99. Thus, radiation incident on the inner surface will be diffusely reflected throughout the sphere, producing a uniform irradiation of the inner surface. Around the surface are placed radiation detectors to sense the reflected radiation. The general measurement technique consists of a comparison of the detector signal for direct impingement of the reference beam with that which results from reflected radiation from the sample. The ratio of these two signals is the reflectivity of the sample. It must be noted that a chopped (ac) input beam is essential for this measurement technique because the detectors in the side of the sphere sense all radiation incident upon them, including the continuous signal resulting from thermal emission of the wall of the sphere. Appropriate electronic circuits filter out all but the chopped signals so that the output will be indicative of the input chopped radiation.

A measurement of transmissivity is made by placing a sample over the input port to the sphere. After transmission through the specimen, the signal is then compared to that obtained for direct impingement on the inner surface of the sphere.

The integrating-sphere technique is widely regarded as the most reliable method for obtaining reflectance and transmissivity measurements and may properly be described as an absolute measurement standard. It is applicable only up to about 3-μm wavelength because of the spurious properties of magnesium oxide at higher wavelengths. The use of the integrating sphere for radiation measurements has been given a very complete exposition in the literature, and the interested reader should consult Refs. [13], [15], and [5] for additional information. A good discussion of radiation-property measurements in both gases and solids is given in Refs. [14], [16] to [20], [30], and [31]. A complete survey of thermal-radiation-property data is available in Refs. [21] and [24].

12-5 SOLAR RADIATION MEASUREMENTS

The sun does not radiate as a blackbody. Estimates of the temperature of the sun vary widely but are generally in the billions of kelvins. Measurements indicate that the radiation spectrum of the sun has a shape similar to that of a blackbody, as shown in Fig. 12-9. Most of the radiation is contained between 0.2 and 2.0 μm, and the peak occurs at about 0.5 μm. If this were a true blackbody spectrum and Wien's law [Eq. (8-21)] applied, the equivalent blackbody temperature would be

$$T \approx \frac{5215.6}{0.5} = 10{,}431°\text{R} = 5795 \text{ K}$$

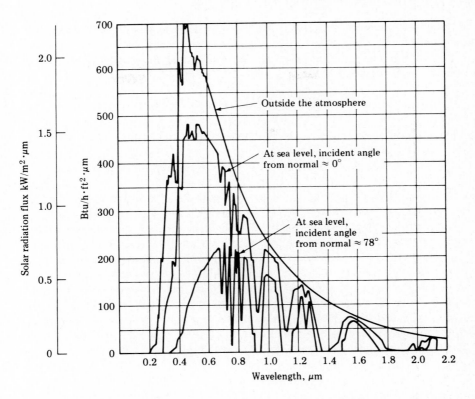

Figure 12-9 Spectral distribution of solar radiation as a function of atmospheric conditions and angle of incidence, according to Ref. [32].

Figure 12-9 indicates the strong absorption of solar radiation in the atmosphere because of carbon dioxide and water-vapor concentrations.

Measurements of solar radiation are important because of the increasing numbers of solar heating and cooling applications and the need for accurate solar irradiation data to predict performance.

Two basic types of instruments are employed for solar radiation measurement: (1) a *pyrheliometer*, which collimates the radiation to determine the beam intensity as a function of incidence angle, and (2) a *pyranometer*, which measures the total hemispheric solar irradiation. The pyranometer measurements are the most common. A typical instrument is the Eppley pyranometer,† which employs two concentric rings serving as hot and cold junctions for a 50-element thermopile. The rings are coated black and white to produce a temperature difference when exposed to solar radiation. Barium sulfate is used as the whitening agent,

---

† The Eppley Laboratory Inc., Newport, R.I. 02840.

and the entire assembly is covered with a hemispherical glass cover which transmits radiation from about 280 to 2800 nm. A typical sensitivity for the device is 7.5 mV · cm² · min/cal or 107.5 mV · cm²/W. For a solar irradiation of 800 W/m², the output of the device would be

$$E = (107.5)(800 \times 10^{-4}) = 8.6 \text{ mV}$$

The *langley* unit is frequently used in solar measurements and is defined as

$$1 \text{ langley} = 1 \text{ cal/cm}^2 = 4.186 \text{ J/cm}^2 = 3.687 \text{ Btu/ft}^2$$

The total solar radiation arriving at the outer edge of the atmosphere is called the *solar constant* and has a value of approximately 1395 W/m².

The Robitsch pyranometer operates on the principle of differential expansion of bimetallic strips exposed to solar radiation. Because the operation is mechanical, readout can be accomplished with a simple mechanical linkage, and no external electric power source is required. Photovoltaic cells may also be employed for solar measurements. Silicon cells, cadmium sulfide cells, and selenium cells have all been used in various applications. Calibration of solar pyranometers is made with the Abbot water flow and silver disk pyrheliometers as primary and secondary standards [26], [28]. Rather complete reviews of solar measurements are given in Refs. [23], [27], and [29].

12-6 NUCLEAR RADIATION

In the following sections we shall be concerned with methods for detecting nuclear radiation. A detailed consideration of the origins of such radiation is quite beyond the scope of this discussion, but a few introductory remarks are appropriate in order to categorize the types of nuclear radiation and indicate some of their particular characteristics. A detailed discussion of the principles of nuclear radiation is given by Kaplan [8]; Price [10] presents a comprehensive appraisal of the subject of nuclear-radiation detection.

We shall be concerned with the detection of four kinds of nuclear radiation: (1) alpha (α) particles, (2) beta (β) particles, (3) gamma (γ) rays, and (4) neutrons. An alpha particle is a helium nucleus that has a positive charge of 2 and a relative mass of 4. A beta particle is a negatively charged electron having a relative mass of 0.000549. The neutron has a zero charge and a relative mass of unity. Gamma rays are high-energy electromagnetic waves that result from nuclear transformations and, as such, do not have mass or charge in the classical sense. They may, however, produce ionizing effects in their interaction with matter. Alpha particles are absorbed rather readily in many materials, beta particles are usually more penetrating, and neutrons and gamma rays are the most penetrating types of radiation because they usually have higher energies and do not interact with coulomb force fields when entering a material. The interaction of these nuclear radiations with particular materials is a very complicated subject and forms the basis for nuclear-shielding applications. Our concern is with the detection of these types of radiation.

12-7 DETECTION OF NUCLEAR RADIATION

Nuclear radiation is detected through an interaction of the radiation with the detecting device, which produces an ionization process. The degree of ionization may be measured with appropriate electronic circuitry. Two types of detection operations are normally performed: (1) a measurement of the *number* of interactions of nuclear radiation with the detector and (2) a measurement of the total effect of the radiation. The first type of operation is a counting process, while the second operation may be characterized as a mean-level measurement. The counting operation frequently ignores the energy level of the radiation, while the mean-level measurement is used for determining the energy level of the irradiation. The popular Geiger-Müller counter is typically used for nuclear counting operations, while ionization chambers and photographic plates are used for energy-level measurements. Scintillation detectors may be used for both counting and energy-level measurements. We shall discuss these different types of detectors and indicate their range of applicability.

12-8 THE GEIGER-MÜLLER COUNTER

A typical cylindrical-tube arrangement for a Geiger-Müller tube is shown in Fig. 12-10. The anode is a tungsten or platinum wire, while the cylindrical tube forms the cathode for the circuit. The tube is filled with argon with perhaps a small concentration of alcohol or some other hydrocarbon gas. The gas pressure is slightly below atmospheric. The ionizing particle or radiation is transmitted through the cathode material, or some window material installed therein, and

Figure 12-10 Typical cylindrical-tube arrangement for a Geiger-Müller counter.

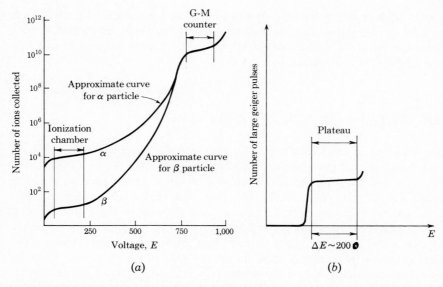

Figure 12-11 (a) Counting performance of the system in Fig. 12-9 as a function of impressed voltage; (b) detail of Geiger-counter region.

through interaction with the gas molecules produces an ionization of the gas. If the voltage E is sufficiently high, each particle will produce a voltage pulse. The counting performance of the tube is indicated in Fig. 12-11. The plateau region typically slopes slightly upward at a rate of from 1 to 10 percent per 100 V. The tube must be operated in the plateau region, which has a width of approximately 200 V for commercial tubes. When the particle causes a discharge or pulse, there is a time delay before the tube can detect another particle and register another pulse. This delay is roughly the time required to recharge the anode and cathode system, i.e., to establish a new space charge in the gas. The counting rate of the G-M tube is thus limited by this delay time. The maximum counting rates are of the order of 10^4 counts/s.

An end-type Geiger-counter tube may be constructed as shown in Fig. 12-12. This type of tube is used for counting α and β particles and low-energy γ rays. For these low-energy radiations the window is usually a thin sheet of mica or Mylar.

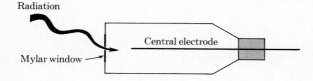

Figure 12-12 Construction of an end-type Geiger counter.

Figure 12-13 Parallel-plate ionization chamber, according to Ref. [10].

12-9 IONIZATION CHAMBERS

An ionization chamber may be constructed in basically the same way as the Geiger counter shown in Fig. 12-10 except that the tube is operated at a much lower voltage. This region of operation is indicated in Fig. 12-11. The arrangement may be modified, however, to accommodate specific applications. A schematic of a typical parallel-plate ionization chamber is shown in Fig. 12-13 according to Price [10]. In operation, the chamber is charged with a voltage that is high enough to ensure that electrons produced by the ionizing radiation will be collected on the anode. The voltage is not so high that it triggers excitation of other molecules by electrons produced in the ionization process. This is in contrast to the operation of the G-M counter, which has a sufficiently high voltage that once ionization is effected by the incoming radiation, subsequent ionization of other molecules also takes place in an avalanche fashion, which produces the voltage pulse used for counting. To measure the energy level of the incoming radiation, a measurement is made of the output current of the device or a determination is made of the charge released in the chamber over a period of time. The ionization chamber may also be used as a pulse-type device for the measurement of the number and energies of high-energy alpha particles. In this application a measurement of the pulse height (strength) is made in order to determine the energy levels of the alpha particles.

12-10 PHOTOGRAPHIC-DETECTION METHODS

When certain types of photographic films are exposed to nuclear radiation and subsequently developed, the opacity of the print may be taken as an indication of the total amount of radiation incident on the film during the time of exposure.

Many specialized films are available for nuclear-radiation measurement, and this method is the primary one used for measuring the total radiation exposure for workers in atomic-energy installations. The photographic film badge may be used for detecting α and β particles, γ rays, and neutrons. In order to use a single badge for measurement of the exposure rates to the different types of radiation, several openings, or windows, may be constructed in a single badge. A different type of filter is placed over each window so that only one type of radiation is permitted to strike the photographic emulsion under each window. Thus, the opacity of the developed film under each of the windows gives an indication of the different radiation-exposure rates.

12-11 THE SCINTILLATION COUNTER

The scintillation counter is the most versatile of the conventional devices for the measurement of nuclear radiation. A schematic of the device is shown in Fig. 12-14. The ionizing radiation strikes the scintillation crystal, which is enclosed in the thin aluminum housing, and produces a flash of light. The light flashes are detected by the photomultiplier tube, the output of which is fed to appropriate electronic circuits. If the number of pulses is to be counted, the output circuit will include appropriate electronic counters. The scintillation counter may also be used for measuring the energy of the incoming radiation since, for certain crystals, the intensity of the flash of light is proportional to the energy of the radiation.

A number of scintillation materials are used, including both solids and liquids. The selection of the proper type of material for detection of specific radiations is discussed by Price [19] and Birks [3].

12-12 NEUTRON DETECTION

The three detection methods described above depend on an ionization process caused by an interaction of the α, β, or γ radiation with a gas or scintillation material. These detectors may not be used for neutrons because neutrons do not produce an ionizing effect. For this reason, the measurement of neutron flux is usually an indirect process which involves the utilization of an intermediate

Figure 12-14 Schematic of a scintillation counter.

Figure 12-15 Diagram of a B^{10}-lined neutron counter. Filling gas is helium plus 5 percent ether at 10 cmHg pressure. (*From Ref. [12].*)

reaction to produce some type of ionizing radiation (α, β, or γ rays), which may then be used to indicate the incoming flux. A variety of neutron-measurement devices are available, and the type which is used in practice depends strongly on the energy level of the neutrons to be measured and the total neutron flux. We shall discuss the principles of operation of only one type of detector which is applicable for the measurement of thermal neutron fluxes ($E \sim 0.025$ eV).

A typical reaction which is used for neutron detection is the interaction of neutrons with B^{10} to produce Li^7 and an alpha particle. The alpha particles may subsequently produce ionizations in a gas and corresponding voltage pulses. The reaction is utilized in two ways: (1) the boron is present in the form of boron-trifluoride (BF_3) gas and placed in a chamber very similar to the G-M counter shown in Fig. 12-10. The potential difference between the central wire and the cylindrical shell is about 1300 V. The wire diameter is about 0.002 in, while the cylinder diameter is about 1 in, with a length of about 6 in. (2) An alternative method of utilizing the B^{10} (n, α) reaction is shown in Fig. 12-15. The inside surface of the cylinder is coated with B^{10}, while the volume is filled with helium and 5 percent ether at 10 cmHg pressure. The sensitivity of a typical BF_3 detector is about 1.0 counts/s per unit neutron flux. The sensitivity of the boron-coated chamber is about 5 to 10 counts/s per unit neutron flux. Neuron flux is measured in neutrons per square centimeter per second.

If sufficiently high neutron fluxes are encountered, the detector may be operated as a current-sensitive device. The current output is then proportional to the incoming flux. Currents of about 0.1 mA may be obtained for neutron fluxes of 10^{10} neutrons/cm$^2 \cdot$ s.

12-13 STATISTICS OF COUNTING

We have already seen that many nuclear-radiation measurements rely on a counting operation. In this section we wish to consider the statistics associated with such operations and indicate the possible errors involved.

In a given sample of radioactive material there are a large number of atoms, and the probability that each of these atoms will decay is quite small. The probability that n atoms will decay is given by the Poisson distribution [Eq. (3-10)] as

$$p(n) = \frac{n^n e^{-n}}{n!} \tag{12-3}$$

where $Np = n$
N = total number of atoms
p = probability that each atom will decay

Equation (12-3) is generally valid for $N > 100$ and $p < 0.01$, and the standard deviation of the distribution may be shown to be†

$$\sigma = \sqrt{n} \tag{12-4}$$

The quantity n represents the average number of radioactive decays recorded in a series of observations. The radioactivity of a certain sample of material might be studied by observing the number of particles or γ rays given off in specified increments of time. The counting operation could be performed with a G-M counter or other device. In general, the number of counts in a fixed interval of time, say, 1 min, might vary considerably. Several counts would be taken, and the arithmetic average of the number of counts in each time interval is given by $\bar{n}$. The standard deviation of the measurement would be given by Eq. (12-4). In the above and following equations, n is taken to represent the average number of counts.

The parameter which is of interest in an experimental determination is the *counting rate*, defined by

$$r = \frac{n}{t} \tag{12-5}$$

where t is the elapsed time for the average number of counts n. The accuracy of the determination of the counting rate depends on the number of counts and the time t.

When a counting measurement is made to determine the counting rate, the detector senses not only the signal from the radioactive source but also some background signal which is the result of spurious radioactivity of the environment. A separate measurement must be made of the background counting rate and this information used to determine the counting rate resulting from the

† See Ref. [10], p. 56.

radioactive source. The counting rate and its associated uncertainty are written as

$$r = r \pm w_r \tag{12-6}$$

The counting rate for the source is

$$r_s = r_T - r_b \tag{12-7}$$

where r_T is the total counting rate and r_b is the counting rate due to the background. The uncertainty in the source counting rate is thus

$$w_s = (w_T^2 + w_b^2)^{1/2} \tag{12-8}$$

where w_T and w_b are the uncertainties in the total and background counting rates, respectively. The uncertainties for these counting rates may be estimated through the use of the standard deviation for the Poisson distribution given by Eq. (12-4) if the nuclear radiation satisfies the conditions for this distribution. The accuracy of the counting-rate measurement is clearly dependent on the length of time available for the measurement. A longer time affords the opportunity for observation of a large number of counts and hence the establishment of a better average counting rate. Example 12-5 illustrates the use of the above relations for determining counting rates.

Example 12-3 A certain radioactive sample is observed, and the total and background counting rates are approximately 800 and 35 counts/min, respectively. If a total time of 45 min is available for both counting measurements, estimate the apportionment of this total time such that there is minimum uncertainty in the source counting rate.

SOLUTION We may assume that both the source and background counts follow the Poisson distribution, so that the uncertainty is proportional to the standard deviation; i.e.,

$$\sigma_s = n_s^{1/2}$$

$$\sigma_b = n_b^{1/2}$$

$$\sigma_T = n_T^{1/2}$$

Thus,
$$w_s = \frac{n_s^{1/2}}{t_s} \qquad w_b = \frac{n_b^{1/2}}{t_b} \qquad w_T = \frac{n_T^{1/2}}{t_T} \tag{a}$$

where t_s, t_b, and t_T are the counting times for the source, background, and total, respectively. We have also

$$n_s = r_s t_s \qquad n_b = r_b t_b \qquad n_T = r_T t_T \tag{b}$$

so that

$$w_s^2 = \frac{r_s}{t_s} = \frac{r_b}{t_b} + \frac{r_T}{t_T} \tag{c}$$

in accordance with Eq. (12-8). We wish to optimize w_s by proportioning the times t_b and t_T in accordance with

$$t_b + t_T = \text{const} = 45 \text{ min} \tag{d}$$

Thus, we optimize w_s by setting $dw_s/dt_b = 0$ and obtain

$$2w_s \frac{dw_s}{dt_b} = 0 = -\frac{r_b}{t_b^2} - \frac{r_T}{t_r^2} \frac{dt_T}{dt_b} \tag{e}$$

From Eq. (d),

$$\frac{dt_T}{dt_b} = -1$$

so that the optimum condition is

$$\frac{t_b}{t_T} = \left(\frac{r_b}{r_T}\right)^{1/2} \tag{f}$$

For the given counting rates,

$$t_b + t_T = 45$$

$$\frac{t_b}{t_T} = \left(\frac{35}{800}\right)^{1/2} = 0.2091$$

and $t_T = 37.2$ min; $t_b = 7.8$ min. In the above calculations it has been assumed that the time interval is accurately determined and that the uncertainty for this determination is zero.

Example 12-4 For the estimated counting rates given in Example 12-3, calculate the optimum total time necessary to give a standard deviation of the source counting rate of 1 percent. Repeat for total and background counting rates of 400 and 100, respectively.

SOLUTION The nominal value of the source counting rate is

$$r_s = r_T - r_b = 800 - 35 = 765 \text{ counts/min}$$

We have

$$\sigma_s^2 = \frac{r_b}{t_b} + \frac{r_T}{t_T} = [(0.01)(765)]^2 = 58.5$$

Also, from relation (f) of Example 12-3,

$$t_b = t_T \left(\frac{r_b}{r_T}\right)^{1/2}$$

so that

$$t_T = \frac{1}{\sigma_s^2} (r_b^{1/2} r_T^{1/2} + r_T)$$

$$= \frac{1}{58.5} [(35)^{1/2}(800)^{1/2} + 800] = 16.5 \text{ min}$$

The total counting time is $t_b + t_T$, or

$$t_{\text{tot}} = (1.209)(16.5) = 20 \text{ min}$$

For total and background counting rates of 400 and 100,

$$r_s = 400 - 100 = 300 \text{ counts/min}$$

$$\sigma_s^2 = [(0.01)(300)]^2 = 9$$

$$t_T = \tfrac{1}{9}[(100)^{1/2}(400)^{1/2} + 400] = 66.7 \text{ min}$$

The total counting time is

$$t_{\text{tot}} = 66.7[(\tfrac{100}{400})^{1/2} + 1] = 100 \text{ min}$$

Example 12-5 A radioactive sample is observed over a period of 30 min, and a total of 21,552 counts is recorded. The sample is then removed, and the background radiation is observed for another 30 min. The background produces 1850 counts. Calculate the source counting rate and its uncertainty.

SOLUTION We have

$$r_T = \frac{21,552}{30} = 715.1 \text{ counts/min}$$

$$r_b = \frac{1850}{30} = 61.7 \text{ counts/min}$$

so that

$$r_s = 715 - 62 = 653 \text{ counts/min}$$

The standard deviation of the source counting rate is calculated from Eq. (c) of Example 12-3.

$$w_s = \sigma_s = \left(\frac{r_b}{t_b} + \frac{r_T}{t_T}\right)^{1/2} = 5.08 \text{ counts/min}$$

REVIEW QUESTIONS

12-1 How is thermal radiation distinguished from other EM radiation?

12-2 How is thermal radiation detected?

12-3 Why is it sometimes necessary to chop thermal radiation before it strikes the detector?

12-4 What is an integrating sphere? How is it used for reflectivity and transmissivity measurements?

12-5 What are the four types of nuclear radiation that are normally measured?

12-6 Describe the principle of operation of the Geiger counter.

12-7 How does an ionization chamber work?

12-8 How does a scintillation counter operate?

12-9 What types of instruments are used for detection of (*a*) alpha particles, (*b*) beta particles, (*c*) gamma rays, and (*d*) neutrons?

12-10 Describe the techniques for measuring neutron irradiation.

12-11 Why is a Poisson distribution an adequate representation of nuclear counting statistics?

PROBLEMS

12-1 Consider the thermistor detector and radiation source of Example 12-1. Instead of the bare detector an optical system collects the radiation over a circular area of radius *r* and focuses it on the detector, which is maintained at 20°C. Plot the source temperature necessary to maintain a signal-to-noise ratio of 200 as a function of the radius of the collecting optical system. Assume that the distance between the source and detector is 1.0 m, as in Example 12-1. Neglect losses in the optical system for this calculation.

12-2 Repeat Prob. 12-1 for distances of 6 and 12 ft between the source and detector.

12-3 A thermocouple detector has characteristics such that the minimum input power necessary to produce a signal-to-noise ratio of unity is 3×10^{-6} W. A 2-in-diameter optical system collects the input thermal radiation and focuses it on the thermocouple. The optical-thermocouple system is to be used to measure the temperature of a 2-in black sphere 6 ft away. The thermocouple bead is maintained at a temperature of 70°F. Estimate and plot the uncertainty in the temperature measurement as a function of the temperature of the sphere.

12-4 The apparatus of Example 12-2 is used for an emissivity determination between the temperature limits of 150 and 540°C. Plot the percent uncertainty in emissivity as a function of T_s for nominal emissivities of 0.2 and 0.8. Take the uncertainties in radiometer temperature and output voltage as

$$w_{T_R} = \pm 0.3°C$$

$$w_E = 0.004 \text{ mV}$$

The uncertainty in the source temperature is taken as $\pm 0.5°C$.

12-5 A radioactive sample is observed, and the total and background counting rates are 500 and 100 counts/min, respectively. Estimate the total time necessary to produce a standard deviation of 2 percent in the nominal source counting rate. Repeat for counting rates of 100 and 22.

12-6 The following counting data are collected on a radioactive sample. Each of the number of counts is taken in an exact time interval of 1 min. The background counting rate is known accurately as 23 counts/min. Calculate the nominal value of the source counting rate and the uncertainty in this rate. Apply Chauvenet's criterion to eliminate some of the data if necessary. Assume that the Poisson distribution applies.

Interval number	No. of counts
1	523
2	410
3	342
4	595
5	490
6	611
7	547
8	512

12-7 After a set of radioactive counting measurements has been made, it is found that the timing device has been spurious in operation. A set of calibration measurements on the timer has given the standard deviation of the time-interval measurements as σ_t. Using the approach taken in Example 12-3, derive an expression for the standard deviation of the source counting rate in terms of the standard deviations of the number of background and total counts and the standard deviation of the time interval.

12-8 Derive an expression for the optimum time t_T under the conditions that $r_b = r_s$. Express in terms of σ_s and r_s.

12-9 Plot the optimum counting time from Prob. 12-8 as

$$\frac{(t_T)_{opt}\,\sigma_s^2}{r_s} \qquad \text{versus} \qquad \frac{r_b}{r_s}$$

12-10 Under the condition that the total number of counts for both the background and sample measurements is limited, i.e.,

$$n_b + n_T = K = \text{const}$$

show that the minimum error condition is

$$\frac{n_b}{n_T} = \frac{r_b}{r_T}$$

and the total counting time for the background and sample measurements is

$$t_b + t_T = \frac{2K}{2r_b + r_s}$$

Plot $\qquad\qquad \dfrac{(t_b + t_T)r_s}{K} \qquad \text{versus} \qquad \dfrac{r_b}{r_s}$

REFERENCES

1. Ballard, S., and W. Wolfe: Optical Materials in Equipment Design, A Critique, ONR, *Proc. Infrared Inform, Symposia*, vol. 4, no. 1, p. 185, March 1959.
2. Billings, B. H., E. E. Barr, and W. L. Hyde: An Investigation of the Properties of Evaporated Metal Bolometers, *J. Opt. Soc. Am.*, vol. 37, p. 123, 1947.
3. Birks, J. B.: "Scintillation Counters," McGraw-Hill Book Company, New York, 1953.
4. Fuson, N.: The Infrared Sensitivity of Superconducting Bolometers, *J. Opt. Soc. Am.*, vol. 38, p. 845, 1948.
5. Gier, J. T., R. V. Dunkle, and J. T. Bevans: Measurement of Absolute Spectral Reflectivity from 1.0 to 15 Microns, *J. Opt. Soc. Am.*, vol. 44, p. 558, 1954.
6. Glaser, P. E., and H. H. Blau: A New Technique for Measuring the Spectral Emissivity of Solids at High Temperatures, *Trans. ASME*, vol. 81C, p. 92, 1959.
7. Hackforth, H. L.: "Infrared Radiation," McGraw-Hill Book Company, New York, 1960.
8. Kaplan, I.: "Nuclear Physics," 2d ed., Addison-Wesley Publishing Company, Inc., Reading, Mass., 1963.
9. Marcus, N.: A Blackbody Standard, *Instr. and Automation*, vol. 28, March 1955.
10. Price, W. J.: "Nuclear Radiation Detection," 2d ed., McGraw-Hill Book Company, New York, 1964.
11. Snyder, N. W., J. T. Gier, and R. V. Dunkle: Total Normal Emissivity Measurements on Aircraft Materials between 100 and 800°F, *Trans. ASME*, vol. 77, p. 1011, 1955.
12. "Reactor Handbook," vol. 2, p. 951, U.S.A.E.C. Document, McGraw-Hill Book Company, New York, 1955.
13. Edwards, D. K., J. T. Gier, K. E. Nelson, and R. D. Roddick: Integrating Sphere for Imperfectly Diffuse Samples, *J. Opt. Soc. Am.*, vol. 51, no. 11, pp. 1279–1288, November 1961.

14. Cox, R. L.: Emittance of Translucent Materials: A Qualitative Evaluation of Governing Fundamental Properties and Preliminary Design of an Experimental Apparatus to Measure These Properties to 6000°F. M.S. thesis, Southern Methodist University, Dallas, Tex., 1961.

15. Birkebak, R. C., and S. H. Cho: Integrating Sphere Reflectometer Center-Mounted Blockage Effects, ASME Paper 67-HT-54, August 1967.

16. "Precision Measurement and Calibration: Optics, Metrology, and Radiation," Natl. Bur. Std., Handbook 77, vol. 3, 1961.

17. Richmond, J. C.: Measurement of Thermal Radiation Properties of Solids, NASA Sp. 31, 1963.

18. Katzoff, S.: Symposium on Thermal Radiation of Solids, NASA Sp. 55, 1965.

19. Smith, R. A., F. E. Jones, and R. P. Chasmar: "The Detection and Measurement of Infra-Red Radiation," Oxford University Press, New York, 1957.

20. Richmond, J. C., S. T. Dunn, D. P. Dewitt, and W. D. Hayes: Procedures for the Precise Determination of Thermal Radiation Properties, *Air Force Materials Lab., Tech. Doc. Rept.* ML-TDR-64-257, pt. 2, April 1965.

21. Gubareff, G. G., J. E. Jansen, and R. H. Torborg: "Thermal Radiation Properties Survey," Honeywell Research Center, Honeywell, Inc., Minneapolis, Minn., 1960.

22. Duffie, J. A., and W. A. Beckman: "Solar Energy Thermal Processes," John Wiley & Sons, Inc., New York, 1974.

23. Yellott, J. I.: Low Temperature Application of Solar Energy, in "The Measurement of Solar Radiation," ASHRAE, New York, 1967.

24. Touloukian, Y. S., et al.: "Thermophysical Properties of Matter," vol. 7, "Thermal Radiative Properties—Metallic Elements and Alloys" (1970), vol. 8, "Thermal Radiative Properties—Non-Metallic Solids" (1972), and vol. 9: "Thermal Radiative Properties—Coatings" (1972), Plenum Data Corporation.

25. Siegel, R., and J. R. Howell: "Thermal Radiation Heat Transfer," 2d ed., McGraw-Hill Book Company, New York, 1982.

26. Abbot, C. G., F. E. Fowle, and L. B. Aldrich: Improvements and Tests of Solar Constant Methods and Apparatus, *Ann. Astrophys. Observ.*, vol. 3, p. 39, 1913.

27. Aldrich, L. B., and W. B. Hoover: Pyrheliometry, *Ann. Astrophys. Observ.*, vol. 7, p. 99, 1954.

28. Abbot, C. G.: "The Silver Disk Pyrheliometer," *Smithsonian Misc. Collections*, vol. 56, No. 19, p. 1, 1911.

29. Morikofer, W.: On the Principles of Solar Radiation Measuring Instruments, *Trans. Conf. on Use of Solar Energy*, vol. 1, p. 60, University of Arizona Press, Tucson, 1958.

30. Nutter, G. D.: Recent Advances and Trends in Radiation Thermometry. *ASME paper 71 WA/Temp-3*, November 1971.

31. Androulakis, J. G.: The Universal High Temperature Emissometer, *ASME paper 72-HT-1*, August 1972.

32. Threlkeld, J. L., and R. C. Jordan: Direct Solar Radiation Available on Clear Days, *ASHAE Trans.* vol. 64, p. 45, 1958.

AIR-POLLUTION SAMPLING
AND MEASUREMENT

13-1 INTRODUCTION

Most citizens are acutely aware of the need for air-pollution control because of continual information from government sources and the press. The lay public, however, is not aware that accurate measurements are needed to bring about effective air-pollution control. Measurements are important, first, to establish acceptable levels of contamination of the atmosphere taking into account meaningful biological data in both human and animal subjects. Second, measurements must be performed in different locales to determine the offending sources and degree of control required. On the basis of such measurements control standards can be formulated and an enforcement mechanism set in motion. Finally, periodic sampling and measurement is required for use in proper legal proceedings against violators.

Our presentation of air-pollution measurement techniques certainly cannot be complete because this is an extensive and complicated field involving many disciplines in engineering and science. We shall, however, be able to discuss general sampling techniques for both particulate and gaseous contaminants. We shall also discuss air-pollution standards in a general way, keeping in mind that specific standards will change from time to time. A number of books discuss air-pollution control techniques in detail, for example Refs. [1] to [6].

13-2 UNITS FOR POLLUTION MEASUREMENT

Quantities of pollutants may be expressed on either a volumetric or a mass basis. For the mass basis, appropriate units would be grams per cubic meter or pounds-mass per cubic foot. The volumetric unit normally employed is parts per million (ppm), which is defined as

$$1 \text{ ppm} = \frac{1 \text{ volume of gaseous pollutant}}{10^6 \text{ volumes (air + pollutant)}} \tag{13-1}$$

or 0.0001 percent by volume = 1 ppm.

To convert the volumetric unit to a mass basis, one must obviously know the molecular weight of the pollutant in order to calculate its volume at a given temperature and pressure. Assuming ideal gas behavior, the conversion can be made with

$$\frac{m_p}{V} = (\text{ppm}) \frac{M_p p}{\mathscr{R} T} \times 10^{-6} \tag{13-2}$$

where m_p/V is the mass concentration of pollutant, M_p is the molecular weight of pollutant, p is the total pressure of the air and pollutant mixture, $\mathscr{R}$ is the universal gas constant, and T is the absolute temperature of the mixture. The units for m_p/V will be kilograms per cubic meter in the SI system. A more common unit is micrograms per cubic meter, in which case we can note

$$1 \text{ kg/m}^3 = 10^9 \text{ } \mu\text{g/m}^3$$

Equivalent volumetric and mass concentrations for some common pollutants are given in Table 13-1.

13-3 AIR-POLLUTION STANDARDS

Local, state, and federal agencies may set air-pollution standards or allowable levels on two bases:

1. Pollutant concentration from the source of pollution
2. Ambient pollution concentration from all sources

Thus, the particulate discharge from a smokestack can be controlled by specifying an allowable discharge concentration level at each stack, or by specifying a maximum allowable concentration level in the particular locale. One can argue for both types of standards. Measuring the concentration at the source clearly marks the emission of a prospective offender and thus provides a ready enforcement mechanism. On the other hand, it is the level of pollution in the ambient air which eventually influences public health and thus may be cited as the most meaningful type of standard.

Table 13-1 Mass concentrations for 1 ppm of common pollutants at 1 atm

Pollutant	Mass concentration ($\mu g/m^3$)	
	0°C	25°C
Carbon monoxide (CO)	1250	1145
Nitric oxide (NO)		1230
Nitrogen dioxide (NO$_2$)		1880
Ozone (O$_3$)	2141	1962
PAN [CH$_3$(CO)O$_2$NO$_2$]	5398	4945
Sulfur dioxide (SO$_2$)	2860	2620

The ambient standard is not as easy to enforce as a source standard. In an industrial area with many smokestacks, how shall the measurement engineer say *which* stack(s) are causing ambient pollution levels that are too high? In Los Angeles it would be impossible to say which automobiles were causing the most ambient pollution without performing measurements on each.

Arguments for both types of standards are discussed by Stern [4], and the interested reader will want to consult this reference for more detailed information. Our purpose in this section is to emphasize the need for both source and ambient measurements so that later sections on specific measurement techniques can be considered in their proper perspective.

It should be clear to the reader that the local environment should be taken into account in formulating allowable air pollution emission levels. Thus, the stringent auto emission standard which may be required for Los Angeles may be totally unnecessary in the wide-open space of Lubbock, Texas. As another example, the ambient standards to be used around coal-burning facilities on the east coast may be unrealistic in the southwest, where natural gas and oil are commonly used as fuels.

Imposition of strict air-pollution standards may cause severe economic dislocations on a local, state, or national level. Local unemployment may result if a business cannot afford to install pollution-control equipment and therefore must close a plant. Energy requirements and availability at all levels may be influenced by the air-pollution-control measures required. These political and economic factors are beyond our consideration, but we mention them because they are important in setting standards, some samples of which we give below. Table 13-2 gives a summary of ambient standards for California and the federal basis. The California standards are generally more restrictive because that state has more severe air-pollution problems than the nation as a whole. Table 13-3 lists the federal source standards for some typical sources. The methods of measurement mentioned in these tables will be discussed in later sections.

We should caution the reader that air-pollution standards are subject to revision *up or down* as better legal and technical experience is gained with enforcement and control measures.

Table 13-2 Ambient air quality standards

Pollutant	Averaging times	California standard		Federal standard			Method of measurement	Reference or comment
		Concentration μg/m³	Method of measurement	Primary μg/m³	Secondary μg/m³			
Particulates	Annual geometric mean	60	High-volume sampler	75	60		High-volume sampler	EPA No. AP-49
	24 h	100		260	150			
Carbon monoxide	1 h	46,000 (40 ppm)	Nondispersive infrared spectroscopy	46,000 (40 ppm)	Same		Nondispersive spectroscopy	EPA No. AP-62
	8 h	—		10,000 (9 ppm)				
Sulfur dioxide	Annual average	—	Conductimetric	80 (0.03 ppm)	60 (0.02 ppm)		Pararosaniline or flame photometric	EPA No. AP-50
	24 h	105 (0.04 ppm)		365 (0.14 ppm)	260 (0.1 ppm)			
	3 h	—		—	1300 (0.5 ppm)			
	1 h	1310 (0.5 ppm)		—	—			
Hydrocarbons (nonmethane)	3 h (6–9 A.M.)	—	—	160 (0.24 ppm)	Same		Flame ionization	EPA No. AP-64
Photochemical oxidants	1 h	200 (0.1 ppm)	Neutral buffered KI	160 (0.08 ppm)	Same		Buffered KI or chemiluminescence	EPA No. AP-63
Nitrogen dioxide	Annual arithmetic mean	—	Saltzman method	100 (0.05 ppm)	Same		Saltzman or colorimetric method with NaOH	EPA No. AP-84
	1 h	470 (0.25 ppm)						

Table 13-3 Typical source emission standards (Federal Register, Dec. 23, 1971)

Incinerators, particulates	0.08 grains/std. ft^3 (1 lbm = 7000 grains)
Cement plants	
Particulates	
Kilns	0.3 lb/ton feed
Clinker cookers	0.1 lb/ton feed
Visible emissions	
Kilns	10% opacity
Others	Less than 10% opacity
Fossil fuel–fired steam generators greater than 250 × 10^6 Btu/h heat input	
Particulates	0.1 lb/10^6 Btu (max. 2 h avg.)
SO$_2$	
Coal-fired	1.2 lb/10^6 Btu (max. 2 h avg.)
Oil-fired	0.8 lb/10^6 Btu
Nitrogen oxides	
Coal-fired	0.7 lb/10^6 Btu (max. 2 h avg. NO$_2$)
Oil-fired	0.3 lb/10^6 Btu
Gas-fired	0.2 lb/10^6 Btu
Visible emissions	Less than 20% opacity, except that 2 min in any 1 h emission can be as great as 40%
Sulfuric acid plants	
SO$_2$	4 lb/ton acid
Acid mist	0.15 lb/ton acid
Visible emissions	Less than 10% opacity

13-4 GENERAL AIR-SAMPLING TRAIN

Figure 13-1 illustrates the general air-sampling train. Not all elements of the train must be employed in each application, but we include them for generality. The sampling train has four basic sections:

1. Sample-conditioning devices
2. Sample-collection equipment
3. Metering devices
4. Vacuum source

Figure 13-1 General air-sampling train.

Detailed discussion of different elements of these sections is the basis for this entire chapter, but we shall indicate the general nature of each in the following paragraphs.

Sample-conditioning devices alter the incoming air sample so that it is easier to collect the specific gaseous or particulate component desired. The conditioning can consist of

Filtration
Drying to remove moisture
Saturation with water vapor to provide a later reference
A variety of chemical reactions to remove unwanted pollutants or gases which may interfere with later measurements

In the sample-collection section the gaseous or particulate pollutant is removed from the air stream by one or more of a variety of methods such as the following:

GAS COLLECTION

Adsorption onto solid surface
Absorption with chemical reaction into a liquid
Condensation in a freeze-out trap

PARTICULATE COLLECTION

Filtration in single or multiple stages
Removal by wet or dry impingers
Removal by sedimentation

Once the sample is collected, it is measured quantitatively by appropriate methods. After sample collection, the flow rate, temperature, and pressure of the air flowstream are measured in accordance with the methods we have described in previous chapters. The vacuum source provides the means to pull the air sample through the train. To protect the vacuum source, filtration or moisture removal may be necessary after the flow-measurement section. We should remark that all connecting tubing must be nonreactive with the pollutant(s) to be measured.

13-5 GAS SAMPLING TECHNIQUES

We now consider specific methods for collecting gaseous samples from the overall air sample. We shall also describe some of the analysis techniques employed with the sample.

Absorption Techniques

In absorption techniques the air sample is bubbled through a liquid so that the pollutant is absorbed by chemical reaction. The absorption column can be an impinger, an atomizing scrubber, a fritted glass scrubber, or some type of packed column. The absorption rate is influenced by the rate of gas flow through the absorber, concentration of the absorbing solution, and overall contact time. After absorption the liquid reactant is analyzed to determine the concentration of the component proportional to the original gaseous pollutant.

Adsorption Techniques

In the adsorption process the gaseous molecules to be analyzed are placed in contact with a solid surface, to which they adhere. Typical solids employed for the process are activated carbon, activated alumina (AlO_2), silica gel (SiO_2), and molecular sieves. The solid can then be washed with a reagent to absorb the pollutant, and an analysis can be performed. Obviously, the adsorption process requires periodic regeneration of the solid.

Condensation and Freeze-out Techniques

If the air sample is passed through cooling chambers which are maintained at very low temperatures, it is possible to cause various gaseous components to condense or freeze out. In such an operation it may be necessary to employ multiple freeze-out sections in series to remove water vapor or other constituents before extracting the desired pollutant. Table 13-4 gives a summary of typical low-temperature solutions.

We shall examine sulfur dioxide and combustion products measurements in more detail later. For now Table 13-5 gives a convenient summary of some of the methods employed for gas sampling measurements.

Table 13-4 Summary of low-temperature solutions

Coolant	Temperature (°C)	Coolant	Temperature (°C)
Distilled water and ice	0	Toluene slush	−95
Ice and NaCl	−21	Carbon disulfide slush	−111.6
Carbon tetrachloride slush	−22.9	Methyl cyclohexane slush	−126.3
Chlorobenzene slush	−45.2	N-Pentane slush	−130
Chloroform slush	−63.5	Liquid air	−147
Dry ice and acetone	−78.5	Isopentane slush	−160.5
Dry ice and Cellosolve	−78.5	Liquid oxygen	−183
Dry ice and isopropanol	−78.5	Liquid nitrogen	−196
Ethyl acetate slush	−83.6		

Note: The various slushes are prepared by adding small quantities of liquid nitrogen to the respective solvents in a Dewar vessel until a thick slush consistency is obtained.

Table 13-5 Summary of gas analysis methods

Gas	Name of method	Sampling method	Chemical reaction	Concentration range	Analysis method
CO	Iodine pentoxide	Heated I_2O_5 tube preceded by purification train	$5CO + I_2O_5 \rightarrow 5CO_2 + I_2$	1–10 ppm	Titrimetric, photometric, or coulometric determination of I_2
CO	Infrared absorption	Pressurized nondispersive IR spectrometer with 40-in cell	—	0–150 ppm	Determined from IR absorption; suitable for continuous monitoring
CO_2	Infrared absorption	Same as CO	—	0–150 ppm	Same as CO (IR method)
CO_2	Titrimetric	Bubbler	$CO_2 + Ba(OH)_2 \xrightarrow{BaCl_2} BaCO_3\downarrow + H_2O$	0–150 ppm	Titration of remaining $Ba(OH)_2$ with standard oxalic acid
H_2S	Cadmium sulfide	Two bubblers in series	H_2S reacts with Cd salt in ammoniacal solution to form insoluble CdS	0–0.5 ppm	CdS dissolved in HCl and sulfide determined by iodimetric titration
H_2S	Methylene blue	Greenburg-Smith impinger	H_2S reacts with alkaline $CdSO_4$; sulfide ion then reacts with ρ-aminodimethylaniline and finic and chloride ions to give methylene blue	$<20 \times 10^{-3}$ ppm	Colorimetric determination of methylene blue

Pollutant	Method	Sampling	Reaction	Range	Determination
NO_2	Saltzman	Fritted glass bubbler or freeze-out	Dye complex formed between sulfanilic acid, nitrite ion, and α-naphthylamine in an acid medium	0–5 ppm	Colorimetric determination of dye complex
NO	Saltzman	Same as NO_2	Oxidize NO to NO_2 by passing through potassium permanganate, then same as NO_2	0–5 ppm	Colorimetric determination
C_2H_2	Cuprous acetylide	Silica-gel detector tube in acetone–dry ice bath	Acetylenic hydrocarbons absorbed on silica gel, then reacted with ammoniacal cuprous chloride to produce red-colored copper acetylide	0–0.01 ppm	Hydrocarbon concentration determined by color intensity of detector tube
CH_2O (formaldehyde)	Chromotropic acid	Glass midget impinger with fritted disk inlet	CH_2O reacts with chromotropic acid–sulfuric acid solution to form purple chromogon	0.01–200 ppm	Spectrophotometric determination of chromogon
SO_2	West-Gaeke	Midget impinger	SO_2 reacts with sodium tetrachloromercurate to form dichlorosulfito-mercurate ion	0.005–5 ppm	Addition of acid-bleached pararosaniline and formaldehyde to form para-rosaniline methylsulfonic acid; colorimetric determination
SO_2	Hydrogen peroxide	Impinger or fritted bubbler	$SO_2 + H_2O_2 \rightarrow H_2SO_4$	0.01–10 ppm	Titration of H_2SO_4 with standard alkali
SO_2	Conductimetric	Sampler probe and scrubber	$SO_2 + H_2O_2 \rightarrow H_2SO_4$	0.01–20 ppm	Change in conductivity of H_2SO_4
SO_2	Lead peroxide candle	PbO_2 paste on cylinder or plate	$PbO_2 + SO_2 \rightarrow PbSO_4$	0.02 mg SO_2/ 100 cm²/day	Gravimetric determination of $PbSO_4$

13-6 PARTICULATE SAMPLING TECHNIQUES

The range of particulate-matter sizes encountered in air-pollution applications is extremely broad, as illustrated in Fig. 13-2. The particles range from about 0.001 to 500 μm in diameter, with most atmospheric particles falling between 0.1 and 10 μm. For very small particles, below 0.1 μm, motion and transport are governed mainly by molecular collisions, and particles above about 20 μm tend to settle out of the atmosphere and accumulate in their local surroundings. The rate at which particles will settle out depends on both size and density. For a particle density about equal to that of water (1.0 g/cm^3), the approximate settling velocities are given in Table 13-6.

Settling and Sedimentation

Because of the settling phenomenon, one mechanism for collecting samples consists of a simple container placed in the appropriate locale and observed over a sufficient length of time. Thus, the particulate pollution from an iron foundry can be measured by placing a container near the foundry and measuring the accumulation over a period of several days. Such simple methods are not without problems, however. Typically, the container is filled with water to trap the particles once they fall inside, and some protective cover is installed to prevent leaves and twigs from invalidating the measurement. The container is also placed at a sufficient height above the ground to minimize the collection of dust raised by people or motor vehicles moving in its immediate vicinity.

Mechanical Collection

When a gaseous sample is taken over a very short time interval compared to the total time of the experiment, it is said to undergo a *grab sampling* procedure. The sample may be collected in a container which has previously been carefully evacuated and sealed in the laboratory. Once collected, the sample can be returned to the lab for analysis. When an evacuated container is not available, a clean glass grab sampler must be purged for a sufficient length of time to remove air contained in it. Consider the sampler shown in Fig. 13-3. The volume of the sampler is V, and the concentration of sample gas in the sampler at any instant of time is C. If we assume complete mixing in the volume, the input minus the output flow rate of the sample must equal the accumulation, or

$$QC_i - QC = d(VC)/d\tau \tag{13-3}$$

where Q is the volume flow rate and C is now the exit concentration, i.e., $C_e = C$. For constant Q, V, and C_i, we write

$$\int_0^C \frac{dC}{C_i - C} = \frac{Q}{V} \int_0^\tau d\tau$$

or

$$\ln \left(\frac{C_i - C}{C} \right) = -\frac{Q\tau}{V}$$

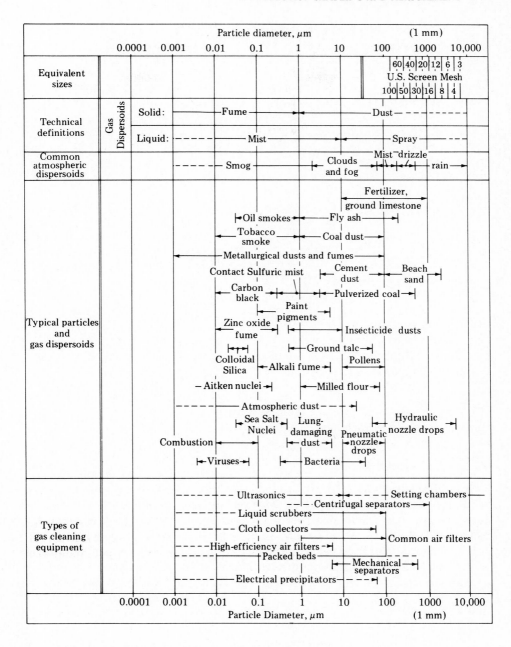

Figure 13-2 Characteristics of particulate matter. (*From Ref.* [*12*].)

Table 13-6 Approximate settling velocities for particles with density of 1.0 g/cm³

Particle size, μm	Settling velocity
0.1	0.4 nm/s
1.0	40 nm/s
10	30 mm/s
100	0.3 m/s

Rearranging,

$$\frac{C}{C_i} = 1 - e^{-Q\tau/V} \tag{13-4}$$

and we see that the concentration in the sampler approaches C_i for τ sufficiently large; the buildup is illustrated in the following tabulation:

$Q\tau/V$	C/C_i
1	0.632
2	0.865
3	0.950
4	0.982
5	0.993
10	0.999955

Thus, by proper monitoring of the grab sampling time, the purity of the sample can be made to approach the desired level.

Filtration Techniques

Using the vacuum source of the sampling train, a volume of air may be drawn across an appropriate filter, which collects the particulate matter. The filter is then removed after a known period of time and taken to the laboratory for examination.

One filtration-collection technique employs a high-volume sampler with a motor similar to that of a vacuum cleaner to pull the sample across a large filter of glass-fiber paper. Such filters typically operate for 24 h with a volume flow rate

Figure 13-3 Purge of a grab sampler.

of about 1.5 m³/min. Ordinary analytical filter paper can be used for particles between 0.5 and 1.0 μm in size.

For smaller-volume samples, porous ceramic, fritted glass, or granular filters can be employed for particles above 1.0 μm in size. Membrane or molecular filters of cellulose esters can be used down to 0.1 μm. For detailed time-varying measurements a tape sampler device is used which has a cellulose tape filter placed over a sampling nozzle connected to the source. The tape is automatically moved at preset intervals so that concentrations can be observed as a function of time. The tape is then examined in the laboratory to determine the percent of light transmitted through each collection spot. The results of such measurements are reported as a *coefficient of haze* (Coh), expressed as Coh/1000 lineal ft of air sample. This measurement is especially valuable in determining the soiling effects of various particulate matter. The following measurements are performed:

$$I_0 = \text{optical transmission of clean tape}$$

$$I = \text{optical transmission of exposed tape}$$

Then, we define

$$\text{Opacity} = \frac{I}{I_0} \tag{13-5}$$

$$\text{Optical density (OD)} = \log_{10}(I/I_0) \tag{13-6}$$

$$\text{Coefficient of haze (Coh)} = \frac{\text{OD}}{0.1} \tag{13-7}$$

$$= \frac{\log_{10}(\text{opacity})}{0.1}$$

$$= 100 \log_{10}(T_{\text{frac}})^{-1} \tag{13-8}$$

where, now, T_{frac} is the fractional transmittance of the dirty filter. For clean air the coefficient of haze is zero.

To express the Coh in terms of lineal feet of air sample, we observe

$$\text{No. of 1000-ft lengths in sample} = \frac{V}{A} \times \frac{1}{1000} \tag{13-9}$$

where V is the sample volume in cubic feet and A is the sample area on the tape in square feet. By combining Eqs. (13-8) and (13-9), we obtain

$$\frac{\text{Coh}}{1000 \text{ ft}} = \frac{10^5 A}{V} \log_{10}(T_{\text{frac}})^{-1} \tag{13-10}$$

$$= \frac{10^5}{u_g \tau} \log_{10}(T_{\text{frac}})^{-1} \tag{13-11}$$

where u_g is the gas velocity in feet per second and τ is the sample time in seconds. Table 13-7 gives values of the Coh for different degrees of air pollution.

Table 13-7 Coefficient of haze

Degree of air pollution	Coh/1000 ft
Light	0–0.9
Moderate	1–1.9
Heavy	2–2.9
Very heavy	3–3.9
Extremely heavy	4–4.9

Example 13-1 A sample is collected at a rate of 0.05 ft^3/s for a period of 45 min through a filter area having a diameter of 1.0 in. As a result of the sample collection, the light transmission of the filter paper is reduced from 80 to 55 percent. Calculate the Coh/1000 ft.

SOLUTION The fractional transmittance is

$$T_{\text{frac}} = \frac{0.55}{0.80} = 0.6875$$

From Eq. (13-8), the Coh is

$$\text{Coh} = 100 \log_{10} (1/0.6875) = 16.272$$

and with Eq. (13-11) we find

$$\text{Coh/1000 ft} = \frac{(10^5)\pi(0.5)^{-2}}{(144)(0.05)(45)(60)} \log_{10} \left(\frac{1}{0.6875}\right)$$

$$= 0.657$$

In this case, only a light degree of air pollution is observed.

Impingement and Precipitator Collectors

In impingement collection devices the air sample is first accelerated to high velocity and then forced to undergo an abrupt change in direction. For particles above about 2 μm, their inertia causes them to impact with the surface causing the change in flow directions. In some devices the surface is wetted and the particles collected in the runoff liquid. In dry devices a cascade arrangement is typically used, with jets of progressively higher velocities to remove progressively smaller particles.

Electrostatic precipitators are very efficient means of particulate collection. The air sample is drawn across a wire grid which is charged to 12 to 30 kV. The particles become charged and are subsequently collected on collector plates carrying an opposite electrical charge. When the device is shut off, the particles can be removed from the plates for laboratory analysis. Precipitators can operate effectively with a broad range of particle sizes, from 0.01 to 10 μm.

Isokinetic Sampling

Up to this point we have described techniques for removing particulate matter from an air sample for subsequent laboratory analysis. But we have not discussed one important sampling technique which must be used to ensure that a representative sample reaches our collection device. With a high-volume sampling device, the particles are collected almost immediately on the filter paper. However, when a sample must be collected from a stack, by either grab or continuous methods, more care must be exerted. Typically, a probe is inserted into the flow at various points and a sample withdrawn with a vacuum source. Because the particles are being transported by the flowstream, they are influenced by the viscosity of the gas, and their motion depends on their size and inertia. If the suction velocity in the probe is much slower than the flow velocity in the stack, much of the flow approaching the probe must be diverted around it. Particles with a large inertia may not be able to follow the change in direction and will enter the probe in a disproportionately high concentration. When the suction velocity is much larger than the flow velocity, large particles again may not be able to make the direction change and thus pass on by the probe. In this case the sample will have too low a concentration of particles. Thus, the most desirable technique is to draw the sample into the probe at a velocity equal to the flow velocity in the stack. This is accomplished by measuring the stack velocity with a Pitot tube and then adjusting the sampling flow rate until a like velocity is produced in the sampling probe. Watson [11] has developed a semiempirical equation to describe this effect.

$$\frac{C}{C_0} = \frac{U_0}{U} \left\{ 1 + f(p) \left[\left(\frac{U}{U_0} \right)^{1/2} - 1 \right] \right\}^2 \qquad (13\text{-}12)$$

where C = indicated particulate concentration
$\quad\quad C_0$ = true particulate concentration in stack
$\quad\quad U_0$ = stack flow velocity, ft/s or m/s
$\quad\quad U$ = mean velocity at sampling orifice, ft/s or m/s
$\quad\quad D$ = diameter of sampling orifice, ft or m
$\quad\quad d$ = particle diameter, ft or m
$\quad\quad \rho$ = density of particles, ft^3/lbm or m^3/kg
$\quad\quad \mu$ = viscosity of gas, lbm/ft $\cdot$ s or kg/m $\cdot$ s
$\quad\quad p = d^2 \rho U_0 / 18\ \mu D$

The function $f(p)$ is shown in Fig. 13-4, and the overall effect on concentration measurement is illustrated in Fig. 13-5. Clearly, large errors can result from anisotropic sampling when large particle sizes are involved. The error is small for small particles.

Figure 13-4 Function $f(p)$ for use in Eq. (13-12), according to Ref. [11].

Figure 13-5 Effect of nonisokinetic sampling as calculated from Eq. (13-12) for spheres with density of 1.0 g/cm^3. (*From Ref. [11].*)

13-7 SULFUR DIOXIDE MEASUREMENTS

Table 13-5 has already indicated some of the techniques employed for SO_2 measurement. Sulfur dioxide arises from the combustion of hydrocarbon fuels with a substantial sulfur content and represents a major source of air pollution. Because of the large number of industries which burn coal and fuel oil with a substantial sulfur content, the measurement techniques are particularly broad in application. We shall discuss three basic measurement techniques and refer the reader to specific manufacturers for details of equipment operation. The control measures which must be employed to remove SO_2 are beyond the scope of our discussion but are examined very thoroughly in Refs. [1] to [6].

Colorimetry Analysis for SO_2

Figure 13-6 illustrates the basic apparatus for colorimetric determinations of SO_2. A vacuum pump draws the air sample through an inlet flow measurement device and then into a rotary disk scrubber. An absorbing agent wets the surface of the disks to provide a large surface for the gas contact area, which allows for rapid absorption and response. The solution containing the pollutant is next

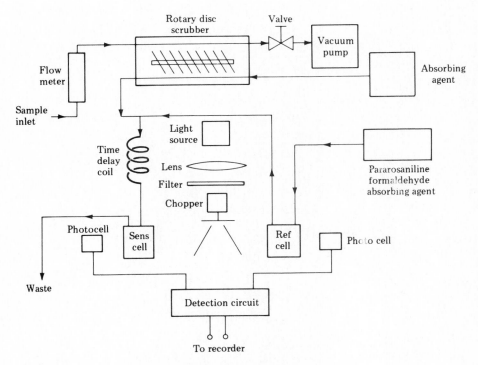

Figure 13-6 Schematic of colorimetric determination of SO_2.

mixed with the reagent, bleached pararosaniline, and allowed to pass vertically through a time-delay coil where the color-development reaction occurs.

A light source is filtered and passed through a motor-driven chopper to provide an alternating source which is adaptable to electronic amplification. Two matched cadmium sulfide photocells are employed to sense the light transmission from a sensing cell containing the sample which has undergone the colorimetric reaction and also from a reference cell containing only the reagent. The electronic detection circuit compares the outputs of the two cells and sends an output voltage to an appropriate recorder. Practical systems may be programmed to take samples and perform the analysis at predetermined time intervals.

Electroconductivity Analysis for SO_2

The electroconductivity measurement relies on oxidation of SO_2 to produce sulfate ions, which change the electrical conductivity of the solution in proportion to the amount of SO_2 present. With water as the reagent, we have the reactions

$$H_2O + SO_2 \rightarrow H_2SO_3$$

$$H_2SO_3 + \tfrac{1}{2}O_2 \rightarrow H_2SO_4 \tag{13-13}$$

As an alternative we can use H_2O_2; we will then experience less interference from acid gases like CO_2. Thus,

$$H_2O_2 + SO_2 \rightarrow H_2SO_4 \tag{13-14}$$

Figure 13-7 Schematic of electric conductivity analysis of SO_2.

Figure 13-8 Schematic of coulometric analysis of SO_2.

The electrical conductivity is a function of temperature, so that the experimental apparatus must be constructed with a temperature control system. The apparatus is indicated in the schematic of Fig. 13-7.

The reagent is passed first through a reference cell and then into an absorbing column, where the air sample is mixed in a counter-current-flow process. The SO_2 is absorbed in the reagent, and the new solution is then passed through a measuring conductivity cell. An ac voltage is impressed on the conductivity cells and the current measured to enable a calculation of the resistances. Zero-level calibration is achieved by passing the air through a soda-lime absorber to remove all SO_2. For this condition the conductivities in the reference and measuring cells should be the same and thus may be used to set the zero-output level.

Coulometric Analysis for SO_2

To perform a coulometric analysis, a detector cell is constructed as illustrated in the schematic of Fig. 13-8. The cell contains a buffered solution of KI, which generates I_2 at the anode according to

$$2I^- \rightarrow I_2 + 2e^-$$

When the air sample containing SO_2 is drawn into the cell, it reacts with I_2 to remove it from the cell. The unreacted portion of the I_2 is then reduced to I^- at the cathode. As a result of the reaction, a reference electrode will detect a difference in the anode and cathode currents as shown:

$$i_r = i_a - i_c$$

This current may subsequently be amplified to provide an output proportional to the SO_2 concentration.

13-8 COMBUSTION PRODUCTS MEASUREMENTS

The analysis of products of combustion is important not only for air-pollution-control applications but also for maintenance of most efficient burning rates and energy utilization. We shall discuss three methods applicable to such measurements.

Figure 13-9 Schematic of Orsat apparatus.

Orsat Apparatus

Figure 13-9 illustrates the simple Orsat apparatus, which is employed to analyze products of combustion. It consists of a measuring burette and three reagent pipettes which are used to successively absorb carbon dioxide, oxygen, and carbon monoxide from the mixture. First, a sample of the combustion products is drawn into the measuring burette. Next, the sampling manifold is shut off and the sample is forced into the first reagent pipette, where carbon dioxide is absorbed. The sample is then brought back into the measuring burette, and the reduction in volume is noted. The procedure is then repeated for the other two pipettes, which absorb the O_2 and CO successively. In the measuring process the combustion products are contained over water in the burette and remain saturated with water vapor. As a result, the volumetric proportions of the products are obtained on a so-called dry basis, i.e., exclusive of the water vapor which is present.

Potassium hydroxide is normally used as the reagent to absorb carbon dioxide. A mixture of pyrogallic acid and solution of potassium hydroxide is employed as the reagent for absorption of oxygen. Cuprous chloride is used to absorb carbon monoxide. Precautions must be taken to ensure that the reagents are fresh in order to minimize errors. The cuprous chloride must be changed after 10 absorptions or so if the carbon monoxide content is above 1 percent.

Example 13-2 An operator reports the following raw volume measurements on an Orsat device in measuring the combustion products for methane fuel (CH_4). Based on these data, calculate the air-fuel ratio and percent excess air for the combustion process.

	Volume (cm^3)
Initial sample	95
After CO_2 absorption	83
After O_2 absorption	79
After CO absorption	78

SOLUTION The analysis of the "dry products" is

$$n_{CO_2} = 95 - 83 = 12$$

$$n_{O_2} = 83 - 79 = 4$$

$$n_{CO} = 79 - 78 = 1$$

On a mole fraction basis we obtain

$$x_{CO_2} = \frac{12}{95} = 12.6\%$$

$$x_{O_2} = \frac{4}{95} = 4.2\%$$

$$x_{CO} = \frac{1}{95} = 1.05\%$$

We now write the combustion equation for 95 moles of dry products.

$$a CH_4 + x O_2 + 3.76x N_2 \rightarrow 12 CO_2 + 4 O_2 + (1)CO + b H_2O + 3.76x N_2$$

The carbon balance yields

$$a = 12 + 1 = 13$$

For the hydrogen $4a = 2b$, so that $b = 26$, and the oxygen balance gives

$$2x = (12)(2) + (4)(2) + 1 + 26 = 59$$

$$x = 29.5$$

The air-fuel ratio for dry air is

$$\text{AF ratio} = \frac{m_a}{m_{CH_4}}$$

$$= \frac{(29.5)[32 + (3.76)(28)]}{(13)(16)}$$

$$= 19.47$$

For stoichiometric combustion (no excess oxygen) the reaction would be

$$12 CH_4 + 24 O_2 + (3.76)(24) N_2 \rightarrow 12 CO_2 + 24 H_2O + (3.76)(24) N_2$$

The percent excess air is therefore

$$\frac{29.5 - 24}{24} \times 100 = 22.9\%$$

Gas Chromatograph

Another technique for measuring products of combustion, and other gas samples as well, is *gas chromatography*, indicated in the schematic diagram of Fig. 13-10. The sample gas is typically collected in a small glass syringe and introduced into

Figure 13-10 Schematic of gas chromatograph.

the chromatographic column. The column can vary in length from 1 to 20 m and in diameter from 0.25 to 50 mm. Inside the column either a solid adsorbant, like granular silica, alumina, or carbon, or a liquid adsorbant is contained to retard the flow of different constituents of the gas sample. The liquid can either coat the wall when the column is a thin capillary or be distributed over an inert granular material like diatomaceous earth in larger packed columns. Hundreds of liquids are available as adsorbants, depending on the particular application.

Once the gas sample is impressed on the column, an inert carrier gas (typically helium) transports the sample through the column. Different components will be retained in the column for different lengths of time, so that they will appear in the discharge stream from the column at different times. The device may be calibrated to predict the retention time for various gas-liquid combinations. The components in the output flowstream may be detected in various ways. A thermal-conductivity detector is most common, but ultrasonic and electron-capture methods may also be used. Special detectors are available for specific elements.

The output of the chromatographic detector can be processed electronically and displayed on a chart recorder. The display takes the form of a series of spikes. The time of appearance of the spike indicates the particular component, while the height of the spike indicates the quantity present. The chromatographic column may require heating or cooling to achieve the proper separation times, so that a temperature control system is required. A flow regulator is also employed to maintain a constant flow rate of the carrier gas.

Nondispersive Infrared Absorption

Nondispersive infrared (IR) measurements may be used to determine concentration of gaseous pollutants as illustrated in the schematic diagram of Fig. 13-11. The absorption of IR radiation occurs in narrow-wavelength bands, with each gas exhibiting its own peculiar characteristics. Two infrared sources are provided, as shown, with a chopper to impose an ac signal, which is handled more easily by subsequent electronic detection circuits. Filters may be placed in the path of the sources so that only the absorption-wavelength band for the particular gas to be studied is investigated. Two cells are exposed to the IR radiation: (1) a reference cell containing an inert gas (usually nitrogen), and (2) a cell which admits passage of the sample gas containing the pollutant to be investigated. No IR absorption occurs in the reference cell, while the absorption in the sample cell is proportional to the concentration of the component of interest.

Beyond the reference and sample cells are two detector chambers which absorb the IR radiation transmitted. These chambers are sealed so that absorption of the radiation causes the temperature of the gas inside to rise. The two

Figure 13-11 Schematic of nondispersive infrared analyzer.

chambers are separated by a thin diaphragm. Because more radiation is transmitted through the reference cell, there will be a greater temperature rise in the reference detector chamber than in the sample detector. As a result, a pressure differential is created which causes movement of the diaphragm; this can be detected with a capacitance pickup. The resulting signal is amplified and transmitted to an appropriate readout device.

13-9 OPACITY MEASUREMENTS

Most laypeople measure air pollution by visual observation. If a discharge stack at a plant emits heavy black smoke, this is judged to be harmful pollution. The accumulation of smog around expressways on still days is evidenced by decreased visibility.

The opacity of the gas stream discharging from a stack is indeed a measure of the concentration of particulate matter in the gas. The *color* of the discharge gas, however, depends on the specific particulate matter involved in the process. Unburned carbon is black, while very harmful asbestos or lead particles might be gray in color. Air-pollution-control officials cannot provide continuous stack gas monitoring of all plants because this would involve excessive costs in instrumentation and personnel, and so they frequently rely on visual observation of the discharge gases to indicate probable air-pollution offenders. At the present time there is no instrument that can be used to sight on a stack and measure the opacity of the discharge gases. One must rely on human visual observations.

To aid in standardizing visual observations, the Ringleman smoke chart system has been devised. Five charts are constructed in even increments of blackness as indicated in Table 13-8. Observers are then trained to compare gas emissions with the charts. In the training process the chart is placed at a distance of 50 ft from the observer and in a line of sight with the smoke. At this distance the charts take on an appearance of different degrees of gray. In observations the observer should stand facing away from the sun. For training purposes optical transmission measurements may also be performed on the smoke to correlate with the Ringleman charts. A person who has undergone an intensive training period is then permitted to make observations in the field strictly on the basis of

Table 13-8 Spacing of lines for construction of Ringleman chart

Ringleman chart no.	Width of black lines, mm	Width of white spaces, mm	Percent black
0	All white		0
1	1	9	20
2	2.3	7.7	40
3	3.7	6.3	60
4	5.5	4.5	80
5	All black		100

visual acuity. Such persons are appropriately called "smoke readers"; when they are employed by air-pollution-control groups, they must be retrained about every 6 months. The trained smoke observer can obviously make many measurements in a region at very low cost. Typically, a plant is not allowed to operate at greater than a Ringleman number 2 discharge opacity. The trained observer can read the opacity within $\pm 1/2$ Ringleman number [14].

The opacity of a stack discharge can be easily changed without changing the total emission of pollutants. Because opacity is a function of optical path length, one simple way to decrease the opacity is to reduce the stack diameter and increase the flow velocity. Another method is to increase the number of stacks employed for the discharge. Air-pollution-control regulations frequently do not allow the use of such techniques once a citation has been issued for excess discharge opacity.

All visual observations can be in serious error if the stack gases contain significant quantities of water vapor which may condense out. In such cases a violator may be excused if engineering data clearly show that the opacity could be the result of the water discharge in the process.

13-10 ODOR MEASUREMENT

We have all encountered odor problems at one time or another. Because odors must be detected by human beings, quantitative measurements and standards are very difficult to establish. Different people react to various odors in different ways. A person who eats garlic likes the taste and odor, while a person sitting nearby may find the garlic smell quite objectionable. We can become so accustomed to odors that they are no longer objectionable. A person who works at a stockyard does not notice the smell, but a newcomer visiting such a facility may find the odor very offensive.

Ideally, we would like to specify acceptable odor levels in terms of odorant concentration in parts per million or micrograms per cubic meter. The technique normally employed is to expose a group of individuals to varying concentration levels of an odorant and ask them to make a rating on the following scale:

5 overpowering perception
4 strong odor
3 easily perceived odor
2 faint perception
1 very faint perception
0 no perception

This rating scale is applied to the intensity of the odor, its pervasiveness, and the acceptability or degree of like or dislike of the odor. When these data are collected, it is generally found that they follow a relationship of the form

$$P = K \log S \tag{13-15}$$

Table 13-9 Odor thresholds in air, from Ref. [10]

Odorant	Odor threshold, ppm
Acetic acid	1.0
Acetone	100
Amine trimethyl	0.0021
Ammonia	46.8
Aniline	1.0
Carbon disulfide	0.21
Chlorine	0.31
Diethyl formanide	100
Diphenyl sulfide	0.21
Formaldehyde	1.0
Hydrogen sulfide	0.00047
Methanol	100
Methylene chloride	214
Nitrobenzene	0.0047
Phenol	0.047
Trimethyl amine	0.00021

where P is the human sensory response, or intensity (from 0 to 5), and S is the ratio of the odor concentration in parts per million to that for 0 response. K is a constant which varies from 0.3 to 0.6, depending on the odorant. Using such techniques it is possible to determine odor thresholds for a number of substances, as shown in Table 13-9. Other values for petrochemicals are given in Ref. [9]. A cursory examination of this table indicates the extremely broad range of odor thresholds (over six orders of magnitude) and therefore the wide variety of instrumentation which may be required for measurements.

In some air-pollution problems it is convenient to define an *odor unit* as:

1 odor unit = quantity of odorant that will contaminate 1 ft³ of
clean air to the mean threshold level

From Table 13-9 we see that 1 odor unit of acetone would be the quantity necessary to produce a concentration of 100 ppm, while one odor unit of formaldehyde would be that necessary to produce a concentration of 1.0 ppm.

REVIEW QUESTIONS

13-1 Distinguish between source and ambient air-pollution standards. What are the advantages of each?

13-2 Name three techniques for particulate measurement.

13-3 What is the range of sizes of most atmospheric particulate matter?

13-4 What is meant by "grab sampling"?

13-5 Define the coefficient of haze (Coh).

13-6 What is the principle of operation of an electrostatic precipitator?

13-7 What is meant by "isokinetic sampling"?

13-8 Describe the colorimetry analysis process for SO_2.

13-9 How does an Orsat apparatus work?

13-10 How does a gas chromatograph work?

13-11 What is a Ringleman chart?

13-12 What easy measures can be taken to reduce the opacity of a stack discharge?

PROBLEMS

13-1 Verify the mass concentrations of carbon monoxide and sulfur dioxide given in Table 13-1.

13-2 A glass grab sampler has a volume of 100 cm^3. Calculate the time required to effect a 90 percent purge of the sampler with a flow rate of 125 cm^3/min.

13-3 A sample is collected at a rate of 10 ft^3/min for a period of 1 h through a filter area having a diameter of 1.4 in. As a result of the sample collection, the light transmission of the filter paper is reduced from 80 to 40 percent. Calculate the Coh/1000 ft.

13-4 An analysis of the dry products of combustion for CH_4 yields

$$CO_2 = 10\%$$

$$CO = 0.7\%$$

$$O_2 = 2.0\%$$

Calculate the air-fuel ratio and percent excess air.

REFERENCES

1. Wark, K., and C. F. Warner: "Air Pollution," IEP Publishers, 1976.
2. Seinfeld, J. H.: "Air Pollution," McGraw-Hill Book Company, New York, 1975.
3. Perkins, H. C.: "Air Pollution," McGraw-Hill Book Company, New York, 1974.
4. Stern, A. C.: "Air Pollution," vols. I, II, III, Academic Press, Inc., New York, 1968.
5. Magill, P. L., F. R. Holden, and C. Ackley: "Air Pollution Handbook," McGraw-Hill Book Company, New York, 1956.
6. "Air Pollution Engineering Manual," EPA, NAPCA, AP-40, 1967.
7. Moncrieff, R. W.: "The Chemical Senses," Leonard Hill Books, London, 1967.
8. Duffee, R. A.: Appraisal of Odor Measurement Techniques, *J. Air Pollution Cont. Assoc.*, vol. 18, p. 472, 1968.
9. Hellman, T. M., and F. H. Small: Characterization of the Odor Properties of 101 Petrochemicals Using Sensory Methods, *J. Air Pollution Cont. Assoc.*, vol. 24, p. 979, 1974.
10. Leonardos, G., D. Kendall, and N. Bernard: Odor Threshold Determinations of 53 Odorant Chemicals, *J. Air Pollution Cont. Assoc.*, vol. 19, p. 91, 1969.
11. Watson, H. H.: Errors Due to Anisokinetic Sampling of Aerosols, *Am. Ind. Hyg. Assoc. Quart.*, vol. 15, p. 21, 1954.
12. Lapple, C. E.: Characteristics of Particles and Particle Dispersoids, *Stanford Research Inst. Jour.*, vol. 5, 1961.
13. Ringleman, M.: Methode d'Estimation des Fumes Produites par les Foyer Industriels, *La Revue Technique*, 268, June 1898.
14. Conner, W. D., and J. R. Hodkinson: "Optical Properties and Visual Effects of Smoke Stack Plumes," Publication 999-AP-30, USHEW, 1967.

FOURTEEN

DATA ACQUISITION AND PROCESSING

14-1 INTRODUCTION

All the previous chapters have illustrated, in one way or another, the manner in which an experiment can be planned to produce meaningful data. Selection of proper instrumentation and consideration of experimental uncertainties are all necessary parts of the planning process. But very little has been said about the manner of collecting data and the later processing of these data to produce the desired results of the experiments. The acquisition of data might consist simply of several people (or perhaps only one person) reading a number of instruments and recording the observations on a data sheet. The subsequent processing of data could be done in many ways, from simple hand calculations to complicated computer routines.

There are systems available today for rapidly collecting a great bulk of data, processing it, and displaying the desired results in printed form. The purpose of this chapter is to present a brief qualitative description of such systems and the functions of the elements that go together to make up an overall data acquisition and processing installation. These systems change rapidly so this discussion is only approximate.

14-2 THE GENERAL DATA ACQUISITION SYSTEM

The essential element in a modern data acquisition system is the instrument transducer, which furnishes an electrical signal that is indicative of the physical variable being measured. The signal may be an analog voltage, current, resistance, or frequency, or a digital representation of any of these quantities in the form of a series of electric pulses. The present discussion assumes that suitable transducers are available to convert the physical variables of interest into electrical signals. As examples of such transducers we may recall that a thermocouple gives a voltage representation of temperature, a strain gage gives a resistance representation of strain, etc.

The object of any data acquisition and processing system is to collect the data, process them in the desired fashion, and record the results in a form suitable for storage, presentation, or additional subsequent processing. Thus, a recording potentiometer is a simple data acquisition system that may be used for collecting temperature data from thermocouples. In this case, the data points must be read from the recorder chart. A more complicated system would convert the analog voltage signal from the thermocouple to an equivalent digital signal, which could be used to operate a printing recorder, so that the numerical value of the temperature is printed on a sheet of paper. Such a system is much more complicated than the simple recorder because of the analog-to-digital conversion process. It is easy to see, however, that the digital output has many advantages.

The major elements of any data acquisition and processing system are shown in block diagram form in Fig. 14-1. This is a redrawing of Fig. 4-18 with an additional subdivision of the system into three major parts. The first is the input stage, which consists of appropriate transducers and an input circuit and additional signal conditioning circuits as needed (amplifiers, filters, etc.). The second is the signal-conversion stage, in which the information is readied for transmission, if such is required (as in situations in which the transducers are physically remote from the location at which data display is desired) as well as the transmission and receiving equipment and any necessary data processors. An example of the latter would be the conversion of a signal from analog to digital form. The last is an output stage which provides two primary functions: data display and data storage. Examples include data display and storage in printed form on a sheet of paper or in graphical form on suitable paper and storage on magnetic or punched tape. The output stage must include suitable coupling circuits to convert

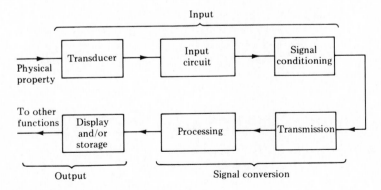

Figure 14-1 General data acquisition system.

the data to a form which may be used to drive a printer, card punch, magnetic tape writing unit, etc.

It is a rare circumstance when data involving only one experimental variable are to be collected. The data acquisition and processing system, then, should include provision for collecting and analyzing multiple channels of data input. This collection process could be accomplished by having a channel like the one shown in Fig. 14-1 for each variable to be studied. The cost of such a system, however, would be quite high because of equipment duplication, so that a scanner/programmer is normally employed for multiple-channel work. The scanner is a device that samples the data channels in rapid sequence so that only one conversion and output stage is necessary. Equipment available today makes possible any particular sequencing of a large number of data channels at the discretion of the laboratory personnel. Thus, the system may be programmed to collect any desired range of variables in any order, and the term *scanner/programmer* is quite appropriate. We will discuss the electronic realization of such systems, in general terms, later in this chapter.

Many experimental situations require the collection of data at regular time intervals or in some particular time sequence. The acquisition system may perform this timing function automatically by incorporating a digital clock and time standard in the scanner and/or conversion stages.

Finally, it may be advantageous to apply signal conditioning to the output of the device functioning as the scanner/programmer. This conditioning might be amplification, analog-to-digital conversion for only some of the channels, filtering, distortion or harmonic analysis of the waveform of some of the signals, etc. When all the above elements are combined, a very flexible data acquisition and processing system results, as shown in Fig. 14-2. It should be noted that the programmable feature of the scanner is normally used in the converter stage as well. This is essential, since some channels may require signal conditioning while others may not.

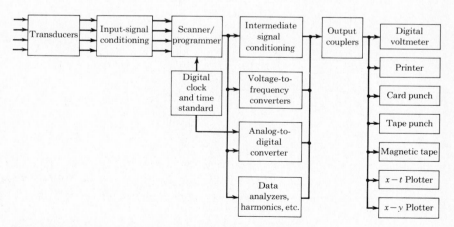

Figure 14-2 Schematic of a multichannel data acquisition system.

The use of a flexible system like the one described above depends on many factors, not the least of which is cost. Progress in the semiconductor industry, particularly the development of microprocessors (simple computers on one small chip of silicon), has greatly reduced the cost of such systems. Data acquisition processing systems which use microprocessors to perform the scanner/programmer function are very widespread. We will devote the remaining sections of this chapter to a careful analysis of the elements of data acquisition systems which were not covered in earlier chapters of this text.

14-3 SIGNAL CONDITIONING REVISITED

As was pointed out in Chap. 4, noise is present in all physical situations in which measurements are attempted or information is conveyed. A consideration of noise and its effects on experiments led us, in Sec. 4-12, to consider filtering electrical signals. We considered circuits which would transmit or "pass" only certain bands or ranges of frequencies which are part of an input signal. It was pointed out that some filters are composed only of passive elements (resistors, capacitors, and inductors), while others contain amplifiers. Circuits of the first type are called *passive filters*, while those in the second category are termed *active filters*.

Table 4-3 displayed four very simple passive filters. In order to illustrate active filters and the methods of analysis applied to such filters, we will analyze an active version of the first filter shown in this table, a lowpass RC filter. For the passive form, the maximum ratio of output voltage to input voltage is unity. As we shall see, the addition of an operational amplifier allows the realization of output-to-input-voltage ratios which are larger than 1. The gain of the operational amplifier serves to improve the overall performance of the filter. It should be mentioned that we will be analyzing only a very simple circuit; there are commercial design manuals which present the analysis of more complex active filters.

The basic ideal operational amplifier is presented in Fig. 14-3 along with the relationship between the output voltage v_o and the two input voltages v_1 and v_2. The amplifier gain A will be taken to be infinite, and the input resistances of both the positively and the negatively labeled terminals of the amplifier will be assumed to be infinite. These approximations greatly simplify the analysis.

The lowpass, single-section active filter to be analyzed is shown in Fig. 14-4. As was the case for each circuit containing a operational amplifier which was

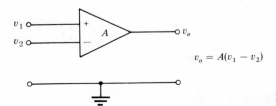

$$v_o = A(v_1 - v_2)$$

Figure 14-3 The ideal operational amplifier.

Figure 14-4 Lowpass, single-section active filter.

analyzed in Chap. 4, we are seeking a relation between the output voltage v_o and the input voltage v_i. It is simplest to assume that all voltages, and consequently all currents, in the circuit are single-frequency sinusoids, i.e.,

$$v_1 = V_1 e^{j\omega t} \qquad v_i = V_i e^{j\omega t}$$
$$i_1 = I_1 e^{j\omega t} \qquad i_2 = I_2 e^{j\omega t} \tag{14-1}$$

Using the reference directions shown in Fig. 14-4, it is possible to write the following four expressions for the currents shown:

$$i_1 = \frac{v_i - v_1}{R} \qquad i_2 = \frac{v_2}{R}$$

$$i_3 = -C \frac{dv_1}{dt} \qquad i_4 = \frac{v_o - v_2}{(K - 1)R} \tag{14-2}$$

In these equations, the ground voltage has been taken to be zero.

Since the input resistances of the two operational amplifier terminals $(+, -)$ are infinite, it follows that the sum of the two currents entering each terminal must be zero. We write

$$i_1 + i_3 = 0$$
$$-i_2 + i_4 = 0$$

Substituting the sinusoidal forms given in Eqs. (14-1) and the current expression in Eqs. (14-2) results in

$$\frac{V_i - V_1}{R} e^{j\omega t} - C \frac{d}{dt} (V_1 e^{j\omega t}) = 0 \tag{14-3a}$$

$$-\frac{V_2}{R} e^{j\omega t} + \frac{V_o - V_2}{(K - 1)R} e^{j\omega t} = 0 \tag{14-3b}$$

After carrying out the derivative with respect to time and canceling common factors, we arrive at

$$V_i - V_1 - j\omega RC V_1 = 0 \tag{14-4a}$$

$$-V_2 + \frac{V_o - V_2}{K - 1} = 0 \tag{14-4b}$$

The relationship between the output voltage and the two input voltages of the operational amplifier is

$$v_0 = A(v_1 - v_2)$$

or, upon substituting the sinusoidal time dependencies,

$$V_0 = A(V_1 - V_2)$$

Since the amplifier gain is approximated by infinity, it is necessary for V_1 to equal V_2 if the output voltage V_o is to be finite.

Solving for V_1 in equation (14-4a) and V_2 in equation (14-4b) and equating the results yields

$$\frac{V_i}{1 + j\omega RC} = \frac{V_o}{K}$$

The ratio between the output and input voltages is given by the expression

$$\frac{V_o}{V_i} = \frac{K}{1 + j\omega RC}$$

or, in terms of amplitudes,

$$\left| \frac{V_o}{V_i} \right| = \frac{K}{\sqrt{1 + (\omega RC)^2}}$$

This result should be compared with the equation given for the lowpass RC passive filter in Table 4-3. The two results differ only by the factor of K in the numerator. Since this may be selected by the designer (refer to Fig. 14-4), the effect of the operational amplifier in the active filter is to allow a choice of the overall gain displayed by the active filter compared with the passive filter. The frequency response of the two circuits is identical. The only difference is the constant K appearing in the relation between output and input voltages for the active filter.

This example indicates that using active filters rather than passive filters often makes adding additional amplifier circuits to a given data acquisition and processing system unnecessary.

The other major advantage of active filters is based on a unique property of integrated circuits. Simply stated, there are no integrated-circuit forms of inductors. To this day, the inductor remains a large element. In order to fabricate passive filters containing inductors, it is necessary to sacrifice some degree of compactness. Active filters can be designed to totally avoid the need for inductors. That is, the combination of resistors, capacitors, and operational amplifiers will replace and outperform any passive filter containing inductors. This means that circuit compactness and integration can always be obtained. The proof of these assertions is complex and well beyond the scope of this chapter. It is mentioned because this avoidance of inductors is a major advantage of active filters.

Although we have analyzed only a lowpass filter, both bandpass and high-pass active filters exist and can often be used in place of passive filters. In many situations, operational amplifiers allow the circuit designer to use smaller values of resistance and capacitance than would be needed for passive filters. This implies an associated reduction in cost and circuit size. There are also available designs for filters containing more than one operational amplifier. Such multi-stage circuits exhibit sharper cutoffs, and hence more useful filtering behavior, than the single-stage circuit analyzed in this chapter.

Once the signal has been filtered, it is necessary to consider the means available to transmit it if the transducer is at a remote location. This is the subject of the next section. It must be mentioned that in some situations, there are advantages to filtering the signals *both* before and after transmission, while in other cases, filtering *either* before *or* after transmission will be adequate. More information on active filters is given in Refs. [16] to [19] of Chap. 4.

14-4 DATA TRANSMISSION

Given a transducer in one location and the need to display the transducer's output in another location, one is immediately faced with choosing the channel over which the necessary transmission will occur. There are many possible choices, ranging from current and voltage on copper wires to the propagation of electromagnetic waves through space to the transmission of light signals through an optical fiber. In selecting a communication channel, the system designer must consider two paramount issues, cost and noise immunity. A detailed examination of these two issues would fill several additional books, but we can consider some of the important characteristics of various transmission channels.

Coaxial cable provides a very secure means of transmitting a signal from one location to another. It consists of a central conductor through which the signals of interest are sent surrounded by another conductor which is held at zero voltage. Any stray electromagnetic noise is attenuated by this outer shield before it can corrupt the signal present on the inner conductor. Coaxial cable is often the communication medium of choice in laboratory situations where low voltage levels are being transmitted. Of course, if the transducer is on a probe heading toward Venus, it is not possible to string a coaxial cable between the transducer and the earth. In these situations, it is necessary to consider communication channels which do not depend upon a material link between transmitter and receiver. This is, of course, necessary not only in communicating from space, but also in any situation in which the cost of a physical link between transmitter and receiver is prohibitive.

It is obvious that empty space is free and available as a channel over which signals can be transmitted. Of course, there may be additional expense in the receiver and transmitter, depending on the frequencies used, but the communication link itself is quite inexpensive. The major drawback is the susceptibility of signals propagating through space to noise. For example, a lightning discharge generates a large amount of noise over a wide range of frequencies. If a signal is propagating in the vicinity of a lightning strike, it will be mixed with the noise produced by the lightning. This illustrates the basic tradeoff between cost and noise immunity. Of course, the effect noise has on a given signal depends upon the frequency being used and the degree of protection the designer builds in. The details, again, are beyond our immediate interest, but a common example of such designed protection is given by AM and FM radio systems. A comparison of the quality of reception of an AM signal and an FM signal during a thunderstorm will immediately indicate that noise is much less significant on the FM channel. This is a result of both the frequency used and the more complex electronics involved in FM radio transmission.

We shall discuss the relative advantages of digital signals in the next section, but it is convenient to examine a simple technique which can be used to provide some degree of noise protection when the signals being transmitted are in digital form. The science of information theory has developed around the topics of transmitting information over channels and protecting the information against noise. Such systematic attempts to build noise immunity into a data transmission link are called *codes*. We present the simplest possible example of such a code to conclude this section.

Consider a case in which a value of temperature is being transmitted from a remote location in the form of a four-bit binary word. The set of possible four-bit words and the associated temperature ranges are shown below.

0000	0–5°C	1000	40–45°C
0001	5–10	1001	45–50
0010	10–15	1010	50–55
0011	15–20	1011	55–60
0100	20–25	1100	60–65
0101	25–30	1101	65–70
0110	30–35	1110	70–75
0111	35–40	1111	75–80

The meaning of this table is quite simple. If, for example, the voltage produced by the transducer indicates that the temperature is between 15 and 20°C, the binary signal 0011 will be sent by the transmitter. This process of transmitting temperatures could occur as frequently as desired.

The receiver would then "see" a series of four-bit binary words. If the experiment is to produce valid results, the four-bit binary word received must be identical to the one sent. This may or may not be the case in a given sending-receiving event, since the signal received depends not only on the signal sent but on the noise present along the channel while transmission was occurring.

A very simple technique can be used to provide a limited degree of protection against noise. It allows the receiver to determine whether one of the bits in the received four-bit word is incorrect. In such a case, the receiver can automatically request a second transmission. The technique is called *parity checking*. As each four-bit word is readied for transmission, the signal-conditioning circuits add a fifth binary bit such that the total number of 1s in the five-bit word is even.

Imagine that the temperatures at three successive times are 57, 52, and 48°. This set of temperatures generates 3 four-bit binary words, 1011 1010 1001.

The parity check bit (the fifth bit) would be added to each of these words so that the total number of 1s would be even. The resulting set of three five-bit words would be 10111 10100 10010.

If the following set of words arrived at the receiving station, 10111 10110 10010, the presence of an odd number of 1s in the second word would initiate a request for retransmission. This request could be generated by part of the receiving circuit, or it could depend on an individual monitoring the received signals.

It should be noted that this simple coding technique will not protect against an even number of errors (i.e., 10111 sent and 10001 received), since in such cases the number of 1s in the received word remains even. In addition, the technique will not be able to discriminate between one error and three errors occurring in a single word. In short, if an even number of errors occurs (and this includes zero errors), the parity check will be satisfied. If an odd number of errors occurs, the parity check will generate a request for retransmission.

Much more sophisticated coding techniques exist. In fact, codes exist which will not only indicate the presence of errors in received data but contain, within the received data, the means to correct the errors without requiring a retransmission. This is an important result and serves as a tribute to the ingenuity of those individuals who have developed so-called error-correcting codes.

Data in digital form are a prerequisite if coding techniques are to be used. Transducers, of course, produce signals which are necessarily analog in nature. In the next section, we shall discuss the means used to convert signals from analog to digital form and from digital to analog form.

14-5 ANALOG-TO-DIGITAL AND DIGITAL-TO-ANALOG CONVERSION

The basic concepts of digital and analog representations of electrical signals were discussed in Sec. 4-2. In analog representations, the physical variables of interest are treated as continuous quantities, while in digital representations, the quantities are restricted to discrete, noncontinuous "chunks." A further refinement of the digital representation was presented in the last section, where a range of temperatures of interest was broken into separate segments, each 5° wide. Then a separate four-bit binary word was used to represent each segment. This is a two-step digital representation; the first step involves discretizing the analog quantity of interest (the temperature in this case), and the second assigns a different four-bit word to each range. It is clear that 2^n different ranges can be designated by means of an n-bit binary word. For example, the selection of an eight-bit word allows the physical signal of interest to be discretized into 2^8 or 256 segments. The phrase *analog-to-digital conversion* is used to describe this kind of signal processing. The immediate question of interest concerns the value of such conversions.

The first reason for converting signals from analog to digital form involves noise immunity during transmission. The problem of determining whether or not a single digital pulse (either a binary 1 or a binary 0) is present at a given time is relatively simple compared with the task of determining whether a voltage is, for example, 4.8 V or 4.81 V or 4.79 V, etc. In crude terms, it is easier to "electrically see" the presence or absence of a pulse than to discern the precise value of an analog signal in the presence of noise induced along the transmission path.

In addition, the techniques of coding, which were discussed in the last section, have been developed only for digital signals. This means that in order to take advantage of the error-detection and error-correction capabilities of these codes, it is essential to convert all the data signals of interest to digital form.

The third major reason for the widespread use of digital representations centers on the data processing capabilities of the digital computer. After all, the digital computer, by its nature, manipulates only data in digital form. Therefore, an analog-to-digital conversion must be made if one wants to carry out signal processing using digital computers. One example should suffice to illustrate the attractiveness of digital computers for many types of signal analysis.

In statistical studies, it is often necessary to evaluate expressions of the form

$$\overline{y(t)y(t-\tau)}$$

where $y(t)$ is some physical signal of interest, the overscore indicates a time average, and τ is a time delay. Such so-called correlation functions are important in analyzing the statistical properties of physical signals. It is necessary that the time average be calculated for different values of the time delay. In some cases, such correlation functions are evaluated between two different physical signals, i.e.,

$$\overline{x(t)y(t-\tau)}$$

In order to carry out the evaluation of the time average in analog form, it is necessary to take the signal $y(t)$ and delay it by the quantity τ. How can this be accomplished for a typical electromagnetic signal? The most obvious and direct method is to physically delay the signal by propagating it over a sufficiently long path. Suppose, for concreteness, that a time delay of 1 s was required in some application. Since electromagnetic signals travel with the velocity of light, a time delay of 1 s can be obtained only by using a signal path which is 300 million m in length (a distance not easily available in the normal laboratory!). Even a time delay of 1 μs requires a rather cumbersome 300 m. The reason for this difficulty is the high velocity with which electromagnetic waves travel. There are two basic ways out of this dilemma.

The first involves converting the analog signal of interest, $y(t)$, to an equivalent digital form. The digital values of the signal are then stored in a large computer memory, with a value for each discrete time stored in a different memory location. Once this storage has been completed, the digital computer can be programmed to carry out the appropriate time averaging. For example, it could "fetch" $y(0)$ and $y(1)$, multiply these two numbers together, and continue this "fetching" and multiplying process until the time average had been calculated. What we have done is replace a very long delay line with a very large digital computer. However, as computer memory technology advances, the cost and size of the digital computer required to do a given calculation are continually decreasing. In order to take advantage of the digital computer, a signal conversion from analog to digital form must be carried out. Although we cannot go into other applications, it must be mentioned that the calculation of correlation functions is only one of the useful processing steps which can be accomplished using digital representations of signals and a computer.

The second way of producing time delays involves converting a signal from a physical form which has a large propagation velocity to a form with a much smaller velocity. Within the past decade, surface acoustic wave (SAW) devices have been developed for this purpose. The essential physical phenomenon underlying their operation is the much smaller velocity of acoustic (pressure) waves in solids. The velocity of sound in a solid is about six orders of magnitude less than the velocity of electromagnetic waves in insulating solids. Figure 14-5 illustrates a simple surface acoustic wave delay line. The electrical signal of interest $y(t)$ is applied to a piezoelectric solid through a pair of metal electrodes. This is shown at the left side of the figure. Piezoelectric solids have the property of producing stress in response to applied electrical voltages.

In some solids, the stress resulting from the application of a voltage in the fashion indicated in Fig. 14-5 produces a surface wave of stress which propagates from left to right across the surface of the device. If the structure is properly designed and made of the correct materials, this acoustic stress wave will propagate in a relatively undistorted fashion at low velocities.

As the acoustic wave passes beneath the metal electrodes on the right side of the device, it will induce a voltage in these electrodes since the piezoelectric effect is reversible, i.e., an elastic stress generates an induced voltage. If the output voltage is to be an exact replica of the input voltage with the desired time delay, it is essential to eliminate "echoes" of acoustic waves from the sample edges. This is accomplished by ensuring that the edges absorb elastic waves without reflection.

The reduction in delay line sizes made possible by this conversion of signals

Figure 14-5 Surface acoustic wave delay.

from electromagnetic to acoustic is illustrated by a problem at the end of this chapter.

There is also an analogous electrical technology based on charge-coupled devices (CCDs). In such devices, the input analog signal is digitized into a sequential series of packets of electric charge. These packets are moved from left to right across a silicon sample by means of an external clock. The frequency of this clock can be reduced, thereby reducing the propagation velocity of the signal. In this way, long delays can be realized in physically small structures. The input and output elements to this device are simple semiconductor diodes.

Both surface acoustic and CCD devices may be fabricated in forms which allow important amounts of digital filtering. Such filters are the digital equivalent of the passive and active analog filters discussed earlier. Digital filtering can also be carried out by computer programs operating on digital signals stored in computer memory. Digital filtering is now quite important because some types of filtering can be carried out more efficiently in the digital domain. In the past, the only way to do this necessarily involved a large computer. As surface acoustic wave and charged-coupled devices have become more available and as computer costs and size have shrunk, the relative drawbacks of digital filters have become less significant. We cannot take the time to discuss these techniques in any detail. They require a significant amount of mathematical modeling and analysis, but they are available in commercial equipment.

We have now discussed all the major elements shown in the block diagram of a general data acquisition system (Fig. 14-1) except the last element, output. This is the subject matter of the next section.

14-6 DATA STORAGE AND DISPLAY

Data storage and data display do not always require different apparatus. The relatively simple methods of data display often provide simultaneous storage. Examples include the xy plotter, the strip chart recorder, paper tape output from a calculator or a computer, and even the recording of experimental data by pen in a notebook. The data are both displayed and stored in all these cases. A storage oscilloscope also provides a dual function because in such a device there is at least temporary storage of the data for as long as the instrument is left on.

There are also devices which are intended to provide only a storage function. Such storage media are most useful if the data stored are easily retrievable for display or subsequent processing. As an example, let us consider the problem NASA has in transmitting television pictures to earth from deep space probes. Normally, the images "seen" by the cameras are stored in digital form on magnetic tape on board the craft. When a transmission is made to earth, the stored data are sent, bit by bit, using appropriate coding techniques, to earth, where the received information is again stored on magnetic tape. This information is digitally filtered to reduce the effects of noise, and the resulting image is printed. In addition to magnetic tape, other forms of more or less permanent storage are punched paper tapes, punched computer cards, magnetic cores, magnetic disks, and various kinds of semiconductor device memories. The semiconductor memories are said to be *volatile* because the stored information will normally be lost if the power is turned off. There are some exceptions to this general rule, however.

There are many technologies and associated devices which compete for dominance in the display area. We have space to mention only two. The first is the light-emitting-diode technology, in which semiconductor diodes which emit light under positive voltages are used as light sources. These devices have the significant advantage of having no filaments or other parts that can burn out; as a consequence, they have very impressive operating lives. They do require significant amounts of power compared with other technologies and tend to "wash out" in bright ambient lighting. The power requirement is the reason that digital watches with light-emitting-diode displays are not constantly on. The user must press a button to cause a brief time display. If the displays were left on constantly, batteries would last only a few days or weeks.

The second technology is based on liquid crystals. These are materials which display different light-reflecting properties when voltages are applied to them. By sandwiching a liquid crystal between two thin transparent metal electrodes, a display can be easily fabricated. This display requires less power than a light-emitting-diode array, but since it depends on light reflection, it is not easily visible in dimly lit surroundings and not at all visible in the dark.

Rather complete packages for display purposes are sold by different manufacturers. In addition, the electronic circuits required to drive these displays are also readily available.

This completes a brief survey of the available display and storage technologies. We have now discussed all the major elements of a data acquisition, processing, display, and/or storage system. To be sure, space did not always allow for sufficient detail to completely prepare the reader for all the experimental difficulties that arise in setting up such a system. Rather, we intended to discuss the major elements in general terms. Any specific advice we could have given would be made obsolete rapidly by advancing technology. In general, the manufacturers of the various instruments and systems described above will be valuable sources of detailed information.

In the next section, we provide an overview of the entire data acquisition and processing topic by considering a significant technological advance—the microprocessor.

14-7 THE PROGRAM AS A SUBSTITUTE FOR WIRED LOGIC

The general data acquisition system which we have been discussing often serves many functions. In addition to gathering, processing, storing, and displaying data, it often is required to perform a control function. In the case of a nuclear

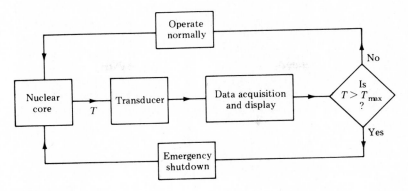

Figure 14-6

power plant, an indication in the data acquisition system that a nuclear core temperature is exceeding some operational limit will result in an automatic shutdown of the nuclear reaction. The entire data acquisition and control system must be able to make logical judgments. It is expected that this system will be backed up by human judgment as well, but for maximum safety, the electronic system itself must be capable of such decisions. How is the control function fulfilled? A simple schematic answer is presented in Fig. 14-6. This sketch shows the familiar elements of a transducer and the subsequent data acquisition and display functions. In addition, there is a new element in which a decision is made electronically. In this module, the question, "Is the temperature T greater than the safe operating temperature T_{max}?" is electronically answered. If the answer is "yes," the system generates an electronic signal which begins an automatic emergency shutdown of the nuclear reaction. If "no," the system indicates that all is well and allows the core to continue to operate.

In order to implement this logical decision-making process, a single integrated circuit called a Schmitt trigger may be used. The output-input characteristic of such a device is shown in Fig. 14-7. The threshold voltage V_T can be

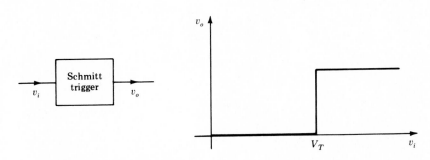

Figure 14-7 The Schmitt trigger.

electrically controlled. This voltage is set to be the voltage produced by the transducer in Fig. 14-6 when the nuclear core temperature is equal to T_{max}. As long as the actual temperature is below this value, the voltage output of the Schmitt trigger will be zero. Once the safe operating temperature limit is exceeded, the Schmitt trigger will have a positive voltage at its output, and this voltage may be fed, through appropriate interface circuitry, to the equipment which initiates the emergency shutdown procedure. This example illustrates a segment of electrical engineering known as logic design, the purpose of which is to develop circuits which can perform logical operations and comparisons. Subsequent action, or the lack of it, depends on the outcome of these logical operations. Although the example given above is analog in nature, circuit designers have tended to concentrate in the area of digital logic circuits. Again, this preference for digital rather than analog representations results from the higher noise immunity of digital circuits. After all, in many industrial situations, logic circuits must operate in quite noisy electrical environments.

As an example of a reasonably simple logic design problem, let us consider the task of preparing a circuit which will monitor a traffic control signal and generate an error message when an illegal combination of lights appears. This signal could be routed to maintenance headquarters to alert the staff to the existence of a traffic signal problem. In the state in which this is being written, legal light combinations are red only, yellow only, green only, and red-yellow together. All others are illegal. Figure 14-8 illustrates the solution to this problem in the form of a block diagram. It should be noted that error messages occur whenever three or no lamps are lit as well as when any two lamps other than red-yellow are lit. In any of these cases, an error signal must be generated as indicated in the diagram. The logical processing shown in this problem can be carried out by means of digital logic circuits. To briefly illustrate this, imagine

Figure 14-8

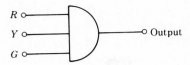

Figure 14-9

that we have available three electrical wires; the first of these will be at a positive voltage if the red lamp is lit and zero volts if the red lamp is out, the second line is tied to the yellow lamp, and the third is tied to the green lamp in analogous fashion. In order to decide if all three lamps are on, a logical AND gate is used as sketched in Fig. 14-9. This gate will produce a positive output voltage only if each of the inputs is positive. In this case, an output will occur only if all three of the lamps are simultaneously lit. Since this is an error condition, this output signal could be used to generate the necessary error message. In an analogous fashion, other gates are used to check if any other error conditions exist. In addition to AND gates, there are other classes of digital logic circuits, and, indeed, the designing of circuits containing these decision-making gates is a separate area of study.

Our purpose in discussing this topic here is not to educate the reader thoroughly in the area of logic design. This would be impossible in the space of one section within one chapter of a text devoted to experimental methods. The salient point is that logical decisions can be made by wiring together various gates and connecting them in the appropriate way to the input and output circuits. Such systems are called *hard-wired* logic arrays. In order to change the function of a given logic system, it is necessary to physically rearrange and reconnect the gates. In situations in which many different logical functions must be performed, this leads to a great proliferation in the types of logic arrays which are manufactured. Semiconductor companies have spent large sums of money attempting to automate the processes whereby custom-designed logic arrays could be manufactured. This was, of course, in response to cost pressures and, to a large extent, did not succeed. Integrated-circuit manufacturing techniques do not lend themselves to a high degree of customization at low cost. It was necessary to develop a universal logic circuit, and this is given the name *microprocessor*.

This device consists of an almost complete small computer on a single piece of silicon. Microprocessors which are available today are in the form of one circuit containing 40 leads. Some of these leads provide interconnection to the necessary dc power supplies and ground, while others serve as interfaces over which data can be fed into the system and results fed out of the system. An external digital clock must also be connected to the unit to provide the timing necessary for the step-by-step digital processing. The chip itself contains a memory in which data *and* a user-generated program may be stored. In addition, the microprocessor has available a set of registers in which arithmetic operations may be performed.

Within limits set by the available instructions, the size of the memory, the interface circuits, and the cleverness of the designer, the microprocessor is truly a universal logic circuit. It may be programmed to perform a particular desired set of logical procedures. Rather than having to manufacture a wide array of custom-designed circuits, the semiconductor manufacturers are free to concentrate on producing only one microprocessor design. The resulting economies have forced the prices of these systems down rapidly. As these prices continue to decrease, microprocessors will replace more and more hard-wired logic circuits. It is anticipated that custom logic circuits will continue to be used only in the simplest applications.

It is important to recognize how pervasive these devices have become. The advent of the microprocessor is responsible for the enormous power now available in the personal computer. These computers now allow direct programming of data input to produce a desired output in either printed or magnetic storage form.

14-8 SUMMARY

This chapter began by considering the entire data acquisition, conditioning, transmitting, storage, and display process. We developed some additional concepts in the areas of filtering and signal conditioning and surveyed some of the technologies available to carry out signal processing in digital form.

The basic technologies used for information storage and display were then considered, and the chapter concluded with a brief introduction to the key concepts underlying the operation and usefulness of microprocessors.

REVIEW QUESTIONS

14-1 What are the major elements of a data acquisition and processing system?

14-2 In what kinds of situations would data transmission be necessary?

14-3 What are the differences between active and passive filters?

14-4 What are the functions of a lowpass active filter?

14-5 Why are digital representations of signals preferred for data transmission?

14-6 Why does coaxial cable exhibit higher noise immunity than signal propagation through free space?

14-7 What is coding?

14-8 How are time delays conveniently realized in systems where the calculation of correlation functions is being carried out?

14-9 What physical principle is responsible for the operation of surface acoustic wave devices?

14-10 What is the function of a microprocessor?

PROBLEMS

14-1 Find the ratio of V_o to V_i for the following circuit, where

$$v_o = V_o e^{j\omega t} \qquad v_i = V_i e^{j\omega t}$$

Figure Prob. 14-1

14-2 At what frequency ω_0 will the ratio of output to input (V_o/V_i) in the above circuit be 50 percent of the low-frequency $(\omega \to 0)$ value?

14-3 Determine V_o/V_i for the following circuit.

Figure Prob. 14-3

14-4 How long must an SAW device be if it is composed of a material with a sound velocity of 10^3 cm/s and it is to produce a time delay of 10 μs?

APPENDIX

Table A-1a Conversion factors

Length
  12 in = 1 ft
  2.54 cm = 1 in
  1 μm = 10^{-6} m = 10^{-4} cm

Mass
  1 kg = 2.205 lbm
  1 slug = 32.17 lbm
  454 g = 1 lbm

Force
  1 dyn = 2.248 × 10^{-6} lbf
  1 lbf = 4.448 N
  10^5 dyn = 1 N

Energy
  1 ft · lbf = 1.356 J
  1 kW · h = 3413 Btu
  1 hp · h = 2545 Btu
  1 Btu = 252 cal
  1 Btu = 778 ft · lbf

Pressure
  1 atm = 14.696 lbf/in^2 = 2116 lbf/ft^2
  1 atm = 1.0132 × 10^5 N/m^2
  1 inHg = 70.73 lbf/ft^2

Viscosity
  1 cP = 2.42 lbm/h · ft
  1 lbf · s/ft^2 = 32.16 lbm/s · ft

Thermal conductivity
  1 cal/s · cm · °C = 242 Btu/h · ft · °F
  1 W/cm · °C = 57.79 Btu/h · ft · °F

Table A-1*b* Useful conversion factors between SI and English units

Physical quality	Symbol	SI to English conversion	English to SI conversion
Length	L	1 m = 3.2808 ft	1 ft = 0.3048 m
Area	A	1 m² = 10.7639 ft²	1 ft² = 0.092903 m²
Volume	V	1 m³ = 35.3134 ft³	1 ft³ = 0.028317 m³
Velocity	v	1 m/s = 3.2808 ft/s	1 ft/s = 0.3048 m/s
Density	ρ	1 kg/m³ = 0.06243 lbm/ft³	1 lbm/ft³ = 16.018 kg/m³
Force	F	1 N = 0.2248 lbf	1 lbf = 4.4482 N
Mass	m	1 kg = 2.20462 lbm	1 lbm = 0.45359637 kg
Pressure	p	1 N/m² = 1.45038 × 10⁻⁴ lbf/in²	1 lbf/in² = 6894.76 N/m²
Energy, heat	q	1 kJ = 0.94783 Btu	1 Btu = 1.05504 kJ
Heat flow	q	1 W = 3.4121 Btu/h	1 Btu/h = 0.29307 W
Heat flux per unit area	q/A	1 W/m² = 0.317 Btu/h·ft²	1 Btu/h·ft² = 3.154 W/m²
Heat flux per unit length	q/L	1 W/m = 1.0403 Btu/h·ft	1 Btu/h·ft = 0.9613 W/m
Heat generation per unit volume	$\dot{q}$	1 W/m³ = 0.096623 Btu/h·ft³	1 Btu/h·ft³ = 10.35 W/m³
Energy per unit mass	q/m	1 kJ/kg = 0.4299 Btu/lbm	1 Btu/lbm = 2.326 kJ/kg
Specific heat	C	1 kJ/kg·°C = 0.23884 Btu/lbm·°F	1 Btu/lbm·°F = 4.1869 kJ/kg·°C
Thermal conductivity	k	1 W/m·°C = 0.5778 Btu/h·ft·°F	1 Btu/h·ft·°F = 1.7307 W/m·°C
Convection heat-transfer coefficient	h	1 W/m²·°C = 0.1761 Btu/h·ft²·°F	1 Btu/h·ft²·°F = 5.6782 W/m²·°C
Dynamic viscosity	μ	1 kg/m·s = 0.672 lbm/ft·s = 2419.2 lbm/ft·h	1 lbm/ft·s = 1.4881 kg/m·s
Kinematic viscosity and thermal diffusivity	v, α	1 m²/s = 10.7639 ft²/s	1 ft²/s = 0.092903 m²/s

Table A-2 Properties of metals at 70°F (21.1°C)

Metal	Density		Specific heat		Linear coefficient of expansion		Electrical conductivity relative to copper	Thermal conductivity	
	lbm/ft^3	kg/m^3	Btu/lbm·°F	kJ/kg·°C	°F$^{-1} \times 10^4$	°C$^{-1} \times 10^4$		Btu/h·ft·°F	W/m·°C
Aluminum, pure	167	2675	0.22	0.921	0.13	0.23	0.66	118	204
Red brass: 85% Cu, 9% Sn, 6% Zn	544	8714	0.092	0.385	0.10	0.18		35	61
Copper, pure	555	8890	0.095	0.398	0.09	0.16	1.00	223	386
Iron, pure	493	7897	0.108	0.452	0.065	0.12	0.18	42	73
Lead	710	11,370	0.031	0.130	0.16	0.29	0.08	20	35
Magnesium	109	1746	0.242	1.013	0.145	0.26	0.40	99	171
Nickel	556	8906	0.106	0.444	0.07	0.126	0.25	52	90
Plantinum	1335	21,380	0.032	0.134	0.05	0.09	0.18	40	69
Silver	657	10,520	0.056	0.234	0.107	0.193	1.15	235	407
Stainless steel, 18% Cr, 8% Ni	488	7817	0.11	0.46	0.003	0.005	0.0234	9.4	16.3
Steel, structural	485	7769	0.11	0.46	0.07	0.126	0.12	33	57
Tin	456	7304	0.054	0.226	0.11	0.198	0.16	37	64
Tungsten	1208	19,350	0.032	0.134	0.02	0.036	0.33	94	163
Zinc	446	7144	0.092	0.385	0.18	0.32	0.31	37	64

Table A-3 Thermal properties of some nonmetals at 70°F (21.1°C)

Material	Density		Specific heat		Thermal conductivity	
	lbm/ft^3	kg/m^3	$Btu/lbm \cdot °F$	$kJ/kg \cdot °C$	$Btu/h \cdot ft \cdot °F$	$W/m \cdot °C$
Asbestos sheet	60	961	0.2	0.84	0.083	0.144
Brick, common	110	1762	0.22	0.92	0.042	0.073
Concrete	140	2243	0.16	0.67	1.0	1.73
Fiberboard	15	240	0.5	2.09	0.027	0.047
Fiberglas	6	96	0.5	2.09	0.022	0.038
Glass, window	160	2563	0.16	0.67	0.50	0.87
85% magnesia pipe covering	15	240	0.2	0.84	0.042	0.073
Plaster	95	1522	0.25	1.05	0.27	0.47

Table A-4 Properties of some saturated liquids at 68°F (20°C)

Fluid	Chemical formula	Density		Specific heat		Kinematic viscosity		Thermal conductivity		Prandtl number
		lbm/ft^3	kg/m^3	$Btu/lbm \cdot °F$	$kJ/kg \cdot °C$	$ft^2/s \times 10^5$	$m^2/s \times 10^5$	$Btu/h \cdot ft \cdot °F$	$W/m \cdot °C$	
Ammonia	NH_3	38.2	612	1.15	4.81	0.386	0.036	0.301	0.521	2.02
Carbon dioxide	CO_2	48.2	772	1.2	5.02	0.98	0.091	0.0504	0.087	4.10
Freon-12	CCl_2F_2	83.0	1329	0.23	0.96	0.213	0.020	0.042	0.073	2.17
Glycerine	$C_2H_3(OH)_3$	78.9	1264	0.57	2.39	1270	118	0.165	0.286	12.5
Mercury	Hg	847.7	13,580	0.033	0.138	0.123	0.011	5.02	8.69	0.025
Motor oil (typ.)		55.0	881	0.45	1.88	1000	92.9	0.08	0.138	10.0
Water	H_2O	62.4	1000	1.0	4.19	1.08	0.10	0.345	0.597	7.02

Table A-5 Properties of gases at atmospheric pressure and 68°F (20°C)

Gas	Formula	Molecular weight	c_p		$\gamma = c_p/c_v$	Thermal conductivity			Viscosity
			Btu/lbm·°F	kJ/kg·°C		Btu/h·ft·°F	W/m·°C	lbm/h·ft	kg/m·s × 10⁵
Air		28.95	0.24	1.005	1.40	0.025	0.043	0.044	1.82
Acetylene	C_2H_2	26.02	0.38	1.591	1.25	0.0124	0.0215	0.024	0.99
Ammonia	NH_3	17.03	0.52	2.177	1.30	0.0142	0.0246	0.024	0.99
Carbon dioxide	CO_2	44.00	0.20	0.837	1.30	0.0095	0.0164	0.036	1.49
Freon-12	CCl_2F_2	120.9	0.15	0.628	1.14	0.0053	0.0091	0.031	1.28
Hydrogen	H_2	2.01	3.4	14.24	1.40	0.103	0.178	0.022	0.91
Methane	CH_4	16.03	0.54	2.26	1.30	0.0196	0.0339	0.027	1.12
Nitrogen	N_2	28.02	0.245	1.03	1.40	0.0150	0.026	0.044	1.82
Oxygen	O_2	32.00	0.215	0.90	1.40	0.0153	0.0265	0.048	1.98

Table A-6 Properties of air at atmospheric pressure (SI units)†
The values of μ, k, c_p, and Pr are not strongly pressure-dependent and may be used over a fairly wide range of pressures

T, K	ρ, kg/m^3	c_p, kJ/kg $\cdot$ °C	μ, kg/m $\cdot$ s $\times 10^5$	ν, m^2/s $\times 10^6$	k, W/m $\cdot$ °C	α, m^2/s $\times 10^4$	Pr
100	3.6010	1.0266	0.6924	1.923	0.009246	0.02501	0.770
150	2.3675	1.0099	1.0283	4.343	0.013735	0.05745	0.753
200	1.7684	1.0061	1.3289	7.490	0.01809	0.10165	0.739
250	1.4128	1.0053	1.599	11.31	0.02227	0.15675	0.722
300	1.1774	1.0057	1.8462	15.69	0.02624	0.22160	0.708
350	0.9980	1.0090	2.075	20.76	0.03003	0.2983	0.697
400	0.8826	1.0140	2.286	25.90	0.03365	0.3760	0.689
450	0.7833	1.0207	2.484	31.71	0.03707	0.4222	0.683
500	0.7048	1.0295	2.671	37.90	0.04038	0.5564	0.680
550	0.6423	1.0392	2.848	44.34	0.04360	0.6532	0.680
600	0.5879	1.0551	3.018	51.34	0.04659	0.7512	0.680
650	0.5430	1.0635	3.177	58.51	0.04953	0.8578	0.682
700	0.5030	1.0752	3.332	66.25	0.05230	0.9672	0.684
750	0.4709	1.0856	3.481	73.91	0.05509	1.0774	0.686
800	0.4405	1.0978	3.625	82.29	0.05779	1.1951	0.689
850	0.4149	1.1095	3.765	90.75	0.06028	1.3097	0.692
900	0.3925	1.1212	3.899	99.3	0.06279	1.4271	0.696
950	0.3716	1.1321	4.023	108.2	0.06525	1.5510	0.699
1000	0.3524	1.1417	4.152	117.8	0.06752	1.6779	0.702
1100	0.3204	1.160	4.44	138.6	0.0732	1.969	0.704
1200	0.2947	1.179	4.69	159.1	0.0782	2.251	0.707
1300	0.2707	1.197	4.93	182.1	0.0837	2.583	0.705
1400	0.2515	1.214	5.17	205.5	0.0891	2.920	0.705
1500	0.2355	1.230	5.40	229.1	0.0946	3.262	0.705
1600	0.2211	1.248	5.63	254.5	0.100	3.609	0.705
1700	0.2082	1.267	5.85	280.5	0.105	3.977	0.705
1800	0.1970	1.287	6.07	308.1	0.111	4.379	0.704
1900	0.1858	1.309	6.29	338.5	0.117	4.811	0.704
2000	0.1762	1.338	6.50	369.0	0.124	5.260	0.702
2100	0.1682	1.372	6.72	399.6	0.131	5.715	0.700
2200	0.1602	1.419	6.93	432.6	0.139	6.120	0.707
2300	0.1538	1.482	7.14	464.0	0.149	6.540	0.710
2400	0.1458	1.574	7.35	504.0	0.161	7.020	0.718
2500	0.1394	1.688	7.57	543.5	0.175	7.441	0.730

† From *Natl. Bur. Std. (U.S.), Circ.* 564, 1955.

Table A-7 Properties of dry air at atmospheric pressure (English units)†

T, °F	μ, lbm/h · ft	k, Btu/h · ft · °F	c_p, Btu/lbm · °F	Pr
− 100	0.0319	0.0104	0.239	0.739
− 50	0.0358	0.0118	0.239	0.729
0	0.0394	0.0131	0.240	0.718
50	0.0427	0.0143	0.240	0.712
100	0.0459	0.0157	0.240	0.706
150	0.0484	0.0167	0.241	0.699
200	0.0519	0.0181	0.241	0.693
250	0.0547	0.0192	0.242	0.690
300	0.0574	0.0203	0.243	0.686
400	0.0626	0.0225	0.245	0.681
500	0.0675	0.0246	0.248	0.680
600	0.0721	0.0265	0.250	0.680
700	0.0765	0.0284	0.254	0.682
800	0.0806	0.0303	0.257	0.684
900	0.0846	0.0320	0.260	0.687
1000	0.0884	0.0337	0.263	0.690

† From: *Natl. Bur. Std. (U.S.), Circ.* 564, 1955.

Table A-8 Properties of water [saturated liquid] (SI units)†

°F	°C	c_p, kJ/kg · °C	ρ, kg/m³	μ, kg/m · s	k, W/m · °C	Pr	$g\beta\rho^2 c_p/\mu k$ 1/m³ · °C
32	0	4.225	999.8	1.79×10^{-3}	0.566	13.25	
40	4.44	4.208	999.8	1.55	0.575	11.35	1.91×10^9
50	10	4.195	999.2	1.31	0.585	9.40	6.34×10^9
60	15.56	4.186	998.6	1.12	0.595	7.88	1.08×10^{10}
70	21.11	4.179	997.4	9.8×10^{-4}	0.604	6.78	1.46×10^{10}
80	26.67	4.179	995.8	8.6	0.614	5.85	1.91×10^{10}
90	32.22	4.174	994.9	7.65	0.623	5.12	2.48×10^{10}
100	37.78	4.174	993.0	6.82	0.630	4.53	3.3×10^{10}
110	43.33	4.174	990.6	6.16	0.637	4.04	4.19×10^{10}
120	48.89	4.174	988.8	5.62	0.644	3.64	4.89×10^{10}
130	54.44	4.179	985.7	5.13	0.649	3.30	5.66×10^{10}
140	60	4.179	983.3	4.71	0.654	3.01	6.48×10^{10}
150	65.55	4.183	980.3	4.3	0.659	2.73	7.62×10^{10}
160	71.11	4.186	977.3	4.01	0.665	2.53	8.84×10^{10}
170	76.67	4.191	973.7	3.72	0.668	2.33	9.85×10^{10}
180	82.22	4.195	970.2	3.47	0.673	2.16	1.09×10^{11}
190	87.78	4.199	966.7	3.27	0.675	2.03	
200	93.33	4.204	963.2	3.06	0.678	1.90	
220	104.4	4.216	955.1	2.67	0.684	1.66	
240	115.6	4.229	946.7	2.44	0.685	1.51	
260	126.7	4.250	937.2	2.19	0.685	1.36	
280	137.8	4.271	928.1	1.98	0.685	1.24	
300	148.9	4.296	918.0	1.86	0.684	1.17	
350	176.7	4.371	890.4	1.57	0.677	1.02	
400	204.4	4.467	859.4	1.36	0.665	1.00	
450	232.2	4.585	825.7	1.20	0.646	0.85	
500	260	4.731	785.2	1.07	0.616	0.83	
550	287.7	5.024	735.5	9.51×10^{-5}			
600	315.6	5.703	678.7	8.68			

† Adapted from A. I. Brown and S. M. Marco, "Introduction to Heat Transfer," 3d ed., McGraw-Hill Book Company, New York, 1958.

Table A-9 Properties of water [saturated liquid] (English units)†

°F	c_p, Btu/lbm · °F	ρ, lbm/ft^3	μ, lbm/ft · h	k, Btu/h · ft · °F	Pr, $c_p/\mu k$
32	1.009	62.42	4.33	0.327	13.35
40	1.005	62.42	3.75	0.332	11.35
50	1.002	62.38	3.17	0.338	9.40
60	1.000	62.34	2.71	0.344	7.88
70	0.998	62.27	2.37	0.349	6.78
80	0.998	62.17	2.08	0.355	5.85
90	0.997	62.11	1.85	0.360	5.12
100	0.997	61.99	1.65	0.364	4.53
110	0.997	61.84	1.49	0.368	4.04
120	0.997	61.73	1.36	0.372	3.64
130	0.998	61.54	1.24	0.375	3.30
140	0.998	61.39	1.14	0.378	3.01
150	0.999	61.20	1.04	0.381	2.73
160	1.000	61.01	0.97	0.384	2.53
170	1.001	60.79	0.90	0.386	2.33
180	1.002	60.57	0.84	0.389	2.16
190	1.003	60.35	0.79	0.390	2.03
200	1.004	60.13	0.74	0.392	1.90
220	1.007	59.63	0.65	0.395	1.66
240	1.010	59.10	0.59	0.396	1.51
260	1.015	58.51	0.53	0.396	1.36
280	1.020	57.94	0.48	0.396	1.24
300	1.026	57.31	0.45	0.395	1.17
350	1.044	55.59	0.38	0.391	1.02
400	1.067	53.65	0.33	0.384	1.00
450	1.095	51.55	0.29	0.373	0.85
500	1.130	49.02	0.26	0.356	0.83
550	1.200	45.92	0.23		
600	1.362	42.37	0.21		

† From A. I. Brown and S. M. Marco: "Introduction to Heat Transfer," 3d ed., McGraw-Hill Book Company, New York, 1958.

Table A-10 Diffusion coefficients of some gases and vapors in air at 25°C and 1 atm†

Substance	D, cm^2/s
Ammonia	0.28
Benzene	0.088
Carbon dioxide	0.164
Hydrogen	0.410
Methanol	0.159
Water	0.256

† From R. H. Perry (ed.): "Chemical Engineers' Handbook," 5th ed., McGraw-Hill Book Company, New York, 1973.

Table A-11 Approximate total normal emissivities of various surfaces at 20°C

Surface	Emissivity
Aluminum, heavily oxidized	0.20–0.30
Aluminum, polished	0.09
Asbestos sheet	0.93–0.96
Brass, polished	0.06
Brick	0.93
Carbon, lampblack	0.96
Copper, heavily oxidized	0.75
Copper, polished	0.03
Enamel, white fused on iron	0.90
Iron, oxidized	0.74
Paint, aluminum	0.27–0.67
Paint, flat black lacquer	0.97
Steel, polished	0.07
Steel plate, rough	0.95
Tin, bright	0.06
Water	0.95

Table A-12 Area moments of inertia for some common geometric shapes

Rectangle

$$A = bh$$

$$I_{y0} = \frac{bh^3}{12}$$

$$I_{y0} = \frac{hb^3}{12}$$

Right triangle

$$A = \frac{bh}{2}$$

$$I_{x0} = \frac{bh^3}{36}$$

$$I_{y0} = \frac{hb^3}{36}$$

Quadrant of circle

$$A = \frac{\pi a^2}{4}$$

$$I_x = I_y = \frac{\pi a^4}{16}$$

$$I_{x0} = I_{y0} = 0.054a^4$$

Circle

$$A = \pi a^2$$

$$I_{x0} = I_{y0} = \frac{\pi a^4}{4} = \frac{\pi d^4}{64}$$

$$\text{(Polar)} \quad J = \frac{\pi a^4}{4} = \frac{\pi d^4}{32}$$

Quadrant of ellipse

$$A = \frac{\pi ab^2}{4}$$

$$I_x = \frac{\pi ab^3}{16}$$

$$I_y = \frac{\pi a^3 b}{16}$$

$$I_{x0} = 0.054ab^3$$

$$I_{y0} = 0.054a^3 b$$

Ellipse

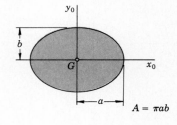

$$A = \pi ab$$

$$I_{x0} = \frac{\pi ab^3}{4}$$

$$I_{y0} = \frac{\pi a^3 b}{4}$$

INDEX